THE CONCISE
SCIENCE
ENCYCLOPEDIA

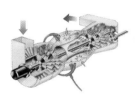

KINGFISHER
Kingfisher Publications Plc
New Penderel House
283-288 High Holborn
London WC1V 7HZ
www.kingfisherpub.com

First published as *The Kingfisher Science Encyclopedia*
by Kingfisher Publications Plc in 2000
Reprinted in a revised format by Kingfisher Publications Plc 2001

2 4 6 8 10 9 7 5 3 1
1TR/0701/SF/UNV(RNB)/157MA

A CIP catalogue record for this book is available
from the British Library.

ISBN 0 7534 0640 3

Printed in China

This edition produced in 2001 by PAGE*One*,
Cairn House, Elgiva Lane, Cheshire, Bucks HP5 2JD

PROJECT TEAM
Project Director and Art Editor Julian Holland
Editorial Team Martin Clowes, Leon Gray, Julian Holland,
Rachel Hutchings, Mike McGuire
Designers Julian Holland, Jeffrey Farrow, Nigel White
Commissioned artwork Julian Baker
Picture Research Wendy Brown

FOR KINGFISHER
Managing Editor Miranda Smith
Art Director Mike Davis
DTP Co-ordinator Nicky Studdart
Editorial team Julie Ferris, Sheila Clewley
Artwork Research Wendy Allison, Steve Robinson,
Christopher Cowlin
Production Manager Caroline Jackson

GENERAL EDITOR
Professor Charles Taylor, D. Sc., F.Inst.P.,
Emeritus Professor of Physics in the University of Wales,
former Professor of Experimental Physics at the
Royal Institution of Britain

CONTRIBUTORS
Clive Gifford, Peter Mellett, Martin Redfern, Carole Stott,
Richard Walker, Brian Williams

THE CONCISE
SCIENCE
ENCYCLOPEDIA

KING*f*ISHER

CONTENTS

CHAPTER 6
LIGHT AND ENERGY

CHAPTER 7
FORCES AND MOVEMENT

CHAPTER 8
ELECTRICITY AND ELECTRONICS

CHAPTER 9
SPACE AND TIME

INTRODUCTION

In the 21st century, science and technology will increasingly dominate our lives. There will be many challenges, not only environmental, but ethical and moral as well. Science is the key subject for all children, and they need easy access to the scientific knowledge that will help them live in an increasingly demanding world.

The Kingfisher Science Encyclopedia is divided into nine thematic sections, each of them tackling a specific area of scientific interest and study. **Planet Earth** examines geological time, how oceans and mountains are formed, and Earth's atmosphere and weather systems. **Living Things** surveys life on our planet from the tiniest bacteria to the largest mammals, while **Human Biology** explores every part of the fantastic group of cells that is the human body. **Chemistry and the Elements** explains how solids, liquids and gases relate to each other, and interact, and **Materials and Technology** examines everyday materials and how they are used. **Light and Energy** looks at light, heat and colour; while human and machine power, sound and pressure are explored in the section called **Forces and Movement**. **Electricity and Electronics** delves into the increasingly technological world of power stations, telecommunications and information technology. In **Space and Time**, Earth is shown as a small part of an incredible Universe that we are only just beginning to explore.

The encyclopedia has been written by a team of specialist science authors and consultants led by the eminent Professor Charles Taylor, the first holder of the Royal Society's Michael Faraday Award for Contributions to the Understanding of Science in 1986. Whether *The Kingfisher Science Encyclopedia* is used for school work, or simply dipped into at random, it will add to knowledge, stimulate natural curiosity and creativity, and prepare the enquiring mind for an exciting future world.

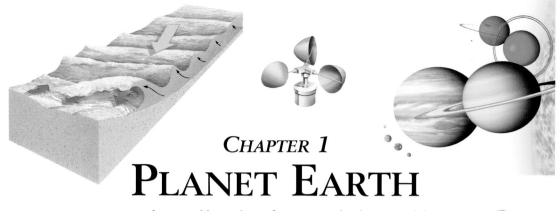

Chapter 1

PLANET EARTH

T he ground beneath our feet seems to be the most solid and unchanging thing we know. It forms foundations for our cities and an environment in which we can live. Yet in reality Planet Earth is spinning on its axis and hurtling through hostile space, as it orbits the nuclear furnace of our Sun. It is an active, dynamic, living planet.

The rock-solid surface is not as solid as it appears. It is cracked like giant slabs of crazy paving. Earthquakes shake our cities and volcanoes erupt, giving clues to the fiery movements beneath the ground. From above, the planet is buffeted by radiation and particles streaming at it through space. But in between is an atmosphere, and oceans with liquid water, and temperatures which are just right for life.

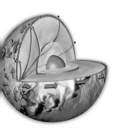

To an alien spacecraft travelling past the solar system, Planet Earth would immediately stand out as something special. The composition of the atmosphere, with free oxygen and traces of gases like methane can only be sustained by life. The aliens might detect the characteristic colours of chlorophyll, the pigment used by plants on land and algae in the sea to trap sunlight. If our aliens were not careful they would pick up the radio cacophony of our broadcasts, revealing that life here is at least moderately intelligent.

Life has transformed Earth and Earth continues to support life. Buried in the rocks are minerals, gems and precious metals. Using energy from Earth in the form of coal and oil, we have transformed them into the artefacts of civilization, from books and buildings to cars and computers. We all depend on Earth.

EARTH AND THE SOLAR SYSTEM

A solar system consists of a star and the planets and other bodies that orbit around it. Earth is the third planet from the Sun in our solar system.

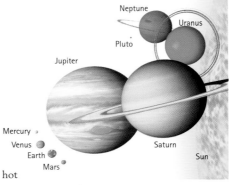

The Sun (right) dwarfs everything else in the solar system, even the gas giants Jupiter and Saturn. Earth and its closest neighbours (left) are minute in comparison.

When the Universe began, around twelve billion years ago, the first elements to form were hydrogen and helium. Nuclear reactions in early generations of stars produced the other elements and spat them into space as clouds of dust and gases. Five billion years ago, one of these clouds began to contract. A spinning ball of dust and gas formed at its centre, and gravity compressed this ball until it was hot enough to form a star – the Sun.

THE ORIGINS OF EARTH

Radiation from the young Sun blew away much of the remaining dust cloud. What was left formed a disc of dust around the Sun. Over time, the dust grains clumped together to form rocky lumps. These lumps bumped into each other, sometimes joining together in a process called accretion. Slowly, the dust disc was transformed into a few planets, one of which was to become Earth.

As Earth gathered mass, its gravitational field increased. The force of gravity pulled the dust into a ball and compressed it

Earth formed from a disc of dust that surrounded the Sun when it was only a few hundred millions of years old. The dust clumped together in a process called accretion.

until it started to melt. A dense core of molten iron formed, surrounded by a solid mantle of silicate rock. Volcanoes and colliding debris helped form the surface features of the new planet. When Earth was almost complete, an object the size of Mars crashed into it, throwing a cloud of material into orbit. This dust condensed and formed the Moon.

OTHER PLANETS

Mercury is closest to the Sun. It has a barren, rocky surface and practically no atmosphere. The outer planets are frozen balls of gas. The three planets in between are Earth, Mars and Venus. Venus is about the same size as Earth and is nearer to the Sun. Mars is slightly smaller and further away. Carbon dioxide in Venus's atmosphere led to a runaway greenhouse effect that boiled away any water. On Mars, water froze or escaped into space leaving a cold desert. If life began on Mars or Venus, it did not survive. On Earth, algae consumed carbon dioxide from the atmosphere, keeping the climate in balance and producing oxygen.

This satellite image of Earth was taken by a camera on a Meteosat weather satellite 35,800 kilometres above the equator as it crosses the Americas. It shows the rich blues of the oceans, and the swirling cloud patterns in the atmosphere of a planet where conditions are just right for life.

SEE ALSO
272–3 The Solar System, 274–5 Earth and the Moon

EARTH'S ROTATION

Earth spins like a top as it orbits the Sun. These rotations cause daily, annual and seasonal variations in sunlight and temperature on Earth's surface.

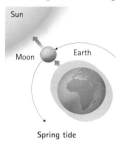

Spring tide

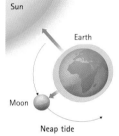

Neap tide

Viewed from Earth's surface, the Sun seems to rise in the East, cross the sky and set in the West. Stars do the same by night. Until the 16th century, people believed that Earth was fixed and the Sun and stars moved around it. We now know that the Sun and stars seem to cross the sky because Earth rotates about its axis every day. Earth also orbits the Sun in 365¼ days, which gives us our year.

THE LUNAR MONTH

The Moon orbits Earth in around 27 days. It takes about 29 days – a lunar month – to pass through all its phases. After each new moon, the Moon is completely dark. It then appears as a thin crescent, lit from one side. The lit area then grows, or waxes, until it becomes full. Finally, the Moon wanes until it becomes a thin crescent again.

TIDES AND ECLIPSES

As the Moon orbits Earth, its gravity pulls water in the oceans towards it. The changes in water level that result are called tides. The Sun also affects tides, and the greatest tidal variations, called spring tides, occur when the Sun and Moon pull in the same direction.

Occasionally, Earth comes between the Sun and the Moon and casts a shadow on the Moon. This event is called a lunar eclipse. A solar eclipse is when the Moon comes between Earth and the Sun. While the radius of the Moon is only ¹⁄₄₀₀th the radius of the Sun, the Moon is ¹⁄₄₀₀th as far away, so total eclipses can happen.

OTHER CYCLES

Sometimes, Earth's orbit around the Sun is less circular and more elliptical. Also Earth's rotation axis slowly wobbles like an out-of-balance spinning-top. These variations add up over tens or hundreds of thousands of years. Some scientists believe they may cause the ice ages that affect Earth every few million years.

▲ Spring tides result when the Sun and Moon pull the waters of the oceans in the same direction. The much weaker neap tides happen when the Moon pulls at right angles to the Sun.

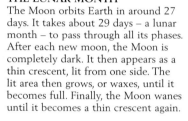

21st March
Equinox. Twelve hours of daylight in all parts of the world.

21st June
Solstice. Longest day of the year in the Northern Hemisphere; shortest day in the Southern Hemisphere.

Sun

22nd December
Solstice. Longest day of the year in the Southern Hemisphere; shortest day in the Northern Hemisphere.

23rd September
Equinox. Twelve hours of daylight in all parts of the world.

Earth rotates around an axis inclined at 23 degrees to the planet's orbit around the Sun. This tilt causes seasonal variations of daylight hours and climate. In March and September, the Sun is directly over the equator. In June, the Northern Hemisphere is angled towards the Sun and becomes warmer. In December, the Southern Hemisphere is angled towards the Sun and experiences summer. The Northern Hemisphere has winter at this time. Near the poles, there are weeks in the summer when the Sun does not set and weeks in the winter when it does not rise.

SEE ALSO

268–9 The Sun, 274–5 Earth and the Moon, 276 Eclipses

FOSSILS AND GEOLOGICAL TIME

Fossils are the preserved remains of once-living organisms. Dating back around 3.5 billion years, they provide vital clues about periods in Earth's history.

This fern frond grew around 300 million years ago in a swamp during the Carboniferous period of Earth's history. When the frond died and was buried, it did not rot away. Instead, it was gradually transformed into coal. The impression of the frond is preserved in the sample.

The origins of life have long posed a dilemma for scientists and theologians. Most cultures have stories relating to the creation of life on Earth – humans are nearly always seen as a pinnacle of the process. Some theologians have estimated dates for creation going back only a few thousand years. In 1650, for example, a bishop from Ireland decided that the world was created in 4004BC. He thought that the shells and bones called fossils found inside rocks were the remains of creatures that perished in the Biblical flood. However, it is hard to see how the planet could have changed so much in such a short space of time.

In the 1800s, geologists realized that slow changes that were still taking place could account for the rise and fall of the mountains and the discovery of fossils. At the time, scientists dated Earth at more than 20 million years old. Today, however, rocks can be dated precisely by measuring the amounts of radioactive elements within them. For example, a radioactive form of carbon is known to decay at a

This fossil of a marine reptile called an ichthyosaurus was preserved in shale dating from the lower Jurassic period. The specimen was found near Lyme Regis in Dorset, England. Ichthyosaurus was a fast swimmer. It fed on fish, using its sharp teeth to tear its prey apart.

fixed rate, and this is used to date charcoal that is up to 50,000 years old. Other elements can date far older rocks and show that the history of Earth began over 4.5 billion years ago.

CHARTING THE EVIDENCE

Careful study of fossils has revealed that similar life forms existed at the same time in different parts of the world. As a result, fossils have become useful for dating rocks. The different types of fossils found in rocks change with time, plotting the evolutionary history of life. Sometimes the changes were slow and gradual, but sometimes they seem very sudden, with entire groups of plant or animal species disappearing from one layer to the next. A few successful species seem to continue almost unchanged for millions of years, while others perish. Sometimes a sudden

FORMATION OF A FOSSIL

When an organism dies, the remains of its body become buried and will slowly fossilize. Usually only the hard parts, such as shells or bones, survive. Sometimes the remains gradually become stone – the original molecules are replaced by minerals such as calcite or iron pyrites. Often, however, the fossil contains many of the original molecules. A new science called molecular palaeontology compares the chemicals or even the genes of extinct species with species that continue to live on Earth.

1 In the Jurassic period, around 150 million years ago, a type of shellfish called an ammonite dies and falls to the bottom of the ocean.

2 The soft body parts of the ammonite decay or are eaten by predators.

3 The empty shell is then covered by sand and mud.

4 The sand and mud layers are compressed, turn to stone and lift up and tilt into land above sea level.

5 The forces of erosion break up the ground and reveal the remains of the ammonite.

6 A fossil hunter cracks open the stone to reveal the fossil and the mould it has left in the rock.

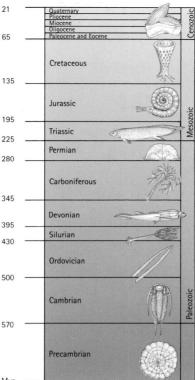

Mya			
21	Quaternary		Cenozoic
	Pliocene		
	Miocene		
	Oligocene		
65	Paleocene and Eocene		
	Cretaceous		Mesozoic
135			
	Jurassic		
195			
	Triassic		
225			
	Permian		Paleozoic
280			
	Carboniferous		
345			
	Devonian		
395			
	Silurian		
430			
	Ordovician		
500			
	Cambrian		
570			
	Precambrian		

◀ Geological time is divided into a series of periods, each of which is characterized by a different range of fossil creatures. The Precambrian period represents 85 per cent of the history of Earth. However, rocks dating from this period are poorly preserved, and there were few large creatures to leave fossils.

In special circumstances, the remains of soft-bodied creatures turn into fossils. Over 40 million years ago, tree resin trapped this fly. The tree resin has turned into amber, and the fly and some of its genetic material have been preserved within it.

diversification populated the world with a whole new range of creatures. The changes define the boundaries between geological periods. The intervals are thought to have occurred as a result of major catastrophes, some of them probably triggered by large asteroids or comets hitting our planet and disrupting the climate. At the end of the Cretaceous period, 65 million years ago, thousands of species, including all the dinosaurs, are known to have become extinct. This boundary also coincides with an enormous impact crater in the Gulf of Mexico. An asteroid, maybe a kilometre across, hit Earth and vaporized, spreading a cloud of dust around the planet, blocking out energy from the Sun, and starting global forest fires. Even more species became extinct at the end of the Permian period 225 million years ago. Indeed, mass extinctions of various degrees mark the boundaries of most of the geological periods.

LIFE ON EARTH
The last 65 million years of the story of life are marked by the rise of the mammals, along with broad-leaved trees and flowering plants. About 200 million years prior to this period, dinosaurs and their relatives ruled the land, and a rich variety of marine life populated the warm seas. In the Carboniferous period, about 300 million years ago, extensive swamps supported a lush growth of primitive plants such as tree ferns and cycads. The remains of these plants have resulted in deposits of coal. Before then, there is not

much evidence for life on land. However, the oceans were teeming with life. Fossils from the Precambrian period, 600 million years ago, are scarce. During this time, there were few large plants and animals on Earth.

A COMMON ANCESTOR
Life on Earth must have begun more than 3.6 billion years ago, its component chemicals seeded from space soon after the new planet cooled. But for three billion years, microscopic bacteria and algae dominated. Then, triggered perhaps by a change in climate and the release of nutrients as a super-continent broke apart, a host of larger, multi-cellular plants and animals appeared. By 600 million years ago, the ancestors of most of the groups of organisms that are around today had emerged. Among them was what may be our own distant ancestor.

Mounds of cyanobacteria, known as stromatolites, are found in the warm tidal waters of Shark Bay in western Australia. Stromatolites are the fossilized remains of some of the earliest-known living organisms on the planet. The Australian stromatolites are more than 3.5 billion years old.

SEE ALSO
38–9 Life: origins and development

EARTH'S STRUCTURE

Planet Earth has a layered structure made up of a central core, a mantle and a surface crust. Earth is constantly changing due to the dynamic forces within.

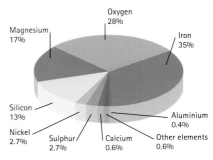

Iron is the largest single component of Earth. This metal is concentrated in Earth's molten core. Compounds called magnesium silicates, which contain magnesium, silicon and oxygen, form the bulk of the mantle. Most of these elements were formed in space millions of years ago.

This dense rock originated in Earth's mantle and contains a green mineral known as olivine. It was brought to the surface during an eruption in the Canary Isles. Geological processes such as volcanic eruptions often reveal clues about the structure of Earth's interior.

➤ Transverse (S) waves

➤ Primary (P) waves

➤ Surface waves

Unlike the rocks at the surface of Earth, the rocks deep within its centre are squeezed to such tremendous pressures and temperatures that, even though they are solid, they can flow slowly like ice in a glacier. Early in the planet's history, the densest material, mostly metallic iron and nickel, settled out to form a molten core. This core is the densest part of Earth, with a radius of over 2,900 km. Above the molten outer core is the mantle. This is made up of dense silicate rocks (containing silicon and oxygen). The ocean crust and continents float on these layers like a film of oil on the surface of water.

INSIDE THE CORE

Conditions in Earth's core are hard to imagine. Pressures are enormous and the temperature exceeds 3,000°C. Geologists can measure the temperature of the boundary between the inner and outer core. Earth's core is made up of iron mixed with certain impurities. Scientists have recreated the pressures within Earth's core and discovered the temperature there to be nearly 4,000°C.

The molten iron in the outer core is slowly circulating. Electrical currents within it generate the Earth's magnetic field. This reaches far into space, and forms a magnetic envelope around the planet's surface, deflecting electrically charged particles from the Sun and shielding us from harmful radiation. The magnetic field generated in the core probably varies enormously, but most of the variations are dampened by the mantle. However, every 100,000 years or so, they become so great that the magnetic field of the planet reverses entirely.

LOOKING INSIDE THE PLANET

The heat generated by the formation of our planet is still cooling, and it is being released from the interior as the inner core freezes and as radioactive elements decay. This heat has to escape, but rock is a good insulator. To let the heat out, the mantle rocks surrounding the inner core must circulate. Heat is carried upwards as hot mantle rock rises. At the surface, where the rock is brittle, the movement causes earthquakes. Seismologists, scientists who

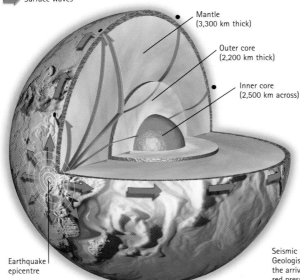

Mantle
(3,300 km thick)

Outer core
(2,200 km thick)

Inner core
(2,500 km across)

Earthquake epicentre

Seismic waves reverberate from an earthquake in eastern Africa. Geologists can work out features of the planet's structure by timing the arrival of the waves at monitoring stations around the world. The red pressure waves can pass through the molten outer core. The blue transverse waves can pass through only the solid mantle and crust.

study earthquakes, operate a network of earthquake monitoring stations. By timing the arrivals of the earthquake (seismic) waves at different stations, powerful computers can build up a picture of the interior of Earth, just as a medical scanner sees inside your body.

ACTIVITY WITHIN

Earth scans reveal plumes of hot mantle material rising towards the surface, often topped by volcanic activity. The seismic waves pass more slowly through this hot, soft material. It contrasts with cold hard rock that descends into the mantle where cold ocean crust disappears beneath continents. By analysing the seismic data, geologists have found a barrier about 670 kilometres down into the mantle. Descending rock seems to accumulate there, leading some geologists to speculate that the entire mantle is not mixing in a single circulation but that there are two layers of rock circulation.

Recent analysis of seismic data suggests that there is another thin layer at the base of the mantle, a few tens of kilometres thick. This layer is not continuous but is more like a series of giant continents on

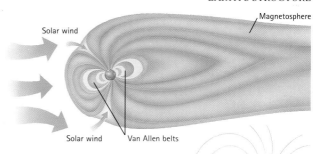

Solar wind

Solar wind   Van Allen belts

Magnetosphere

▲ The magnetic field created by Earth forms an envelope called the magnetosphere, which stretches far above the planet's surface into space. The wind of charged particles streaming outwards from the Sun pushes on the magnetosphere so that it streams out downwind like the tail of a comet.

▶ The shape of the magnetic field makes it look as if there were a huge bar magnet inside Earth. The strong lines of magnetic force are actually created by electric currents circulating within the molten outer core.

Line of magnetic force   Earth

the underside of the mantle. The slabs could have formed by the mixing of silicate rocks in the mantle with iron-rich material from the core. However, another explanation is that this region is where ancient oceans came to rest. After descending to the base of the upper mantle, the cold ocean crust compressed into an extremely dense rock layer. Then this layer could break through the 670 kilometre layer and sink further. This layer continues to spread out at the base of the mantle. As the core slowly heats the dense rock layer, it will rise once more to form new ocean crust.

WORKING OUT THE CLUES

Land compressed by ice during the last ice age, together with the pull of the Moon on the tides, is gradually slowing down our spinning planet. As a result, the lengths of the days and nights are increasing by tiny amounts. However, there are other even smaller variations of a few billionths of a second. These may be the result of atmospheric pressure on mountain ranges. More importantly, the circulation in Earth's outer core pushes on ridges and valleys, similar to upside-down mountains, in the base of the mantle. The changes in day length are a measure of the core's circulation, and provide another clue to the geological processes inside Earth.

Sea

Drill

Crust

Mantle

Drill cores can reveal the layers of rock in Earth's crust. It has not yet been possible to drill through to the mantle.

The aurora borealis fills the night sky above the Arctic Circle. Where Earth's magnetic field converges on the poles, charged particles from the Sun hit atoms in the atmosphere, creating the spectacular display. The aurora australis occurs in regions surrounding the South Pole.

SEE ALSO

14–15 Earth's atmosphere, 22–3 Earthquakes, 274–5 Earth and the Moon

EARTH'S ATMOSPHERE

A gaseous envelope called the atmosphere surrounds
Earth. It protects us from the extremes of space, keeps
us warm and is the cause of our weather systems.

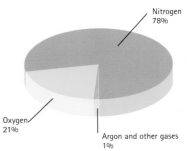

This pie chart reveals the composition of Earth's
atmosphere. Nitrogen and oxygen make up most of the
atmosphere. Other gases include argon, carbon dioxide
and methane. Human activities have increased the levels
of these gases, warming the climate considerably.

Without the atmosphere, living
organisms would not be able to
survive the constant barrage of solar
and cosmic radiation, bombardment by
meteors, and exposure to extremes of
temperature. The atmosphere protects
living things from these potentially lethal
threats by surrounding Earth with a layer
of gases, liquids and other particles,
300 kilometres thick. The force of gravity
holds the atmosphere in place. Near
Earth's surface, the atmosphere is highly
compressed, but it gets thinner with
increasing altitude. In the lower levels
of the atmosphere, winds and storms
distribute the heat from the Sun. In the
upper levels, the molecules that make up
the atmosphere collide with incoming
meteors and radiation.

A jet aircraft accelerates
through the atmosphere,
leaving a visible record
of its passage, called a
vapour trail, behind it.
The trail is supercold air
caused by the deposition
of water vapour in the
engine exhaust as tiny
crystals of ice.

A BRIEF HISTORY

For the first billion years of Earth's life,
the atmosphere was very different from
ours today. Originally, it was a mixture of
nitrogen, carbon dioxide and water vapour.
Carbon dioxide is known as a greenhouse

gas, which means it lets sunlight through
to warm the planet but prevents heat
from escaping. As a result, carbon dioxide
acted like a blanket to keep the young
Earth warm. When the first living things
evolved, they began to use up the carbon
dioxide in the atmosphere. Since the Sun
was growing stronger, a balance was
struck. In addition, the organisms released
a new gas – oxygen – on Earth. This
meant that other animals could survive
by breathing oxygen, first through gills
and eventually through lungs. In the last
billion years or so, oxygen concentrations
have stayed constant.

LAYERS OF THE ATMOSPHERE

The atmosphere does not have a definite
boundary. Space satellites orbit more than
300 kilometres above Earth's surface –
here an atmosphere exists, but it is so thin
it is almost a vacuum. This region is called
the thermosphere, where atoms are very
hot (up to 2,000°C), but so sparse they
would not burn you. The thermosphere

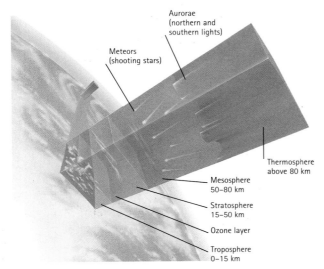

Aurorae
(northern and
southern lights)

Meteors
(shooting stars)

Thermosphere
above 80 km

Mesosphere
50–80 km

Stratosphere
15–50 km

Ozone layer

Troposphere
0–15 km

This is a section through the atmosphere from sea level
up to the beginnings of space. The troposphere makes
up the first 15 kilometres of the atmosphere and
contains the world's weather systems and major aircraft
routes. About 20 kilometres up is the protective ozone
layer, within the stratosphere which is thin and cold.
Weather balloons can rise through the stratosphere and
supersonic aircraft and clouds of volcanic ash reach it.
Above it lies the mesosphere, which includes the radio-
reflective layer of the ionosphere. The thermosphere
stretches up into space and contains the exosphere,
where gas molecules escape into space. Aurorae occur
towards the base of the thermosphere at each pole.

reaches down to about 80 kilometres above Earth's surface. Here a region called the mesosphere begins. Atoms within the mesosphere are ionized. This means that they have lost electrons and can reflect short wavelength radio waves. This area is commonly called the ionosphere and is extremely important for global radio communications. The stratosphere is the next layer, reaching down to about 15 kilometres above Earth's surface. This colder layer contains the ozone layer, a protective screen that blocks out potentially harmful ultraviolet radiation from the Sun, although it has been damaged by chemicals released by human activities. Powerful volcanic eruptions can inject dust and acidic gases into the stratosphere. The troposphere makes up the last 15 kilometres of the atmosphere and contains 80 per cent of its mass. In this part of the atmosphere, the world's weather runs its course.

A DELICATE BALANCE

The atmosphere hangs in a precarious dynamic balance. In the process known as photosynthesis, plants constantly absorb carbon dioxide and produce oxygen. Conversely, animals absorb oxygen in a process called respiration, returning carbon dioxide and other gases, such as methane, to the atmosphere. Today, human activities have transferred much of the carbon stored away in the rocks back into the atmosphere. This process is causing the world's climate to get warmer. Similarly, the ozone layer is rapidly depleting as a result of human activities, allowing harmful solar radiation to reach the surface. If we continue to disrupt the atmosphere, our planet may not be such a comfortable place to live in the future.

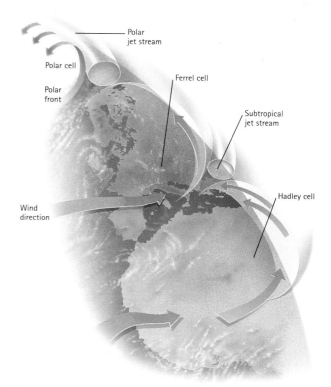

Polar jet stream
Polar cell
Ferrel cell
Polar front
Subtropical jet stream
Hadley cell
Wind direction

▲ Atmospheric circulation transfers heat to and from the equator by a series of convection cells. The first cell, called a Hadley cell, transfers warm air north over the tropics. The temperate latitudes are under the control of the Ferrel cell. Lastly, polar cells, as their name suggests, cover both poles.

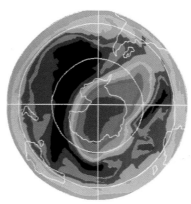

◀ For more than 20 years a hole has appeared in the stratospheric ozone above Antarctica every October. In the cold, still air of the Antarctic winter, chlorine-containing chemicals, called chlorofluorocarbons, break down the ozone. The hole is observed here in satellite data from space.

A scientist studies data at the observatory high on Mauna Loa in Hawaii. Pulses of laser light are used to measure the amount of dust, volcanic gas and ozone in the stratosphere above.

SEE ALSO

28–9 Climate,
44–5 Plant anatomy

THE OCEANS

More than 70 per cent of Earth's surface is covered by water. Around two per cent of that water is ice; less than one per cent is freshwater and water vapour.

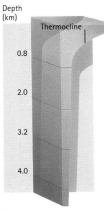

Depth (km)

Thermocline

0.8

2.0

3.2

4.0

Water at the surface of the ocean is warmed by the Sun. Waves mix this warm water with cooler water to a depth of about 100 metres. Beneath this, there is little mixing and the temperature falls rapidly. The boundary between the warm (pink) and cool (blue) layers is called the thermocline. This boundary prevents nutrients in deep water from moving upwards.

⟶ Cold currents
⟹ Warm currents

Four billion years ago, Earth's surface was too hot for water to exist as liquid. Water that was erupted as steam in volcanic gases boiled away to be lost in space. By around 3.85 billion years ago, Earth had cooled to form an atmosphere of volcanic gases, including steam. Then, water started to condense and form oceans in dips in Earth's surface.

Since the oceans formed, rain has been falling on land and washing salt from rocks into the sea. This is why seawater tastes salty. On average, 2.9 per cent by weight of an ocean is salt. Seas such as the Baltic, with plenty of freshwater from rivers and little evaporation, are less salty. The Dead Sea, where evaporation is rapid, is six times more saline (salty) than average.

UNDER THE SURFACE

When we look out across the sea, or even when we sail or swim in it, we are only aware of its surface. But the average depth of the oceans is around 5,000 metres, and the deepest ocean trenches reach down 11,000 metres. Mount Everest is more than two kilometres shorter than these trenches are deep.

The top few metres of the oceans can be as warm as 26°C in the tropics. They take in heat from sunlight during the day and warm the atmosphere at night. This layer of the oceans contains more heat

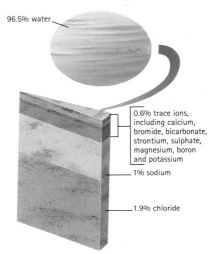

96.5% water

0.6% trace ions, including calcium, bromide, bicarbonate, strontium, sulphate, magnesium, boron and potassium

1% sodium

1.9% chloride

Average seawater contains 96.5 per cent water. Almost three per cent is sodium chloride, which gives seawater its salty taste. There are traces of many other salts.

than the entire atmosphere. Where there are dissolved nutrients in the sunlit waters, extensive blooms of tiny marine algae or phytoplankton flourish. But the warm water floats on the top of the ocean, and very often nutrients are scarce unless they are washed down in rivers or stirred up from deeper water. Beneath the sunlit or photic zone lies a very different world of cold, dark water. Yet it supports a rich diversity of life. The oceans provide food for millions of people. They also conceal rich deposits of oil, gas and minerals.

OCEAN CURRENTS

Heat circulates in the oceans in a series of circular currents or gyres. Blown by the winds, they tend to flow clockwise in the Northern Hemisphere and anticlockwise in the Southern Hemisphere. This pattern is disrupted by the continents.

The currents of water at the surfaces of oceans are driven by prevailing winds. These currents move in swirls, called gyres. Most warm currents start near the equator, and most cool currents start near the poles. The Gulf Stream and North Atlantic Drift carry warm water from the Straits of Florida towards western Europe and Scandinavia. The Antarctic gyre is a circular current of cold water that flows clockwise around the South Pole.

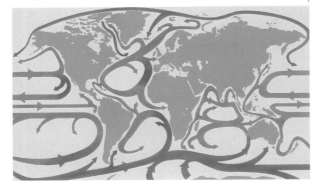

THE POWER OF WAVES

Waves begin in the open ocean as simple up-and-down oscillations, driven by the wind. As the wave advances, the water moves in a circular motion. As the wave enters shallow water near the shore, the lower part of that motion is slowed down and the crests of the waves break or topple over, pounding onto the shore. The forward motion, or swash, drives sand and gravel up the beach, the backwash pulls it back down. Where waves meet the shore at an angle, material drifts along the beach.

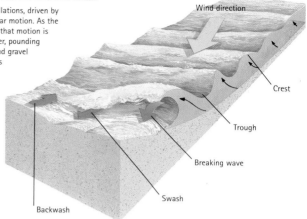

Wind direction

Crest

Trough

Breaking wave

Swash

Backwash

Where large waves from the open ocean meet the shore, spectacular breakers can build up, offering an exciting and sometimes dangerous challenge to surfers.

A map of the main ocean currents at the surface does not reveal the deep circulation that also takes place. The Gulf Stream and the North Atlantic Drift bring warm waters from the Gulf of Mexico northeast across the Atlantic. This keeps western Europe and the British Isles warm. As the warm water flows along, some of it evaporates and the water gradually becomes cooler and more salty. This makes the water denser. Eventually, it becomes too dense to stay on the surface. The water then dives down and turns south to complete the circulation like a conveyor belt. Were this conveyor belt to stop, western Europe would have winters as cold as those in northeast Canada.

EXPLORING THE DEPTHS

The deep oceans are the least explored parts of our planet. Going there in a submarine or even sending robot craft can be as complex as mounting a space mission. The crushing pressures on the ocean floor are far greater than those encountered by any spacecraft.

Strange worms, blind shrimps and giant squids have been found in deep oceans, as well as hydrothermal vents, which eject water at up to 350°C. Living bacteria that have been found in deep-ocean sediments suggest that more than one tenth of life on Earth lies in the mud and rock beneath the ocean floors. It is possible that these life forms are relatives of the first species to colonize the planet.

The Dead Sea lies on the border between Israel and Jordan. Its water comes mainly from the River Jordan. The sea has no exit, so water can only escape by evaporation. This process concentrates salts in the sea, making it a hostile environment for marine species. The salt makes the water so dense that the human body floats easily on its surface.

◀ This satellite image shows the green plant pigment chlorophyll in phytoplankton in the oceans. The concentration of plankton in the oceans (shown as yellow and green here) is highest where warm water and plentiful nutrients mix and produce blooms of plankton.

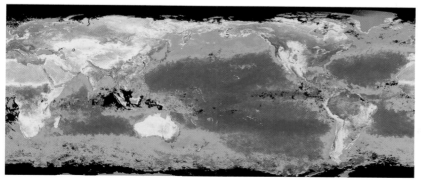

SEE ALSO

14–15 Earth's atmosphere, 18–19 Continental drift

CONTINENTAL DRIFT

Earth's land masses move around the surface of the planet. Over millions of years, they have broken up and joined together to form the present continents.

The solid surface of Earth consists of slabs of ocean crust and continental crust. Both types of crust float on the denser rock of the mantle. Although the mantle rock is solid, the pressure and temperature in the mantle make it flow like a thick paste.

The transfer of heat from Earth's core makes the mantle rock rise from the core towards the surface and sink back again, just as water swirls around a pan when it is placed on a hot stove. This process is called convection. As the surface of the mantle rock moves, it causes the continents to drift slowly around Earth's crust.

CONTINENTAL CRUST

Ocean crust is a layer of dense basalt just six or seven kilometres thick, perhaps with sediments on top. Continental crust is much thicker, averaging 30 kilometres and reaching 60 kilometres in mountain ranges. It is less dense than ocean crust, being mostly silica-rich granite and sediment, and it floats like a scum on Earth's surface. Like a floating iceberg, the higher the mountains, the deeper the continent's roots sink into the mantle.

Ocean crust is continually being created and destroyed, so its sediment is relatively new. But continental material has been accumulating ever ince Earth's surface

▲ 200 million years ago, there was only Panagaea, a supercontinent (1). 100 million years ago, North America was splitting away from Europe. South America from Africa (2). By around 80 million years ago, Africa was about to collide with Europe, and India was joining Asia (3). 50 million years from now, Africa is likely to have joined Europe, and North America will have separated from South America, and become joined to Asia (4).

→ Direction of plate movement

⌒ Divergent boundary

⌢ Convergent boundary

Transform fault

This view of the Andes was taken from a NASA space shuttle. The mountain range runs 8,900 kilometres along the western coast of South America. It formed when the edge of the continental plate buckled under pressure from the Pacific plate, which lies to its west.

solidified. The material at the centre of some continents, such as Australia and North America, is up to four billion years old. A continent grows thicker as sediment piles up around its edges, volcanoes erupt new rocks onto it and molten rock is injected into its base.

Where the base pushes lowest into the mantle, it heats up and can begin to melt. Granite is the result of this melting. Lubricated by moisture in the rocks that formed it, great masses of granite can rise through a continent like giant bubbles of liquid. The granite bakes the rock around it as it solidifies to form great masses of crystalline rock. Where these rocks are exposed by surface erosion, they produce granite moors such as Dartmoor in England.

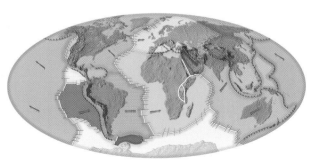

The tectonic plates of Earth's crust are shown in different colours here. The red arrows on this map show in which direction each plate is moving. Some plates are drifting apart, some are moving together and others are grinding past one another. Most volcanoes and earthquakes happen at the edges of tectonic plates.

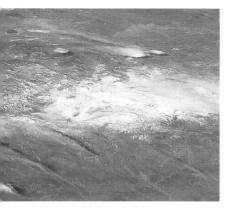

◄ This photograph looks southeast along the San Andreas fault in California, in the United States. The Pacific plate, right, is gradually moving northwest relative to the North-American plate.

Fault line

▲ The stars on this map show the epicentres of past earthquakes along the San Andreas fault. It is difficult to forecast where or when the next earthquake will happen.

PLATE TECTONICS

The slabs of continental and ocean crusts that drift around Earth's surface are called plates. Plate tectonics is the theory of how these plates have split, moved and collided to form Earth's surface as it is now.

Tectonic plates do not move rigidly across Earth's surface. Continents can stretch as they move. When this happens, the crust becomes thinner and the surface level drops. The Great Rift Valley, which runs from Syria to Mozambique, formed when the African continent stretched at a weak point. The North Sea formed when the European continent stretched. If a continental plate stretches too much, it can break up. New ocean crust then forms between the fragments.

When continental plates collide, they buckle at the edges and form mountain chains. The Alps formed when Africa collided with Europe. The Himalayas, which formed when India crashed into Asia, are still rising as the plates push together.

FAULT LINES

A boundary between two tectonic plates is called a fault. When neighbouring plates move in different directions, they grate against each other. This happens at the San Andreas fault, close to the coast of California. If the movement at the fault sticks for months or years, enormous stress can build up in the surrounding rocks. An earthquake results when the fault finally gives and releases stress.

Although continents drift only a few centimetres per year, they can be tracked by lasers that measure movement at fault lines, or by satellite monitoring. Geologists can trace the history of continental drift using the magnetism of volcanic rocks, which record the magnetic field at the time and place where they solidified.

This photograph shows a pile of granite on a tor in Dartmoor, England. Granite solidified from molten rock that had bubbled up from Earth's mantle. With time, wind and rain wore away the hill and left the hard granite exposed.

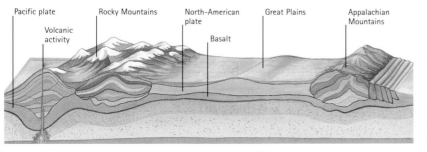

Pacific plate

Volcanic activity

Rocky Mountains

North-American plate

Basalt

Great Plains

Appalachian Mountains

◄ The North-American continental plate. Pressure from neighbouring plates has buckled the edges of the continent, forming vast mountain ranges. There are volcanoes in the west, where the Pacific plate dives under the continental plate.

SEE ALSO

12–13 Earth's structure, 22–3 Earthquakes

VOLCANOES

A volcano is an opening in Earth's crust through which molten lava, ash and gases erupt. In many cases, lava and ash form a mountain around the opening.

Fissure volcanoes erupt through narrow slits.

Shield volcanoes ooze runny lava over wide areas.

Dome volcanoes are formed by sticky lava.

Conical volcanos form from volcanic ash.

Composite volcanoes may have several side vents.

Calderas are formed by explosive eruptions.

▲ The shape of a volcano depends on the type of lava and how it erupts. Cone-shaped volcanoes are formed from layers of ash and cinders that have sprayed from a central crater. Broad calderas are formed when a volcano first bulges then erupts in a violent explosion.

It used to be thought that volcanoes leaked molten rock and gases directly from Earth's core. That is not the case. As hot, solid rock rises in the mantle, the pressure drops and a small part of the rock begins to melt. This liquefied rock, called magma, is less dense than solid rock. It squeezes out from the solid like water from a sponge. The rising magma creates wide channels in the crust as it forces its way to the surface. When it breaks through the surface, the pressure drops. Gases dissolved in the magma force it to erupt through the opening as lava.

TYPES OF VOLCANOES

The behaviour of a volcano depends on the type of magma that fuels it. Volcanoes such as those near Hawaii and Iceland are sitting on top of a rising plume of hot mantle rock, called a hot spot. The lava that erupts from these volcanoes comes from great depths, sometimes more than 150 kilometres into the mantle. Its composition is not the same as the mantle, since only a tiny fraction of the mantle rock melts. This lava is runny when molten and sets as dense, black basalt. Because the lava is so runny, it can pour out through fissures at vast rates and flow across the land at speeds of up to 50 kilometres per hour. Where this type of volcano erupts under water, the lava cools quickly and

This photograph shows Mount Etna in Sicily, erupting at night. Gas eruptions spray lumps of molten rock from a crater, resulting in a spectacular fiery fountain.

builds volcanic islands as it sets. Where gas bubbles through it, the runny lava erupts in spectacular fountains. Since this type of lava flows freely, eruptions are smooth rather than explosive.

A different type of volcano is found where ocean crust dives under the edge of a continent. The ocean crust partly melts to form a sticky lava that is rich in silica and contains some water. During an eruption, the sudden drop in pressure causes the water to turn to steam. This results in an explosion of fine ash and hot gases. This mixture, which can race down the flanks of a volcano at 200 kilometres per hour, is called a *nuée ardente* in French, meaning 'glowing avalanche'.

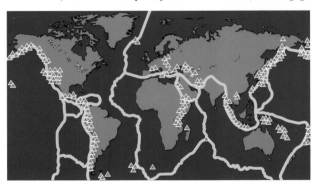

◀ This map shows the sites of active volcanoes as pink triangles. Most volcanoes are located near plate boundaries (yellow). Some are located over hot spots in Earth's mantle.

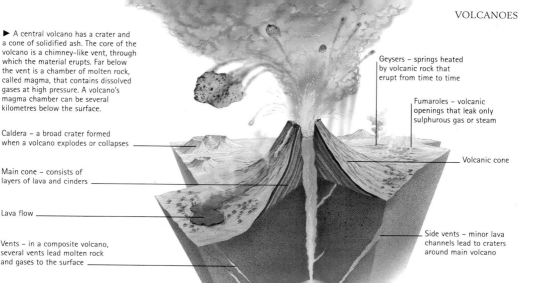

▶ A central volcano has a crater and a cone of solidified ash. The core of the volcano is a chimney-like vent, through which the material erupts. Far below the vent is a chamber of molten rock, called magma, that contains dissolved gases at high pressure. A volcano's magma chamber can be several kilometres below the surface.

Geysers – springs heated by volcanic rock that erupt from time to time

Fumaroles – volcanic openings that leak only sulphurous gas or steam

Caldera – a broad crater formed when a volcano explodes or collapses

Volcanic cone

Main cone – consists of layers of lava and cinders

Lava flow

Side vents – minor lava channels lead to craters around main volcano

Vents – in a composite volcano, several vents lead molten rock and gases to the surface

Magma chamber

LIVING WITH VOLCANOES

With their combinations of red-hot lava, toxic gases and suffocating ash, volcanic eruptions can be deadly phenomena. But people continue to live on the flanks of volcanoes despite the dangers. This is because volcanic soil is often fertile and eruptions can be few and far between, giving a false sense of security. The consequences can be disastrous. In 1902, when Mount Pelée on the Caribbean island of Martinique erupted, a *nuée ardente* raced down the mountain and engulfed the port of San Pierre. More than 29,000 people were killed. The only survivor was a prisoner in an underground cell. In AD79, a similar type of eruption from Mount Vesuvius smothered the Roman towns of Boscoreale, Herculaneum, Pompeii and Stabiae with mud and ash.

It is possible to predict at least some eruptions by monitoring volcanic gases and measuring changes in gravity as molten lava rises inside a volcano. Sometimes, the whole mountain bulges. When Mount St Helens, Washington, USA, started to bulge in 1980, most people were evacuated before the mountain blew its top. A huge landslide removed part of the volcano, exposing the pressurized molten lava. The lava then exploded sideways and upwards. The blast hurled almost a cubic kilometre of rock into the air and flattened trees up to 30 kilometres away.

On 18 May 1980, Mount St Helens in Washington, erupted with enormous force. The blast of fast-moving dust and hot gases devastated the landscape, killing all forms of life and snapping great pine trees like matchsticks.

These photographs show Mount St Helens weeks before the 1980 eruption (far left) and during the eruption (left). Almost one cubic kilometre of rock was pulverized and blasted into the air.

SEE ALSO

12–13 Earth's structure,
18–19 Continental drift

EARTHQUAKES

Earthquakes are caused by the sudden release of stress in rocks deep below Earth's surface. This happens mostly at the borders between tectonic plates.

In a seismometer, a roll of paper turns slowly under a pendulum. If the ground shakes, a pen attached to the pendulum records the intensity of the motion.

The San Andreas fault in California is a typical earthquake zone. It is where the Pacific crust and the North-American plate meet. The Pacific crust is moving northwards at an average of 34 millimetres per year, but the motion at the San Andreas fault is not that smooth. It can get stuck for months or years. When this happens, stresses build up along the fault. The longer the wait, the greater the stress. When the fault gives, the ground on either side of the fault can shift by up to 12 metres in a short time, sending shock waves through the surrounding ground.

The shock waves of an earthquake can shake buildings until they collapse. The ground movements can destroy roads, railway lines and underground pipes. The gas and water that leak from these pipes add fire and flooding to the side effects of an earthquake.

When an earthquake happens at sea, the rapid shift in water level forms tidal waves, tsunamis, that travel quickly over great distances. As they enter shallow water, they slow down and can rise ten or more metres above normal sea level. A tsunami can destroy any building in its path and cause terrible flooding.

MAJOR EARTHQUAKES			
Location	Year	Magnitude	Deaths
Taiwan	1999	7.7	2,400
Turkey	1999	7.8	17,118
Afghanistan	1998	6.1	4,000
North Iran	1997	7.1	1,560
Russia	1995	7.5	2,000
Japan (Kobe)	1995	7.2	6,310
South India	1993	6.4	9,748
Philippines	1990	7.7	1,653
Northwest Iran	1990	7.5	36,000
San Francisco	1989	7.1	275
Armenia	1988	7.0	25,000
Mexico City	1985	8.1	7,200
North Yemen	1982	6.0	2,800
South Italy	1980	7.2	4,500
Northeast Iran	1978	7.7	25,000
Tangshan, China	1976	8.2	242,000
Guatemala City	1976	7.5	22,778
Peru	1970	7.7	66,000
Northeast Iran	1968	7.4	11,600
Nanshan, China	1927	8.3	200,000
Japan	1923	8.3	143,000
Gansu, China	1920	8.6	180,000

The blue dots show the sites of past earthquakes. Most earthquakes occur along the fault lines at the boundaries between tectonic plates, shown here as green lines.

PREPARING FOR EARTHQUAKES

Tiny variations in roughness in a fault or the presence of lubricating water are enough to set off an earthquake. It is almost impossible to predict exactly when and where a major earthquake will strike, but it is possible to deal in probabilities. The regions of the world most at risk of earthquakes are concentrated in thin bands along the major faults between the tectonic plates of Earth's crust. In these areas, seismologists can predict with near certainty that there will be a major earthquake at some time in the future.

If an area is known to be earthquake-prone, shelters can be built for protection in the event of an earthquake. Buildings can be designed to sway rather than shatter, and rubber in their foundations can absorb some of an earthquake's force. But these precautions are expensive, and many buildings – particularly in poorer countries – are built without them. This is why earthquakes of similar magnitudes can kill tens of thousands in one part of the world but very few in another.

The only way to completely avoid the risk of death in an earthquake is to evacuate the area before it strikes. However, the inconvenience and cost of evacuating a city are enormous. This is why scientists search for ways to predict earthquakes. Sometimes, but not always, slight earth tremors can indicate that a

▶ The Richter and Mercalli scales record earthquake magnitudes. The Richter scale measures the energy of a quake, and the Mercalli scale charts destructive effects.

◀ Destruction caused by the 1989 earthquake at Loma Prieta near San Francisco, California. This photograph shows the collapsed double-decker Nimitz freeway on the east side of San Francisco Bay. A total of 275 people died in this quake. Some 25,000 people died in Armenia when a quake of a similar magnitude hit poorly constructed buildings in 1988.

Richter below 3; Mercalli I. Detected by seismographic instruments, but too weak to be felt by people.

Richter 3–3.4; Mercalli II. Detected by instruments and a few people. Delicate objects may shake.

Richter 3.5–4; Mercalli III–IV. Obvious shaking felt indoors. Walls crack, hanging objects swing.

Richter 4.1–4.8; Mercalli V. Felt by most people. Some windows may crack and loose objects fall over.

Richter 4.9–6; Mercalli VI–VII. Felt by all. Furniture moves. Some chimneys topple.

Richter 6.1–7; Mercalli VII–IX. Some houses collapse, roads crack and pipes rupture.

Richter 7.1–8.1; Mercalli X–XI. Large cracks form in the ground. Few buildings remain standing.

Richter more than 8.1; Mercalli XII. Total destruction. The ground rises and falls in waves.

major earthquake is due. Attempts have been made to record other warnings such as the level of water in wells, releases of gas and animal behaviour. In 1975, the city of Haicheng, China, was evacuated hours before a massive earthquake on the basis of such warnings. But a year later, in Tangshan, 240,000 people died in a quake for which there had been no warning.

It is now possible to give warnings of a few tens of seconds. Earthquake detectors along a fault zone can sense the start of a major earthquake and send radio messages to control centres in nearby cities. The warning time is not enough to evacuate a city, but a 30-second warning can allow time for computers to save their data, for lifts to open their doors to release people, and for fire engines to move into the open where they will not be damaged. It also helps operators make industrial processes safe before the shock hits.

In January 1995, more than 6,000 people died in an earthquake and the resulting fires in Kobe in Japan. Poorly enforced building regulations and poor-quality concrete construction were largely responsible for the death toll being so high.

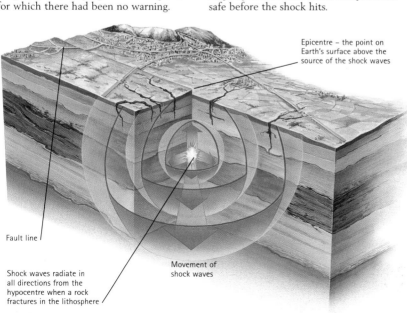

Epicentre – the point on Earth's surface above the source of the shock waves

Fault line

Shock waves radiate in all directions from the hypocentre when a rock fractures in the lithosphere

Movement of shock waves

◀ When an earthquake strikes, shock waves radiate from a point beneath the ground.called the hypocentre. The point on Earth's surface directly above the hypocentre is called the epicentre. Both pressure waves and shock waves can radiate from the hypocentre, causing cracks in the ground and destroying buildings.

IGNEOUS ROCK

Igneous rocks are all born from fire, and most of them originate deep within Earth. Slowly, igneous rocks rise towards the surface in a molten or semi-molten state.

This cliff face contains layers of pumice and obsidian, both of which are extrusive igneous rocks.

A microscope reveals the mineral crystals in this thin section of magnified porphyritic granite.

Black obsidian glass, of volcanic origin, is covered with white 'snowflakes'.

Igneous rocks all form from molten magma. They are classified by their texture, composition and origin. Acidic igneous rocks tend to be light in colour and low in density. They contain plenty of silica or quartz, often with various minerals of the feldspar group. Basic igneous rocks are more dark and dense and contain combinations of olivine, pyroxene and hornblende or amphibole. Ultrabasic rocks are very dense and are close in composition to the upper mantle.

DIFFERENT IGNEOUS ROCKS

Igneous rocks are either intrusive or extrusive. Intrusive rocks are pushed up beneath overlying rocks as large masses that have already solidified. Extrusive rocks erupt from volcanoes or into the dykes and sills associated with them.

The texture of igneous rocks depends on how fast they cool. If they cool slowly in big intrusions, large crystals may grow. Sometimes these settle out to form coarse, crystalline pegmatite or get caught up as large phenocrysts in finer-grained material.

If the magma cools rapidly, the grains will be fine or even glassy. They may trap bubbles of gas or even large fragments of surrounding rock (xenoliths).

Basic igneous rocks may be formed by the melting or partial melting of the upper mantle, particularly where a mantle plume rises under a hot spot. When this magma erupts from a volcano and cools rapidly, it forms finely grained basalts, which contain feldspar, mica and hornblende, and is very dark in colour. The same mixture injected underground, between rock layers, cools more slowly, forming larger crystals called dolerite. Even slower cooling, over millions of years and at greater depths, makes an even coarser form called gabbro. This still had the same overall composition.

When the ocean crust sinks under a continent, parts of both melt to make a magma that is more acidic and richer in silica. Water and carbon dioxide make the magma flow more easily, but silica makes it thick. Eruptions of this material form andesite, a paler form of basalt that is rich in silica. Millions of years later, when the continent has been thickened by volcanic eruptions, the base of the continent may begin to melt. This produces intrusions of silica-rich granite.

▶ This is a section through an intrusion of granite, showing a batholith with dykes and sills around it. Sometimes the intrusion takes the shape of a dome and the surrounding sedimentary rocks produce a laccolith. As the rocks wear away, the granite is exposed.

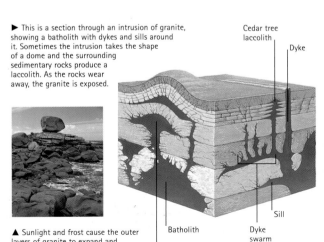

Cedar tree laccolith

Dyke

Batholith

Dyke swarm

Laccolith

Sill

▲ Sunlight and frost cause the outer layers of granite to expand and contract to make rounded boulders.

Devil's Tower in Wyoming in the United States is the plug or vent of a basalt volcano.

SEE ALSO

12–13 Earth's structure, 20–21 Volcanoes, 25 Metamorphic rock

METAMORPHIC ROCK

Metamorphic rocks have been transformed from other rocks by the enormous pressures and temperatures that exist thousands of metres below Earth's surface.

Marble is metamorphosed limestone. It makes a popular decorative building stone. This slab forms part of the exterior of the Stock Exchange building in London.

This sample of gneiss contains coarse crystals and looks rather like granite. However, the texture of this sample is due to the alignment of grains under pressure.

Schist is formed by regional metamorphism. Crystals of mica are aligned by high pressure. Some of the mica in this piece of schist from the Austrian Alps has turned into garnet.

When igneous and sedimentary rocks are subjected to enormous pressures and high temperatures, their structure and sometimes their chemical make-up can be changed. Hot, percolating fluids may add or remove different minerals. These conditions can make existing minerals take on different forms. Metamorphic forces can exceed 100 atmospheres and 400°C. Under these conditions, layered minerals may become distorted, stretched or destroyed. Accordingly, metamorphic rocks are classified according to their texture, composition and source.

CLASSIFYING THE ROCK

There are two types of metamorphism. The first is contact metamorphism, where hot intrusions of igneous rock bake the surrounding country rocks. The second is regional metamorphism, where much larger volumes of rock become squeezed, buried and heated, for example in continental collision zones.

Metamorphism will bind the grains of sandstone together to make quartzite. It will also turn limestone into marble. As the calcium carbonate recrystallizes in a marble, impurities tend to get forced out of the crystals. This often forms bands and makes the marble highly decorative. Some of the world's most famous statues were carved in the famous white Carrara marble, which is quarried in Italy.

Regional metamorphism can take a

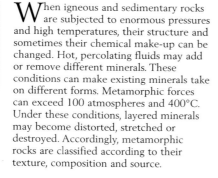

Slate is metamorphosed from shale largely by pressure. The mineral grains are aligned, making it possible to split slate into thin sheets. This quarry in Snowdonia, Wales, used to produce slates used in the roofing industry.

sedimentary rock, such as shale, through a series of changes. A regional compression re-forms many of the clay minerals into thin, flat grains of mica. These tend to align at a perpendicular angle to the direction of compression.

This process can form slate which can be split into flat sheets along the grain, in a direction that may bear no relationship to the original sedimentary bedding. If the compression continues, the slate changes to phyllite, with more micas and other higher-pressure minerals such as garnet. As the pressure and temperature increase, the mineral layer becomes more and more distorted, and this produces foliated schist. As the temperature rises further, the minerals become coarser-grained, producing gneiss. It may be gneiss that melts at the base of continents, rising to produce granite and completing the cycle.

This diagram shows the two different forms of metamorphism. The fold mountains to the left of the fault have piled up a thick layer of rock, squeezing them so that their mineral grains have realigned and turned shale into slate. The deepest, hottest layers of shale and slate have turned into schist and gneiss. On the right, an igneous intrusion has baked the surrounding rocks, causing contact metamorphism and turning limestone into marble.

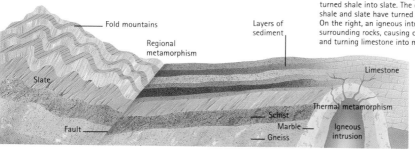

Fold mountains — Layers of sediment

Regional metamorphism

Slate

Fault

Schist

Marble

Gneiss

Igneous intrusion

Limestone

Thermal metamorphism

SEE ALSO

12–13 Earth's structure, 24 Igneous rock, 26–7 Sedimentary rock

SEDIMENTARY ROCK

Earth is carpeted by layers of sedimentary rocks.
These consist of chemical and biological precipitates,
fossils and the weathered fragments of other rocks.

These blocks of calcareous mudstone, also known as marl, were deposited deep beneath the sea millions of years ago.

Soft clay is formed when rocks were broken down into particles and sedimented by the wind, water or glaciers.

▼ This imaginary landscape depicts some of the conditions in which layers of sedimentary rocks form in the earth.

Rocks are forming all the time, and not just in the eruptions deep beneath Earth's surface. Some rocks, called sedimentary rocks, are the result of the consolidation of tiny particles called sediment that has accumulated in layers.

DIFFERENT SEDIMENTS

There are three main types of sedimentary rocks: clastic, biogenic and chemical. Clastic rocks form when pre-existing rocks are broken down into small particles that are redistributed by the wind, water or glaciers. These rocks are classified on the basis of their grain size, from large rocks to the finest clays. The grains can be rounded and worn, or angular and broken. They can be loose or unconsolidated, compressed together or cemented with materials dissolved in ground water such as calcite, silica or iron oxide. Over 75 per cent of all sedimentary rocks are clastic.

Chemical sediments form as a result of physical and chemical processes. They can be precipitated from solution in sea water, as is the case of flint and chert, forms of silica chemically similar to quartz. They can form when salty lakes or shallow seas evaporate; examples include gypsum and rock salt. They can also be formed by

The chalk cliffs of southern England are fine-grained deposits of limestone formed from the shells of tiny marine organisms that lived some 70 million years ago.

leaching, when ground water dissolves and re-deposits them – for example, bauxite.

Limestone can form from the chemical precipitation of calcium carbonate, or it may be biogenic, that is made up of the skeletons of millions of microscopic organisms, as is the case with chalk. Biogenic sediments include fossil fuels such as coal, which is the compressed remains of plants, and oil, made from buried organic material by bacteria.

THE WEATHERING PROCESS

Weathering is a complex process whereby rocks are broken down into sediments. Chemical weathering occurs when rocks are affected by water, carbon dioxide and organic acids and is accelerated by warm temperatures. Physical weathering occurs when rocks are fractured and broken

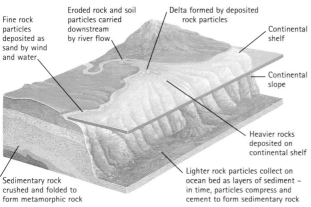

Fine rock particles deposited as sand by wind and water

Eroded rock and soil particles carried downstream by river flow

Delta formed by deposited rock particles

Continental shelf

Continental slope

Heavier rocks deposited on continental shelf

Lighter rock particles collect on ocean bed as layers of sediment – in time, particles compress and cement to form sedimentary rock

Sedimentary rock crushed and folded to form metamorphic rock

apart, for example, by a freeze-and-thaw action. Water that seeps into the cracks of the rock during the warmer daytime temperatures expands as it freezes at night. This shatters the rock.

TRANSPORTING SEDIMENT

Most of the sediments that form rock are transported by rivers. For example, the Mississippi River delivers around 180 million tonnes of sediment each year to the Gulf of Mexico. Some sediment settles on the river bed, while some is deposited at the mouth of a river, forming what is known as a river delta. Yet more sediment is carried to the ocean floor.

Sediment can also be carried by the wind and glaciers. In every transportation process, the sediment gets sorted out by size in a process called sedimentary differentiation. Coarse sediment is difficult to move and is found only in rapid high-energy flows. Very fine-grained muds, however, can be shifted many hundreds of kilometres or deposited in quiet water, such as shallow lakes and deep seas.

RECORDING EARTH'S HISTORY

More than a billion years of Earth's history is recorded in layers of sedimentary rocks. In the Grand Canyon in Arizona there is a spectacular sequence of horizontal sedimentary rock layers, called strata, 1,500 metres deep and nearly as many millions of years old. Fossils within the strata record the development of different forms of life, from the first corals and

worms to fish, dinosaurs and mammals. The rock types reveal the conditions at the time they were formed. Coarse conglomerates of rounded pebbles record high-energy, fast-moving rivers. Sandstones indicate ocean shore and deltas. Mudstones indicate sluggish waters and limestones must have been laid down in warm shallow seas teeming with life.

Linking present-day deposits from different places involves painstaking research, comparing the fossils in rocks and getting date estimates for convenient markers such as lava flows. However, this research has enabled geologists to piece together the history of the development of the land, seas and life itself.

Different colours of sandstone, worn smooth by glacial erosion, have formed an attractive pattern of lines in the smooth sides of hills in Paria Wilderness Area, Arizona. The curves in the rocks are the result of the actions of wind and water.

Sandstone is often found in brown, pink or red layers. The colour is due to varying amounts of iron oxides that bind the sediments together.

◄ The mineral content of the sandstone determines its red or yellow colour .The grey sedimentary rock at the front of this picture is called greywacke, and dates from the late Triassic period – approximately 210 million years ago.

◄ Limestone is made up of calcium carbonate and generally originates as a deposit of marine animal skeletons on the seabed. Acid rain dissolves limestone to produce a pavement-like structure similar to this one near the coast in England.

▶ Layers of sedimentary rocks in a cliff face in Zion National Park, Utah, date from the Triassic period. The red and yellow layers are deposits of sandstone. The grey layer is called greywacke, a deposit formed as a result of underwater landslides.

SEE ALSO

10–11 Fossils and geological time, 12–13 Earth's structure, 28–9 Climate

CLIMATE

The Earth's climate depends on energy from the Sun and the ability of Earth's oceans and atmosphere to circulate the heat efficiently around the planet.

Cold air →
Cool air →
Warm air →

Warm air rises at the equator where sunlight is strongest. From there, it moves towards the poles, drawing cool air behind it.

The rainforest vegetation helps keep global warming under control by absorbing carbon dioxide from the air as it grows.

▼ The eight main climate zones have different mean temperatures and rainfall rates. These differences influence the type of vegetation found in them.

The heat that drives Earth's weather systems comes from the Sun. As sunlight shines onto Earth, some of it is reflected back into space by bright white clouds and ice caps. The rest is absorbed by the land and sea, which get warmer and radiate heat as infrared light. While visible light shines freely through clear air, gases such as carbon dioxide reflect infrared back to Earth. This is called the greenhouse effect. Carbon dioxide in the atmosphere traps heat in the same way that glass in a greenhouse does. Without this effect, the average temperature on Earth's surface would be about −15°C. Ice would cover the planet and human life would not exist.

The amount of sunlight that reaches Earth's surface varies with latitude. Sunlight is most intense at the equator and weakest at the poles. Winds tend to even this out, carrying warm air to higher latitudes. The climate is also affected by the ocean currents. The top two metres of the sea store more heat than the entire atmosphere. Ocean currents carry some of this heat towards the poles, and the Gulf Stream and North-Atlantic Drift give Northern Europe its mild weather. Away from the oceanic heat stores, the interiors of continents have more extreme weather, with cold winters and hot dry summers.

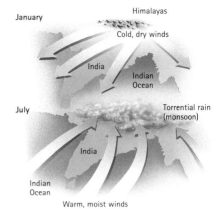

January

Himalayas

Cold, dry winds

India

Indian Ocean

July

Torrential rain (monsoon)

India

Indian Ocean

Warm, moist winds

In winter, the sea is warmer than the land and the air over it rises, drawing cold, dry air south from the Himalayas. In summer, the air over northern India becomes very hot and rises. This causes warm, moist air to move northwards from the ocean, bringing heavy rain to southern Asia – the monsoon season.

CLIMATE CHANGE

For all the variations of weather from day to day and year to year, average world temperatures have not changed by more than half a degree in the last century.

Some climate changes have natural causes. In the 17th century, for example, there were no reported sunspots. Overall, the Sun was cooler and this resulted in what has been called the Little Ice Age. At that time, rivers in Europe froze in winter and frost fairs were held on the ice.

Further back in time there were some much greater variations. There is evidence of this in many forms. Tree rings record good and bad growing seasons, including a series of bad winters around 1450BC, which may have been caused by volcanic eruption blocking out sunlight. The record goes back even further in ice cores from Greenland and Antarctica, and in layers of sediments in lakes. Sediment cores from the oceans provide the longest history. Variations in the ratios of oxygen isotopes can reveal the ocean temperature and the amount of ice at the poles. Carbon isotopes in shells reveal how much carbon dioxide was being drawn down from the atmosphere. Around 50 million years ago

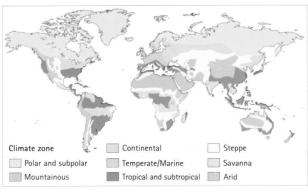

Climate zone

Polar and subpolar
Mountainous

Continental
Temperate/Marine
Tropical and subtropical

Steppe
Savanna
Arid

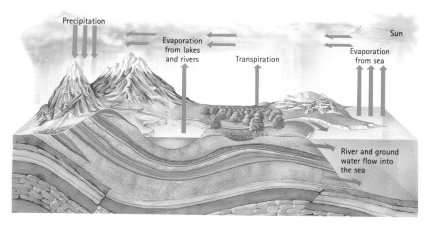

Heat from the Sun drives water in a cycle around the planet. It causes water to evaporate from lakes, rivers and the sea, and plants to transpire, or 'sweat', water into the atmosphere. Warm, moist air cools as it rises over hills and mountains, or when it meets cold air. The cooled air can no longer hold all of its water as vapour, so droplets of water form as clouds and rain. Rain feeds streams and rivers and soaks into the soil as groundwater. In these forms, water returns to lakes, forests and the sea, completing the cycle.

there was probably no ice on Earth apart from on high mountains. Antarctica was covered with vegetation. Before that, for more than 100 million years, Earth's climate was much warmer than it is today and dinosaurs populated the planet.

GLOBAL WARMING

Not all climate changes are natural. Since 1958, scientists have been monitoring the concentration of carbon dioxide in the atmosphere high on a mountain in Hawaii, far away from any sources of pollution. Every year, the concentration has been getting higher. The additional carbon dioxide comes mainly from the burning of fossil fuels, such as coal, natural gas and fuels from oil. Burning releases into the atmosphere carbon that was stored in the tissues of organisms that lived many millions of years ago. Also, the burning

of forests in the Amazon releases carbon dioxide and destroys trees and vegetation that would otherwise consume carbon dioxide. Each year, human activity adds about eight billion tonnes of carbon to the atmosphere in the form of carbon dioxide. During the 20th century, the additional carbon dioxide increased the greenhouse effect enough to raise the average global temperature by 0.5°C.

Meteorologists use supercomputers to model the climate and make predictions of future climate changes. The results suggest that the climate is likely to warm by a further 2.5°C over the next century – more than the change from the ice age to present day. Locally, the changes could be even more dramatic. As polar ice melts, sea levels could rise and cause flooding, deserts would become drier and coastal regions would get more stormy.

Frontal systems are where cold and warm air meet. A cold front is where cool, often dry air drives warm air out of its way.

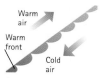

A warm front often brings cloud and rain. Cold air pushes beneath the warm air and lifts it.

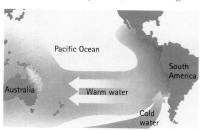

Normal conditions

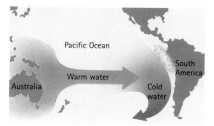

El Niño

An occluded front is where cold and warm air mix. As they do, the front weakens and disappears.

El Niño is the name given to an occasional reversal in the currents of the Pacific. Normally, surface currents move westwards across the South Pacific. This provides moist air for the humid climate of Southeast Asia and allows nutrient-rich water to well up along the coast of Peru. During El Niño, the warm current flows eastwards. This causes floods in the Americas and droughts in Southeast Asia. El Niño also starves fish off the coast of Peru.

SEE ALSO
16–17 The oceans, 34–5 Winds, storms and floods, 268–9 The Sun

RAIN AND SNOW

Rain and snow are two forms in which water falls to Earth from the sky. The water they supply is vital for the survival and growth of animals and plants.

When water freezes, it forms flat, six-sided crystals. Under certain weather conditions, these crystals can join together, building complex and beautiful snowflakes.

Water that falls from clouds is called precipitation. In one form of precipitation, raindrops form in clouds when air currents cause tiny droplets of water to bump into each other. These droplets join together to form larger drops, which fall as rain. The air must be humid for rain to reach the ground without evaporating, so this type of formation happens mainly in the tropical regions.

Most rain starts as crystals of ice that form high in the atmosphere in clouds where the temperature is low. The ice crystals grow as water freezes onto them. Whether these crystals reach the ground as rain or snow depends on the height of the freezing level – the minimum height at which the temperature and pressure cause water to freeze.

If the freezing level is less than 300 metres above ground, ice crystals do not have time to melt before reaching the ground, so they fall as snow. In warmer conditions, the freezing level is higher and the crystals turn to rain before they reach the ground.

HAILSTORMS

Inside a storm cloud, falling ice crystals or raindrops may be carried upwards by currents of warm air called updraughts. At the top of the cloud, they freeze and attract more water. As the developing hailstones then fall to warmer levels, their outside layers melt, but are re-frozen as a clear layer of ice when the hailstones are carried to the top of the cloud.

RAIN AND SNOW

If the base of a stratus cloud is low, small droplets of rain may fall as a fine drizzle. Dry snow falls when the temperature near the ground is below 0°C, the freezing point of water. If snow falls into air that is just above 0°C, some of it will melt. The resulting mixture of snow and rain is called sleet. Wet snow falls when the temperature is exactly 0°C.

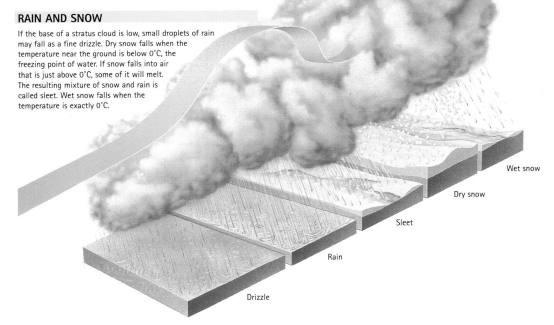

Wet snow

Dry snow

Sleet

Rain

Drizzle

Clouds form and produce rain when warm, moist air rises. This can happen when a warm air front meets a mass of cold air. The warm air cools as it rises over the cold air, forming clouds, light rain and drizzle. When a cold front pushes under warm air, the moist air rises rapidly, forming storm clouds and often giving heavy rain. Clear, cooler weather usually follows.

The hailstones continue to rise and fall, getting larger with each cycle, until they are too heavy for the updraught and fall to the ground. Typical hailstones can be a centimetre or more in diameter, and the record weight is 760 grammes.

RAINFALL LEVELS

Annual rainfalls of 60–150 centimetres are normal for temperate regions such as northern Europe, but there are desert regions where there has been no rain for years. In part of the Atacama Desert in Chile, no rainfall was recorded for more than 400 years. At the other extreme, Waialeale peak, Hawaii, receives more than 11 metres of rain in an average year. The greatest snowfall in 12 months was

31.1 metres, which fell in the winter of 1971–72 at Paradise, Mount Rainer in the state of Washington, USA.

On 16 March 1952, Cilaos, on Reunion in the Indian Ocean, received 1.87 metres of rain. Torrential rain of this type can loosen tree roots and destabilize slopes, causing catastrophic landslides and washing away homes. Prolonged heavy rain and sudden thaws can raise the water levels of rivers, making them burst their banks.

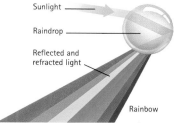

Hailstones start off as tiny seeds of ice at the top of a storm cloud. They fall to the bottom of the cloud but are then swept back to the top by a warm updraught of air. More water freezes onto the seed before it drops and is swept up again. Each hailstone goes through several cycles, building up many layers of ice before it finally falls to the ground.

Sunlight is a mixture of wavelengths of light. When it is reflected by a raindrop, each wavelength is refracted, or bent, through a different angle. As a result, the different wavelengths are seen as separate colours in a rainbow.

◀ Rainbows can be seen when sunlight is refracted and reflected by millions of droplets of rain.

SEE ALSO

CLOUDS AND FOG

Heat from sunlight makes water evaporate from the ground. When the air temperature drops, water vapour in the air condenses to form clouds and fog.

The amount of water that air can hold as invisible vapour depends on its temperature. If moist air cools below a certain temperature, minute droplets of water start to form around tiny grains of dust or smoke in the air. Clouds and fog are made of these droplets. Fog is cloud that has formed at ground level. At high altitude, where the air is cold, the water in clouds forms ice crystals.

CLOUD FORMATION

In sunny weather, warmth and moisture from the ground produce rising currents of warm, moist air. When this rises into cooler air, the vapour begins to condense and form cloud. But the inside of the cloud is still warm and continues to rise, forming tall, fluffy clouds.

A different type of cloud forms when a warm, moist air front meets a body of cold air. The warm air rises over the cool air and starts to cool. Sheets of unbroken cloud can form at the boundary between warm and cool air. Cloud can also form when moist air rises and cools as it passes over hills or mountains.

TYPES OF CLOUDS

Clouds are usually classified by their appearance and height. The height of a cloud is measured to the cloud's base, but some types of clouds can tower many thousands of metres above their bases. Low-level clouds have bases at an altitude of less than 2,000 metres. Medium-level clouds have bases at 2,000–5,000 metres and high-level clouds have bases around 5,000–14,000 metres.

Examples of low-level clouds include stratus cloud, which forms in unbroken sheets, and fluffy white cumulus clouds that resemble balls of cotton wool. Cumulonimbus clouds also have low-level bases, but they can reach 13,000 metres into the atmosphere. These clouds are usually anvil shaped, and they can bring heavy showers and thunderstorms.

Medium-level clouds include altostratus, which forms as fine sheets, and darker nimbostratus clouds, which often bring persistent rain or snow. Altocumulus are spectacular bands of medium-level cloud that resemble ripple marks on a seashore.

High-level clouds are types of cirrus, which take their name from the Latin word meaning 'tuft'. This is because of their wispy appearance. Cirrus clouds contain water as minute crystals of ice.

1 Sunshine warms bare soil faster than grass. On sunny days, humid air rises from these areas.

2 As humid air rises and cools, the water vapour in it starts to condense and form clouds.

3 The clouds grow as further pockets of rising warm air feed them with droplets of water.

Storm clouds produce lightning discharges of 100,000 volts or more. These huge sparks heat air to more than 30,000°C. The rapid expansion of air causes a thunderclap.

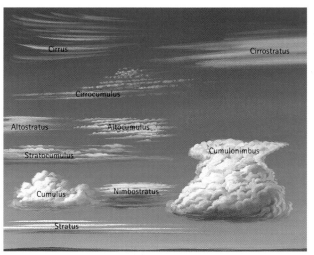

Cirrus
Cirrostratus
Cirrocumulus
Altostratus
Altocumulus
Stratocumulus
Cumulonimbus
Cumulus
Nimbostratus
Stratus

Cirrus clouds are high-altitude strands of cloud. Cirrostratus clouds form a veil. Cirrocumulus clouds are tufts of high-level cloud that can make a regular rippled pattern called mackerel sky. Altostratus is a thin layer of medium-level cloud. Nimbostratus is a thicker sheet of grey cloud that can bring rain or snow. Cloudlets at medium level are called altocumulus. Low-level cloud includes stratus, a layer that often shrouds the tops of hills, and fluffy cumulus clouds. Stratocumulus is a layer of joined cumulus clouds. Strong updraughts of warm air can form anvil-shaped cumulonimbus clouds, bringing the threat of thunderstorms.

FEATURES OF CLOUDS

A trained meteorologist can often predict the weather by looking at the shapes of clouds and watching how they change. Clouds can form amazing and sometimes beautiful features in the sky. Unstable bulges can hang down below altocumulus clouds to form udder-like fingers of cloud. Air currents descending from mountains sometimes produce series of vertical waves that form clouds that resemble a pile of plates. These are the clouds that have been mistaken for flying saucers.

When sunlight passes through holes in cloud but the Sun itself is hidden, beautiful rays of light can be seen as the sunlight catches on particles of dust in the atmosphere. Droplets of very cold water in thin clouds can act as prisms, splitting sunlight into its component wavelengths and making the clouds shine with a variety of colours. At sunset and sunrise, when the Sun is low in the sky, its light is scattered by the atmosphere, so that the undersides of high clouds are illuminated with beautiful pink, orange and red colours. This effect can be particularly spectacular in polluted areas, where dust in the atmosphere adds to the natural scattering effect. High-altitude cirrus clouds can produce the effect of a halo around the Sun or the Moon as light is refracted through the ice crystals.

FOG AND SMOG

When cloud forms at ground level, it is called fog. Because fog reduces the range over which objects can be seen, it is particularly dangerous for drivers and mountaineers. In extreme cases, visibility can be less than one metre.

Smog is a mixture of smoke and fog that occurs in polluted areas. Until the mid-1950s, smoke from coal fires in cities in England, caused such severe smogs that many people died of respiratory illnesses. Pollution control has reduced the severity of smogs, but occurrences of smog still increase death rates, as well as causing eye irritation and asthma attacks.

Sea mist is a common sight in San Francisco Bay, California. Here, the centre of the Golden Gate Bridge is hidden by mist.

Sea mist is a type of advection fog that forms when warm moist air meets a cold ocean current, in this case off the Californian coast.

FOG FORMATION

Advection fog forms when moist, warm air travels over a cool surface on land or at sea. Frontal fog forms where two air masses of different temperatures meet. Radiation fog forms at night when the ground rapidly loses heat by radiation and cools damp air above it. Radiation fog forms in hollows in the ground. Upslope fog can form when moist air cools as it travels up a slope.

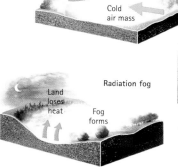

Frontal fog

Warm air mass

Fog

Cold air mass

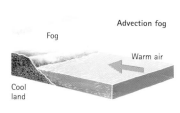

Advection fog

Fog

Warm air

Cool land

Radiation fog

Land loses heat

Fog forms

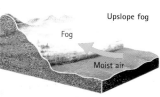

Upslope fog

Fog

Moist air

SEE ALSO

28–9 Climate, 30–1
Rain and snow, 180–1
Refraction

WINDS, STORMS AND FLOODS

Gentle winds can be refreshing and pleasant. Strong winds, storms and floods can range from being a simple inconvenience to a threat to life and property.

Force	Speed (km/h)	Effects
0	<1	Calm. Smoke rises vertically.
1	1–5	Light air. Smoke drifts but flags to not move.
2	6–11	Light breeze. Smoke shows the direction of the wind.
3	12–19	Gentle breeze. Flags move gently, leaves rustle.
4	20–29	Moderate breeze. Loose paper blows around.
5	30–39	Fresh breeze. Small trees sway.
6	40–50	Strong breeze. Umbrellas blown inside out.
7	51–61	Moderate gale. Resistance felt when walking into wind.
8	62–74	Fresh gale. Twigs and branches break.
9	75–87	Strong gale. Chimneys topple, roofs damaged.
10	88–102	Whole gale. Trees blown over but not moved.
11	103–120	Storm. Trees uprooted and moved. Vehicles blown over
12	>120	Hurricane. Buildings destroyed, widespread devastation.

In 1805, British admiral Sir Francis Beaufort (1774–1857) devised the Beaufort wind scale for measuring wind force. His scale is still widely used.

Dark clouds form

Rotating air

Cloud rotation becomes visible in the dark sky.

Rotating air forms funnel

A rotating funnel snakes down from the cloud.

Funnel extends to ground and picks up dust

The funnel touches down and sucks up debris.

▲ A tornado can result from a clash between two wind systems that are travelling in different directions. The first sign of a tornado forming is when a patch of dark cloud starts to rotate. Then, a rotating funnel of warm air descends to the ground. The spiralling updraught sucks up dust and debris, which it hurls around at high speed.

The wind patterns on Earth's surface are part of the three-dimensional pattern of air circulation in Earth's atmosphere. Around the equator, where sunshine is strongest, warm air rises and horizontal wind speeds can be low. Sailors know these regions as the doldrums. The updraught of air at the equator draws in air over the tropics, causing reliable trade winds from the northeast in the Tropic of Cancer and from the southeast in the Tropic of Capricorn. Higher latitudes are dominated by winds from the west.

In some places, strong seasonal winds are so well known they are given names. In southern Europe, the mistral is a cold north wind that funnels down the valleys of the Rhone and other rivers. In western Africa, a dry easterly wind is called the harmattan, meaning 'doctor', because it brings relief from humid conditions.

TORNADOES

As pockets of air become warm, they expand and rise, swirling around and drawing more air in beneath them. A tornado usually begins inside thunder clouds, where currents of warm and cold air meet in a rotating system called a supercell. As the supercell sucks air more and more rapidly from underneath, a funnel reaches down from the base of the cloud until it meets the ground. The suction inside the funnel is immense, and wind speeds more than 480 kilometres per hour, or 133 metres per second, are not unusual. Some tornadoes last for hours, others blow out in a few seconds. The base of a twister can be anything from a few metres across to a kilometre wide.

The Midwest of the United States is sometimes called Tornado Alley. Hundreds of tornadoes strike the area each year as hot humid air from the Gulf of Mexico meets cold, dry air from Canada. On 19 April 1965, a total of 37 tornadoes tore through six states in nine hours, killing 271.

Wind speeds can exceed 130 metres per second

Structures burst apart

Airborne debris is a major cause of casualties

The high-speed wind around a tornado can cause damage in its own right. The damage is worsened by fast-moving debris sucked up through the funnel of the tornado. Added to this is the explosion risk caused by trapped air when a sealed building is caught in the low-pressure funnel.

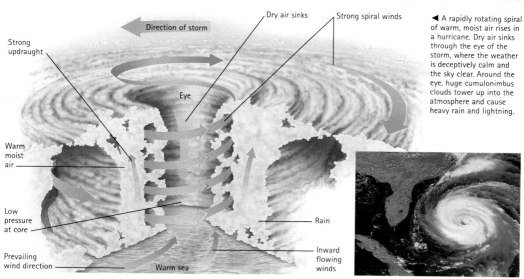

Direction of storm

Dry air sinks

Strong spiral winds

Strong updraught

Eye

Warm moist air

Low pressure at core

Prevailing wind direction

Warm sea

Rain

Inward flowing winds

◄ A rapidly rotating spiral of warm, moist air rises in a hurricane. Dry air sinks through the eye of the storm, where the weather is deceptively calm and the sky clear. Around the eye, huge cumulonimbus clouds tower up into the atmosphere and cause heavy rain and lightning.

This satellite photograph shows Hurricane Fran approaching the North American mainland from the Caribbean Sea in 1996. The eye of the hurricane is clearly visible as a clear patch in the centre of the spiral of clouds. This hurricane, whose wind speeds reached 190 kilometres per hour, killed 34 people.

TROPICAL CYCLONES

Tropical cyclones cause more widespread destruction than tornadoes. Most tropical cyclones form in late summer. This is when sea temperatures are at their highest and hundreds of storm systems can come together and rotate as a single great low pressure system, sometimes hundreds of kilometres across. As the wind speed increases, tropical cyclones tend to move away from the equator. They gather force over warm seas until they hit land.

When they occur in the Gulf of Mexico and the western Atlantic, tropical cyclones are called hurricanes. Hurricanes rotate anticlockwise and are most likely to occur between August and October.

In the Southern Hemisphere, tropical cyclones rotate clockwise and mostly occur between January and April. When they occur off the coast of Southeast Asia, they are called typhoons.

In all cases, tropical cyclones bring heavy rain and wind speeds of up to 200 kilometres per hour or more as warm air spirals around the storm system. Since they are regions of low pressure, tropical cyclones raise the sea level beneath them and can cause devastating floods, called storm surges, if they run into coastlines that do not have strong defences.

FLOODS

Storm surges are not the only cause of flooding. In mountainous areas, flash floods can surge down steep valleys after heavy rain or when snow melts quickly. Rivers can burst their banks and flood wide plains. Also, a combination of low pressure, a high tide and strong onshore winds can flood coastlines. In 1953, storm winds and high tides drove a wedge of water down the North Sea onto the coasts of eastern England, the Netherlands and Belgium. Sea defences failed and the sea swept 60 km inland. If global warming causes sea levels to rise, such floods could become more frequent and low-lying coral islands such as the Maldives and parts of countries such as Bangladesh and Holland could be lost to the seas.

Disastrous floods are an almost yearly event in Bangladesh and many other countries that suffer from tropical cyclones. Tragically, many of these countries are too poor to be able to fund adequate defences against flooding.

SEE ALSO

14–15 Earth's atmosphere,
28–9 Climate

FACTS AND FIGURES

PLANETARY DATA

Equatorial diameter	12,756 km
Polar diameter	12,714 km
Equatorial circumference	40,077 km
Volume	1.083×10^{12} km³
Mass	5.98×10^{24} kg
Density (average)	5.52 (water = 1.0)
Surface gravity	9.78 ms⁻²
Escape velocity	11.18 kms⁻¹
Day length	23 hr 56 min 4.1 sec
Year length	365.24 days
Axial inclination	23.44 degrees
Axial speed at the equator	1,600 km/h
Average temperature	14°C
Age	4,600 million years approximately
Distance from the Sun	149,503,000 km (average) 147,000,000 km (minimum) 152,000,000 km (maximum)
Length of orbit	938,900,000 km
Orbital speed	106,000 km/h
Surface area	510 million km²
Land surface	148 million km²
Oceans cover	71% of surface
Mass of water	1.35×10^{21} kg
Ocean depth	3.8 km
Atmospheric composition	N₂ 78% O₂ 21%
Atmospheric thickness	1,100 km
Atmospheric pressure	101,325 Pa (Nm⁻²) (sea level) 33,440 Pa (Everest summit)

EARTH'S STRUCTURE

Continental crust	35 km thick (av.)
Oceanic crust	7 km thick (av.)
Lithosphere	100 km thick
density	2.7–3
composition	O 46.6%, Si 27.7%, Al 8.1%, Fe 5.0%, Ca 3.6%, K 2.6%, Na 2.8%, Mg 2.1%.
Mantle	2,900 km thick
temperature	3,000°C at base
density	3.3–6.0
composition	silicates of iron and magnesium
Outer core	2,200 km thick
temperature	4,000°C at base
density	10
composition	molten iron, traces of nickel
Inner core	1,300 km radius
temperature	6,500°C at centre
density	13
composition	solid iron, traces of nickel

BRANCHES OF EARTH SCIENCES

Climatologists study the climate: including seasonal and longer-term variations in temperature and moisture in the lower atmosphere across the world.
Geochemists study the chemical composition of Earth's crust, its oceans and its atmosphere.
Geologists study Earth's origin and the structure and composition of its layers.
Geomorphologists study the forms and formation of land features such as ocean basins and mountain ranges.
Meteorologists study and attempt to predict day-to-day variations in the weather. They measure and forecast conditions in the lower atmosphere such as temperature, rainfall and wind speed.
Mineralogists are geologists who study crystalline minerals and ores.
Oceanographers study physical and chemical conditions in the oceans, the ocean floor and marine life.
Palaeontologists study the structure, evolution, environment and distribution of ancient organisms by examining their fossilized remains. Palaeobiologists study animal fossils, palaeobotanists study plant fossils and palaeoclimatologists study the climate of the past.
Petrologists study the origins and structures of rocks.
Planetologists examine and make comparisons between planets.
Sedimentologists are geologists who study rocks formed from silt and sandy deposits.
Stratigraphers study layers of rock and how they relate to one another.

KEY DATES

BC
c.235 Eratosthenes – a Greek astronomer, geographer and mathematician – calculates Earth's circumference from shadows cast at different latitudes at midday.

c.5 Greek geographer Strabo proposes frigid, temperate and tropical climate zones.

AD
c.30 Strabo suggests that Earth is so big that there might be unknown continents.

79 Roman writer Pliny the Younger describes the eruption of Vesuvius that caused the destruction of Pompeii.

132 Chinese invent first seismograph – finely balanced metal balls that fall if the ground shakes.

1086 Chinese engineer Shen Kua outlines the principles of erosion, sedimentation and uplift processes.

1517 Italian scholar Girolamo Fracastoro suggests that fossils are the remains of creatures left by the biblical flood in the story of Noah's ark.

1546 German metallurgist Georgius Agricola first uses the term 'fossil' for the rocklike remains of plants and animals.

1600 British physician William Gilbert proposes that Earth is like a giant magnet.

1735 British meteorologist George Hadley uses mathematics and physics to explain how Earth's spin affects trade winds.

1785 British geologist James Hutton proposes that Earth's features form through processes, such as sedimentation and volcanic activity, that have been acting for a very long time.

1795 French anatomist Georges Cuvier identifies a set of fossil bones as belonging to a giant marine reptile.

1811 British teenager Mary Anning and her family discover and collect fossils of the first known *Ichthyosaurus* sample.

1815 British surveyor William Smith publishes his *Geological Map of England and Wales, with Part of Scotland*.

1822 *Iguanodon* is identified.

1825 Cuvier proposes that species become extinct as a result of catastrophic events.

1830 British geologist Charles Lyell suggests Earth is hundreds of millions of years old.

1840 Swiss-born naturalist Louis Agassiz proposes that most of Earth was once covered by ice.

1859 British scientist Charles Darwin publishes theory of evolution.

1896 Swedish chemist Svante Arrhenius shows that carbon dioxide in air helps trap heat in Earth's atmosphere.

1906 Irish geologist Richard Oldham finds evidence for Earth's core in records of seismic waves.

1915 German meteorologist Alfred Wegener publishes a theory that continents are in motion.

1925 Echo soundings reveal the Mid-Atlantic Ridge.

1935 US seismologist Charles Richter devises a scale for reporting the strengths of earthquakes.

1965 Canadian geophysicist Tuzo Wilson explains how the floors of oceans spread.

1981 Luis Alvarez and his son Walter propose that a giant meteorite impact killed the dinosaurs.

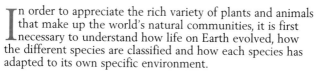

Chapter 2
LIVING THINGS

In order to appreciate the rich variety of plants and animals that make up the world's natural communities, it is first necessary to understand how life on Earth evolved, how the different species are classified and how each species has adapted to its own specific environment.

The story of life on Earth began many millions of years before the appearance of the first human beings. Life began in the oceans over 3,500 million years ago. Today, there are more than 2 million living species on Earth. Many are so tiny that they can only be seen with the aid of microscopes. Others live in such remote and alien habitats, such as the cold and sunless depths of the ocean, that humans know little about them.

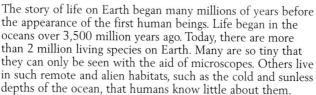

All the species of plants and animals have evolved by gradually adapting to the many different types of environments that Earth offers its inhabitants. Some plants and animals have had their evolution guided by human intervention. Many garden plants, agricultural crops, farm animals and domestic pets are examples of species that arose through selective breeding.

Each wild plant and animal species lives in harmony with its surroundings. The living world is a marvellously intricate system that has taken millions of years to develop, and which is in a delicate and constantly changing balance.

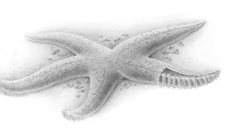

LIFE: ORIGINS AND DEVELOPMENT

Life began on Earth more than 3,000 million years ago. Since then, an astonishing variety of animals and plants has evolved from single-celled organisms.

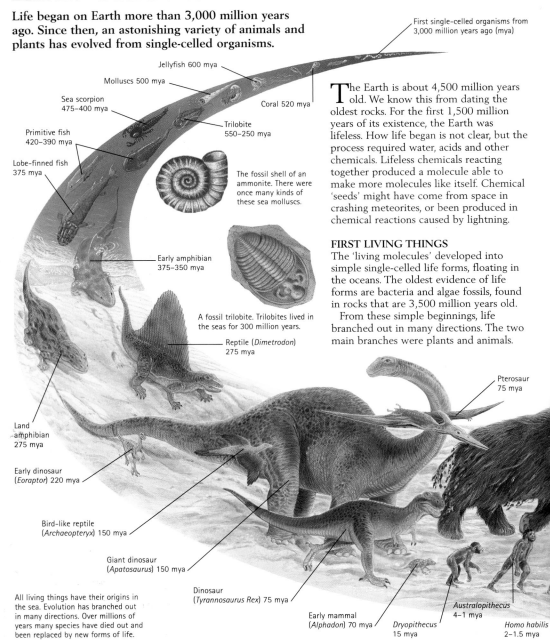

First single-celled organisms from 3,000 million years ago (mya)

Jellyfish 600 mya

Molluscs 500 mya

Sea scorpion 475–400 mya

Coral 520 mya

Primitive fish 420–390 mya

Trilobite 550–250 mya

Lobe-finned fish 375 mya

The fossil shell of an ammonite. There were once many kinds of these sea molluscs.

Early amphibian 375–350 mya

A fossil trilobite. Trilobites lived in the seas for 300 million years.

Reptile (Dimetrodon) 275 mya

Land amphibian 275 mya

Early dinosaur (Eoraptor) 220 mya

Bird-like reptile (Archaeopteryx) 150 mya

Giant dinosaur (Apatosaurus) 150 mya

Pterosaur 75 mya

Dinosaur (Tyrannosaurus Rex) 75 mya

Early mammal (Alphadon) 70 mya

Dryopithecus 15 mya

Australopithecus 4–1 mya

Homo habilis 2–1.5 mya

All living things have their origins in the sea. Evolution has branched out in many directions. Over millions of years many species have died out and been replaced by new forms of life.

The Earth is about 4,500 million years old. We know this from dating the oldest rocks. For the first 1,500 million years of its existence, the Earth was lifeless. How life began is not clear, but the process required water, acids and other chemicals. Lifeless chemicals reacting together produced a molecule able to make more molecules like itself. Chemical 'seeds' might have come from space in crashing meteorites, or been produced in chemical reactions caused by lightning.

FIRST LIVING THINGS

The 'living molecules' developed into simple single-celled life forms, floating in the oceans. The oldest evidence of life forms are bacteria and algae fossils, found in rocks that are 3,500 million years old.

From these simple beginnings, life branched out in many directions. The two main branches were plants and animals.

The land dinosaurs and giant sea reptiles died out 65 million years ago. The Earth may have been hit by a gigantic meteorite. This would have caused a 'long winter' as the Sun was blotted out by dust clouds. Plants died; so did most plant-eating reptiles and the carnivores that preyed on them.

The gingko or maidenhair tree has lived on Earth for more than 300 million years. It is the only survivor of a once common group of plants.

EVOLUTION

All plants and animals have arisen through a process of gradual change known as evolution. The species (kinds) of animals and plants alive today have evolved from much earlier kinds, which are now extinct. Trilobites and ammonites swarmed in the seas millions of years ago. Now only their fossils remain in rocks.

Animals that died out did so because the conditions changed where they lived. They were replaced by other species that were able to adapt to changing conditions.

ADAPTATION

Dinosaurs ruled planet Earth for over 160 million years. There were more than 500 species. Yet they all died out, and today fossils are all that remain of these remarkable animals.

Living things are constantly evolving through adaptation. They can adapt because every individual life form is slightly different, even from members of its own species. This means that when conditions change, such as the climate growing colder, some life forms may be better at surviving than others.

Plants and animals have colonized almost every environment on Earth – freezing polar regions, hot and dry deserts, even the sunless ocean depths.

The DNA molecules in the cells of all living things contain coded instructions (genes). These control the way the cells behave.

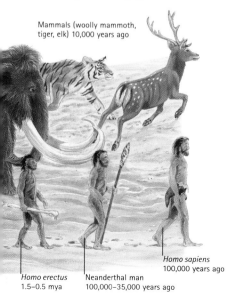

Mammals (woolly mammoth, tiger, elk) 10,000 years ago

Homo sapiens
100,000 years ago

Homo erectus
1.5–0.5 mya

Neanderthal man
100,000–35,000 years ago

Berry-feeding

Seed-feeding

Cactus-feeding

Insect-feeding

On the Galapagos Islands in the Pacific, Darwin found finches that were descended from one species, but that now had different shaped beaks. Evolution had adapted each kind's beak to eat a different type of food.

Charles Darwin (1809-1882) was a British scientist. His studies of living animals and fossils led him to his theory of natural selection, published in 1859.

SEE ALSO

10–11 Fossils and geological time, 107 Genes and chromosomes

CLASSIFICATION OF LIVING THINGS

Classification is the method by which living things are grouped into categories based on their appearance and the natural relationships between them.

No one knows exactly how many different living things there are on Earth today. Scientists have discovered over two million, but there may be four times as many, mostly microscopic, organisms that are yet to be discovered.

HOW THINGS ARE NAMED

The scientific study of the diversity of living things and the relationships between them is called systematics. Taxonomy, which is part of systematics, is the study of the rules and procedures of classifying plants and animals. When classifying a living organism, it is given a scientific name written in Latin so that it can be identified by scientists all over the world. Classification helps us to study and understand the natural world. It also shows how living species are related to species that died out long ago.

The classification system scientists use today is based on one developed in 1758 by Swedish naturalist Carolus Linnaeus (1707–1778). The individual organisms, called species, can be classed in different levels. The highest level is called a kingdom. All animals belong to the kingdom Animalia. There are four other kingdoms, but the animal kingdom is by far the largest. The differences in the structure of an organism's cells partly determines to which division the organism belongs.

The level below the kingdom is called the phylum (plural: phyla). There are more than 20 different phyla within the animal kingdom. All vertebrates (animals with backbones) belong to a phylum called Chordata.

The fossil fish shown above, called *Priscacaria*, is extinct today. However, the species can be fitted into the classification of living things. Taxonomists compare certain features, such as the arrangement of fins, with some of the 22,000 kinds of fish that live on Earth today. The fossil fish is then grouped with existing fish that have similar attributes.

THE FAMILY TREE OF LIFE

The prehistory of Earth is marked by periods of time called eras, each lasting many millions of years. There are four eras: the Precambrian, Palaeozoic, Mesozoic and Cenozoic. During the Palaeozoic Era, there was an enormous increase in the number of different species that lived on Earth. Some organisms left the warm, shallow seas and lived on dry land. The gradual change in the characteristics of living things over time is called evolution. It has created the many branches of the family tree of life. As species have become extinct (died out), others have developed. Evolution has created the amazing diversity of organisms that are alive today.

Eras	Cenozoic	Mesozoic	Palaeozoic	Precambrian
Millions of years ago	0–65	65–245	245–570	570–3000

CLASS, FAMILY AND SPECIES

Organisms are further divided into smaller levels. Below the phylum comes a level called the class. All mammals belong to the class Mammalia. Below the class is the order. All meat-eating mammals, such as foxes, leopards and otters, belong to the order Carnivora.

Then comes a level called the family. Foxes, hyenas and wolves all belong to the family called Canidae. Within the family are subgroups of animals that cannot breed with one another – they can mate, but do not have offspring. Each group is a genus. The genus name is written in italics, so the fox genus is *Vulpes*.

Within a genus are one or more species. Each species has a name that is also written in italics. The little fennec fox from North Africa is *Vulpes zerda*.

EXTINCTION AND CHANGE

Scientists think that the number of organisms today is only a tiny fraction of all the living things that have existed. More than 99 per cent of all the species that have ever existed are now extinct.

Animals, plants and other organisms gradually change as the conditions around them alter. In this way, species may evolve into new species, preserving the diversity of the varied family tree of life.

The graph below shows the relative numbers of living species. Insects far outnumber all other life forms. Next come plants and molluscs. Fish are the most numerous vertebrate species (animals with a backbone). There are comparatively few species of mammals in the world.

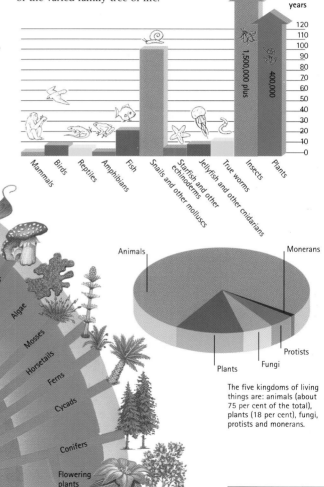

The five kingdoms of living things are: animals (about 75 per cent of the total), plants (18 per cent), fungi, protists and monerans.

Precambrian	Palaeozoic	Mesozoic	Cenozoic
3000–570	570–245	245–65	65–0

SEE ALSO

38–9 Life: origins and development

SINGLE-CELLED ORGANISMS

The simplest living things are tiny single-celled organisms. They were the first living things on the planet, and they are still the most common.

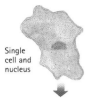

Single cell and nucleus

Nucleus divides

Two new organisms

Cells multiply by splitting in two. Some bacteria can do this once every 15 minutes. A single-celled protist like this amoeba becomes two by a complex process of cell division known as mitosis.

▼ Algae include single-celled diatoms and giant seaweeds. This picture, taken with a powerful electron microscope, shows 'epiphytic' green algae cells that cling to another plant for support.

Cells are the smallest units capable of life. The simplest living things have just one cell, which contains all the information and processes needed to keep that particular unit alive and for it reproduce.

INSIDE THE CELL

A cell has a thin outer wall, which lets chemicals in and waste out. Within the cell wall is a jelly-like fluid that is called cytoplasm, which contains tiny structures each of which has a function. The central structure is the nucleus which contains the genes, which decide the cell's shape and function. Other structures release energy from food, deal with waste, or protect the cell against attack from other organisms.

FIRST LIFE

More than 3,000 million years ago, the first single-celled organisms appeared in the Earth's seas. What kind of chemistry went on to create life is not known. It may have involved 'seeding' of spores from space, but is more likely to have been the result of chemical reactions in the Earth's oceans and atmosphere, which produced complex molecules able to organize themselves into living organisms.

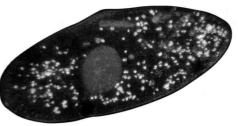

Paramecium, a kind of protozoa. These single-celled organisms use tiny hair-like filaments called cilia, which they wave in order to move and to catch food.

Today, these simplest of all creatures are classified as monerans. There are two main groups, bacteria and cyanobacteria (plant-like blue-green algae). They are so tiny that they can be seen only through a powerful microscope.

Protists are single-celled organisms such as amoebas and tiny algae, but also include many organisms that are made up of cells. Some feed like animals, and others trap the energy in sunlight, like plants. Other protists can feed in both ways. Many protists have one or two tail-like structures called flagella, which they flick to push themselves along. Others are tiny hunters that can engulf their prey. Most protists reproduce by dividing into two new cells.

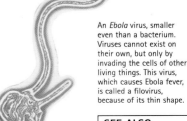

An *Ebola* virus, smaller even than a bacterium. Viruses cannot exist on their own, but only by invading the cells of other living things. This virus, which causes Ebola fever, is called a filovirus, because of its thin shape.

SEE ALSO

38–9 Life: origins and development, 40–1 Class -ification of living things

FUNGI AND LICHENS

Fungi and lichens are simple forms of life that depend for food and survival upon partnerships with plants and with one another.

Amanita muscaria, is one of the toadstools to avoid. It is poisonous and may kill any animals and people that eat it.

Fungi have no chlorophyll and so, unlike plants, cannot make their own food. They feed by making chemicals that rot the bodies of other living things (such as plants) and their dead remains. The fungus absorbs nourishment as the bodies decay. Mushrooms, toadstools, yeasts and slime moulds are fungi.

The toadstool or mushroom seen above ground is the 'fruiting body' of the fungus. Beneath the soil or within the rotting wood of trees is the hidden part of the fungus. This is a mass of thread-like cells, called the mycelium. The fruiting body appears when the fungus is ready to reproduce. It contains spores, which it releases into the wind.

All fungi make sure they can spread spores. The earth star hoists its fruiting body above the soil on star-like 'jacks'.

In lichen, a fungus provides the 'body' for single-celled algae. The algae's function is to make food, and so keep the partnership alive.

LICHENS

A lichen is actually two living things in partnership or symbiosis. The partners are a single-celled alga and a fungus. The alga uses photosynthesis to turn the energy in sunlight into food, and this food supports both the alga and the fungus. The fungus protects the alga from the outside conditions. Some have hard chalky skin.

Lichens are often found on rocks, walls and the barks of trees. They are very hardy, able to withstand even the freezing polar cold and the snow of high mountain peaks. Some types of lichen can live for as long as 4,000 years.

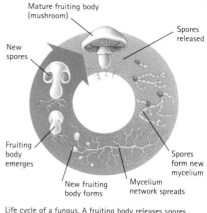

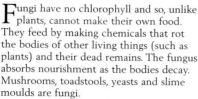

Life cycle of a fungus. A fruiting body releases spores, which form a network of filaments. New fruiting bodies push up to the air and scatter spores of their own.

HARMFUL AND HELPFUL

Lichens are eaten by animals such as caribou. Fungi, too, can be consumed by both animals and people. However, before eating, it is important to identify a fungus, as some kinds are poisonous and can cause stomach upsets and even death.

Fungi are also found in the sea and fresh water, where they can sometimes be seen as foam on the surface. Some fungi live on the skin of animals. Many skin diseases and mouth and ear infections are caused by fungi. But fungi can also be beneficial. The life-saving antibiotic penicillin is made from a mould, while yeast, used in bread-making, is a fungus.

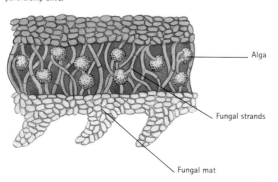

Alga

Fungal strands

Fungal mat

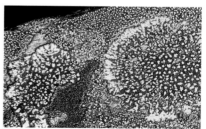

This lichen (*Rhizacarpon geographicum*) is growing in Norway. Lichens can put up with extremes of heat, cold and dryness.

SEE ALSO

40–1 Classification of living things, 54–5 Biomes and habitats

PLANT ANATOMY

Apart from some bacteria, plants are the only living things able to make their own food. Tiny grasses and huge trees share the same anatomy or structure.

There are two main classes of plants. Non-vascular plants, such as mosses and liverworts, do not have tissues to carry food and water from one part of the plant to another. Vascular plants do. They are the larger of the two classes, and include trees and flowers.

WHAT PLANTS NEED

All plants need light, because they use energy from sunlight to make their food. They need water and minerals, which most plants obtain through leaves and roots. To reproduce, many plants have flowers that make seeds. But plants have other means of reproducing too – for example, the strawberry sends out runners. All plants are made of cells. Plant cells differ from animal cells in that they absorb

water and grow larger and stiffer as the plant ages. This is why young vegetables are more tender to eat than old ones. Plant cells build thick walls made of cellulose. Each cell is stuck firmly to the walls of neighbouring cells.

GREEN MAGIC

Plants also differ from animal cells in that they contain a green pigment called chlorophyll. This enables the plants to make their own food with the aid of sunlight. Only plants and some bacteria can perform this chemical trick.

There are many cells in each plant, and each group of cells has its own tasks. The main body parts of a plant are roots, stem, leaves and flowers.

Vacuole (liquid energy store)

Cell wall

Chloroplast (traps Sun's energy)

Plant cells are rigid because they have walls of tough cellulose. The vacuoles store water. The chloroplasts trap energy from sunlight.

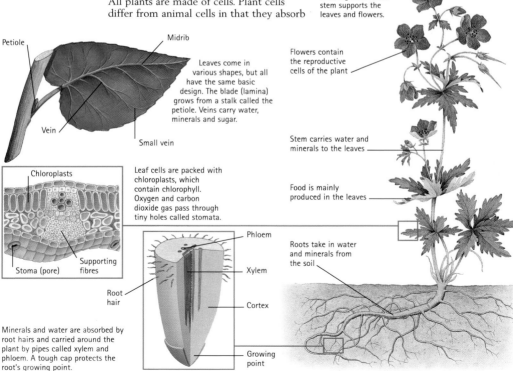

A plant is anchored in the soil by its roots. A strong but flexible stem supports the leaves and flowers.

Petiole

Midrib

Leaves come in various shapes, but all have the same basic design. The blade (lamina) grows from a stalk called the petiole. Veins carry water, minerals and sugar.

Vein

Small vein

Flowers contain the reproductive cells of the plant

Stem carries water and minerals to the leaves

Food is mainly produced in the leaves

Chloroplasts

Leaf cells are packed with chloroplasts, which contain chlorophyll. Oxygen and carbon dioxide gas pass through tiny holes called stomata.

Stoma (pore)

Supporting fibres

Phloem

Xylem

Root hair

Cortex

Roots take in water and minerals from the soil

Minerals and water are absorbed by root hairs and carried around the plant by pipes called xylem and phloem. A tough cap protects the root's growing point.

Growing point

ROOTS

Roots hold the plant fast in the ground, like anchors. They absorb water and mineral salts from the soil through fine hairs. A cap protects the root as it pushes down through the soil in search of water.

STEM

The stem supports the leaves and flowers. Inside are tubes which carry and store water and food. When these tubes are filled with water they are very strong; when they are dry, they become weak and the plant wilts. Tree wood is really a mass of stiffened tubes.

LEAVES

Leaves are the plant's food-making factories. Their cells contain chlorophyll, which uses the Sun's energy to make food from carbon dioxide gas in the air and water. Veins carry water from the roots and move food made in the leaves to the rest of the plant.

FLOWERS

In many plants, flowers are the reproductive parts. Most flowers have male and female organs. The male part makes pollen, which pollinates the female part (usually on another flower). This pollination produces a seed, from which a new plant can grow. Some plants spread their pollen from plant to plant using the wind. Others use insects and other animals to carry it for them. Some flowers grow singly, others form clusters. Many have dazzling colours and strong scents, but some are dull-coloured and do not smell.

Light energy

Carbon dioxide

Oxygen out

In photosynthesis, the plants take in small molecules and build them into large ones, storing the Sun's energy. As part of the process, they give off oxygen gas.

Chlorophyll in leaves

Chloroplast

Grana

Inside a granum

Chlorophyll

Membrane systems

Inside a plant cell's chloroplasts are tiny grana. Each granum contains chlorophyll, which absorbs light rays.

INS AND OUTS OF PLANT LIFE

The food-making process is called photosynthesis. Water from the roots and carbon dioxide gas from the air combine to make glucose (a sugar) and oxygen gas. The plant uses glucose as fuel to make energy, in a process called respiration.

Glucose molecules are joined together into long chains. One chain is called cellulose, which is used for growth and body-building, and the other is starch, which is used as a reserve food store. Plants also make amino acids for protein, enzymes and hormones.

The leaves of *Mimosa pudica*, the 'sensitive plant', recoil when touched. Changes in pressure within some cells cause this defensive movement.

The sharp thorns of a rose are modified leaves. Thorns prevent hungry animals from eating plant leaves, fruit and flowers.

These time-lapse photographs show a lily flower opening. Many flowers are pollinated by insects, which are attracted by brightly coloured petals, tasty nectar and scent. The centre of the flower, where the reproductive organs are, is the 'target area' for visiting insects.

SEE ALSO

46 Non-flowering plants,
47–49 Flowering plants,
50–1 Fruits and seeds

NON-FLOWERING PLANTS

About 400 million years ago, plants began to colonize the land. They were the ancestors of present-day ferns, horsetails, mosses and liverworts.

In prehistoric times, Earth was covered with a dense mat of giant ferns, horsetails, liverworts and club mosses. Today, these plants are still common, but they are some of the most primitive species in the plant world. Unlike the more familiar flowering plants, non-flowering plants have no flowers or seeds. Instead, they reproduce by dispersing tiny reproductive units, called spores, in air currents.

▲ Mosses clump together, often forming thick mats where no other plants can survive. This moss is called *Leucobryum glaucum*.

Horsetails prefer to live in wet places such as bogs. The marsh horsetail has a hollow, jointed stem and looks like a miniature tree.

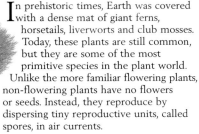

Maidenhair spleenwort is one of the 10,000 kinds of ferns growing on Earth. Along with mosses, horsetails and liverworts, ferns are one of the most primitive species of plants.

MOSSES AND LIVERWORTS

Mosses and liverworts are known as bryophytes. Unlike flowering plants, they have no real roots. Instead they have shallow rhizoids, which look like roots and fix the plant to the ground. Unlike roots, rhizoids cannot take in food or water. Bryophytes have no veins in their leaves, and most species are fairly small.

Mosses and liverworts like damp, shady places, but mosses also cling to exposed areas, such as rocks and walls. Sphagnum mosses form thick mats on swamps and bogs. The squashed remains of these dead plants eventually turn into peat.

HORSETAILS

There are 29 species of horsetail that all belong to one genus, *Equisetum*. Horsetails are jointed, rush-like plants that grow in wet, swampy soil. Their stems contain small quantities of minerals, including gold.

The diagram below shows the stages of growth of a typical fern. Spores from the spore cases are wind-dispersed. Some land many kilometres from the parent plant. The spore grows into a prothallus, from which a young sporophyte develops into the mature fern. The fern develops the typical curly fiddlehead at its tip.

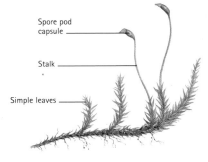

Spore pod capsule

Stalk

Simple leaves

Mosses have rhizoids instead of true roots, and their leaves have no veins. Instead of a flower, mosses grow reproductive structures called spore pods.

FERNS

Most ferns grow in damp, shady places. At first, the leaves, or fronds, are curled up in a structure known as a 'fiddlehead', but the leaves uncurl as the plant grows. Beneath each frond are spore cases. Spores are dispersed by the wind and fall to earth – some grow into a prothallus, with male and female cells. A new fern, called a sporophyte, feeds on the prothallus until it grows roots and can live on its own.

Tropical tree ferns can grow up to 24 metres tall. Tree ferns have woody trunks without branches, topped with clusters of feathery leaves, or fronds.

Tree ferns

Fern

Sporangia

Prothallus

Young sporophyte

New fern grows

SEE ALSO

38–9 Life: origins and development, 44–5 Plant anatomy, 54–5 Biomes and habitats

FLOWERING PLANTS

Flowering plants, or angiosperms, are the most successful of all the plants. Angiosperms reproduce using seeds that develop in the ovaries of their flowers.

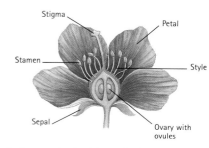

Parts of a flower

The cotyledon is the leafy part of a plant embryo. It provides the embryo with nutrients during germination and growth.

Flowering plants, or angiosperms, are among the most successful group of organisms on Earth. Angiosperms include most of our garden plants, farm crops, and the flowers that are often grown for decoration. They range in size from tiny pondweeds to large trees such as birches and oaks. The name angiosperm is Greek for 'enclosed seed'. The developing embryos of angiosperms are enclosed in special structures, called seeds, within the flower. After fertilization, the seeds are protected inside a fruit. Flowering plants therefore have a lower-risk survival strategy than other plant species, which probably accounts for their success.

A RICHER WORLD

Flowering plants are believed to have evolved from a now extinct group of conifers that lived around 250 million years ago. This period is known as the Permian Period. As they evolved, flowering plants came to have a huge influence on other living organisms. Many animals ate the plants or feasted on the nectar they produced. Others used plants for shelter. For animals, including humans, the world would be less rich without flowers.

Angiosperms advertise their reproductive organs to potential insect pollinators using their flowers, which also serve to protect the organs. The stamens produce male sex cells, called pollen. These are transferred to the female sex cells, called ovules, through the stigma.

TWO MAIN GROUPS

There are two groups of flowering plants, each separated by the way their leaves form as the seed grows. The first group, called monocotyledons, have only one seed leaf. Dicotyledons – the second group – have two. Monocots and dicots differ in many ways. For example, dicot leaves are usually broad and grow from the tip, and can be many different shapes, with smooth or irregular edges. Monocot leaves are long and narrow, like a blade of grass, and usually grow from the base, not the tip. As a result, grasses do not die when animals nibble the tips of the plants. Once nibbled, however, a dicot leaf usually dies. ▶

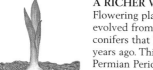

▲ Monocotyledons have smooth leaves that grow from the base of the plant. These plants are also characterized by parallel leaf veins, and flowers with their parts usually in multiples of three.

▶ Gardeners choose specific characteristics when breeding ornamental garden flowers.

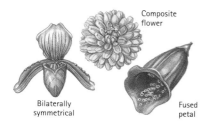

Composite flower

Bilaterally symmetrical

Fused petal

▲ Flowers show many different designs, reflecting their enormous diversity. The daisy is an example of a composite flower, foxgloves have fused petals, and lilies are bilaterally symmetrical (its two halves are alike).

Dicots have complex stems that are either woody or herbaceous. Inside the stem, bundles of tubes are arranged in a ring around a central core. These tubes are important structures that transport water, mineral salts and other nutrients around the plant. As a woody dicot plant grows, the stem adds new rings and gets thicker. As a monocot grows, the stem gets longer without getting thicker and without adding new growth rings.

HOW PLANTS GROW

Unlike animals, plants continue to grow throughout their lives. In favourable conditions, some plants grow extremely quickly. In the humid tropical rainforests, for example, bamboos can shoot up 30 centimetres in just one day.

Many plants are annuals, which means they grow, flower, produce seeds and die within a year. Other plants are biennials, which means they follow a two-year life cycle. Biennials grow their stem and a few leaves during their first summer. These will die after the first hard frost, but the roots survive. During the second growing season, the plant develops a new stem and more leaves, as well as colourful, seed-bearing flowers. Biennials die after the second year. Other plants are perennial. Their stems and leaves die in winter, but the roots survive, and the plant flowers for successive seasons.

REPRODUCTIVE STRATEGIES

Mushroom spores (reproductive units) are dispersed by the wind in their millions.

Some plants have evolved flowers that mimic insects to attract the potential pollinators. The flower of the bee orchid is shaped like a bee. Passing bees try to mate with the flower, collect pollen and pollinate the next plant they visit.

Flowering plants that live in hot deserts, such as the prickly pear shown above, flower quickly after the seasonal rains, setting seeds that may not germinate until the next rainy season.

The Venus flytrap is a carnivorous plant that feeds on insects. The bright red colour of the modified leaves attracts insects, which are trapped as the leaves snap shut.

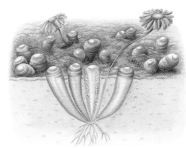

Fenestria is a plant that lives in South Africa. The leaves of this plant shelter beneath the sand. Only the tips emerge above the ground to absorb sunlight.

However, only a few spores fall in the places that are suitable for them to grow into new mushrooms. Flowering plants have a less risky strategy, which does not rely on their being fertilized by accident. Male cells, called pollen, join with female cells, called ovules, to make an embryo contained within a protective structure called a seed. Seeds are then dispersed by a variety of means – they can be carried by the wind or by animals brushing against the plant, or the plants can be eaten by animals that later disperse the seeds in their droppings.

INSIDE A FLOWER

Although an enormous variety of flowers exists, they all flower in a similar way. There are usually four main sets of organs: sepals, petals, stamens (with anthers and filaments) and pistils (stigma, style and ovary). The four organs are collectively called the whorl and they can be arranged

▶ Many flowering plants are cultivated by farmers. Typical examples are the tea and coffee plants, which are used to make drinks. Other flowering plants are harvested for their medicinal properties. For example, foxgloves are harvested for their leaves, which contain a chemical called digitalis. This is used to treat certain heart conditions. These sunflowers in Italy are harvested for their seeds, which are used in cooking and to make sunflower oil.

Meadow foxtail and false oat grass (at left) are two of the huge group of grasses. Grasses are monocotyledons. Meadow cranesbill (at right) is a dicotyledon.

– which makes male sex cells called pollen, and a filament that attaches the anther to the flower.

During pollination, grains of pollen are carried from the anther to the stigma. Once there, the pollen grain forms a tube which grows down through the style and ovary, so that male cells can fertilize the female eggs, or ovules. Following fertilization, the ovules develop into seeds and the flower dies. The plant then puts all its energy into the production of fruits to protect the developing seeds.

SINGLES AND COMPOSITES

Some plants have single flowers, each on its own stem. Composite flowers, such as sunflowers and daisies, are really many small flowers (florets) in a group called an inflorescence. Some flower petals fuse, forming a tube-like shape. Others are not circular but bilaterally symmetrical, with their structures arranged equally on both sides of a line. Some flowers have no petals – the catkins of the hazel tree, for example. Many but not all flowers close at night or in cold weather. Some even close when the Sun goes behind a cloud.

in a spiral pattern or at a single level.

The sepals are the outermost parts of the flower and are green and leaf-like. Their main function is to protect the flower bud and support the delicate petals. The bright colours of the petals make up the second layer. Petals attract insects that are drawn to flowers by their colours and scents, and by the sweet-tasting nectar that they produce.

At the heart of a flower lie the stamens and pistil. The stamens are inside the petals. Each consists of a head – the anther

The world's largest flower belongs to a genus called *Rafflesia*, which is found in the rainforests of southeast Asia. Also called the monster plant, the flowers of *Rafflesia* species can reach a diameter of 1 metre. The flower smells like rotting flesh, which attracts flies.

Wetland plants such as this giant water lily are specially adapted for floating on the surface of lakes and ponds. The roots of this plant cling to the soil at the bottom of the lake or pond, and a white flower blooms at the side of the giant leaf.

THE REPRODUCTIVE CYCLE

Flowering plants, such as the poppy, protect their flowers within tough buds until the pollen grains are ripe. When the flowers open, the anther releases the male sex cells, called pollen. Pollen is transferred from plant to plant by insects (cross-fertilization), or it can be deposited on the stigma of the same plant (self-fertilization). The pollen caught by the stigma triggers the growth of a tiny pollen tube, which grows from the style down to the ovary and the egg cells. When this tube touches an ovule (female sex cell), it fuses with it. This process is called fertilization and results in the development of an embryo contained within a protective structure called a seed capsule.

1 Flower buds stay closed until pollen grains inside are ripe.

2 Buds begin to open, revealing brightly coloured petals and other flower parts.

3 Flowers attract insects looking for nectar. Pollen from carpels is carried by insects to other flowers.

5 Wind shakes ripened seeds out through holes in seed capsule

6 Most poppies are annuals, living for just one year.

4 After pollen is released, the flower petals fall off, leaving the seed capsule.

SEE ALSO
44–5 Plant anatomy, 46 Non-flowering plants, 50–1 Fruits and seeds

FRUITS AND SEEDS

The seeds of flowering plants are protected inside fruits. All fruits have a way of dispersing their seeds that gives the seeds a greater chance of survival.

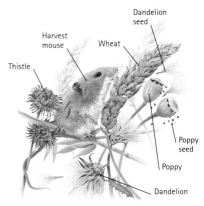

Many seeds are spread by the wind. Others are eaten by animals and pass out with the animal's droppings. Some seeds have hooks which catch on a passing animal's fur.

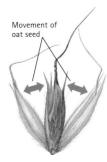

Movement of oat seed

Oat seeds plant themselves by drilling into the soil. Each seed has a long bristle-like structure called an awn. As it dries out, the awn coils and uncoils, twisting the seed into the soil until it reaches the correct depth for germinating.

Seeds are embryos, or potential plants. The seed is self-contained, with a food supply and a tough coat for protection. It needs to get away from its parent to find room to grow in a suitable site. But it can survive for months or even years, waiting for the right conditions.

When the conditions of temperature, light and moisture are right, the seed germinates – it sprouts a tiny root and stem and begins to grow into a new plant.

DRY AND JUICY FRUITS

Seeds are contained inside fruits. There are dry fruits, such as a poppy capsule and a pea pod. These release the seeds inside through holes or by splitting open. An acorn is another dry fruit. Its seed simply forces its way out through the skin.

There are juicy fruits, such as berries, which contain more than one seed. The cherry is a drupe or stone fruit. Its inner layer forms a woody pip or 'stone'. The blackberry is a cluster of small drupes. Its seeds are inside the tiny pips.

Apples and pears are called pomes. They have a fleshy outer layer (the false fruit) around a core (the true fruit), which contains seeds. Pomes are often sweet to the taste to encourage animals to eat them and so help in the seeds' dispersal.

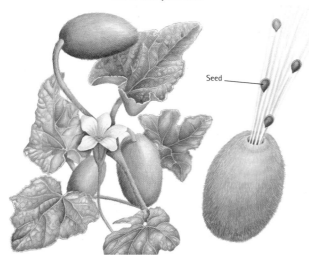

The squirting cucumber is a member of the gourd family. It shoots out its seeds in a jet of liquid. Water pressure builds up inside the fruit until it parts from the stalk and fires its water-jet of seeds.

Seed

Banana buds open into small flowers that form a hand of bananas. At first, the bananas are green. New banana plants can be grown from cuttings, or suckers.

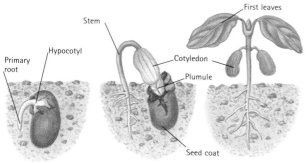

Stem

First leaves

Hypocotyl

Primary root

Cotyledon

Plumule

Seed coat

When a seed starts to germinate, it splits and a primary root is formed, pushing downwards. All other roots grow out from this main root. The stem grows up from the hypocotyl through the soil, and the cotyledons break out of the seed coat. Roots begin to grow and the first leaves form as the stem grows upwards towards the sunlight.

Acorns

Blackberry

Peas

Date

Pear

Ash key

Tomato

Marrow

MALE AND FEMALE

Some flowering plants, such as marrows, produce separate male and female flowers on the same plant. Others, such as willow and holly trees, produce them on separate plants. Holly trees that have only male flowers never have berries. Only female date palms produce dates.

SEEDS BUT NO FRUITS

Some other plants, apart from ferns and mosses, do not have seeds in fruits – conifers, for example.

Conifers are mainly trees or shrubs, and many of them, such as pine trees, have very thin, pointed leaves called needles. Almost all conifers have tough cones. Male cones make pollen, female cones contain ovules. Wind carries pollen from male to female to make seeds.

Two small and ancient plant groups are the gingkoes and cycads. Gingkoes have fan-shaped leaves and fleshy seeds which are not hidden inside cones. Cycads look like large ferns but, unlike ferns, produce seed cones. The Welwitschia, a strange droopy-looking plant that has adapted to life in the desert, has seeds surrounded by modified leaves, which are called bracts.

SEED DISPERSAL

Many plants rely on the wind to scatter their seeds. The poppy has a 'pepper-pot' capsule to scatter its seeds. Dandelion seeds parachute on the breeze. Some gourds shoot out their seeds.

Winged seeds, like those of the sycamore and ash, are really fruits. Seeds eaten by animals have tough coats to protect them as they pass through the animal's digestive

system. Some of these coats are so thick that the seed cannot germinate unless its coat has been weakened by the animal's digestive juices. The droppings of the animal, with which the seeds fall to the ground, can act as fertilizer, helping the young plant to grow.

NEW GROWTH

When a seed germinates, it absorbs water and swells. Its skin splits and the embryo root (radicle) pushes down to find water and anchor the plant firmly in the ground. Now the embryo shoot (plumule) can uncoil and grow upwards. When it reaches daylight, the first leaves grow, and the plant starts making its own food. It no longer needs the food store (cotyledon) in the seed from which it grew.

The seeds of some plants grow very quickly and produce a fully grown plant in a matter of weeks. Plants like these live for only a few months, in which time they flower and make their own seeds for the next year's growing season. Longer-living plants, such as trees, take time to grow from their seeds and do not make their own seeds for several years.

Fruits may not look alike, but they all have the same function. Many of what we call vegetables, such as marrow, tomatoes and runner beans, are in fact fruits because they contain seeds.

◀ A coconut palm produces huge fruits. Palm nuts may drift across the sea and take root on islands that are thousands of kilometres away.

SEE ALSO
44–5 Plant anatomy,
47–49 Flowering plants,
74–5 Animal reproduction

TREES

There are two main groups of trees: conifers and broad-leaved trees. Trees are among the largest and longest-living of all life forms on Earth.

This cross-section through a tree trunk shows heartwood surrounded by lighter sapwood. A new growth ring is added every year, so this section shows the tree's age. The tree grows more slowly during winter leaving a dark ring. The fast summer growth causes a light-coloured ring.

Of the two main types of tree, conifers have been on Earth the longest. Conifer trees are also known as softwoods. They include the pines, spruces, cedars, firs, junipers, cypresses and, largest of all, redwoods. Common broad-leaved trees include the oak, ash and willow.

CONIFER LEAVES

Most conifers have small, needlelike leaves. All except the larch are evergreen. They shed old leaves and grow new ones all year round, not all at once in autumn, like many broad-leaved trees. The stiff needles of conifers lose much less water than the leaves of other trees. So conifers are able to grow in cold regions, on mountains and on the fringes of deserts. In Canada, northern Europe and Russia, they form vast evergreen forests. Many conifers have a triangular shape, which stops snow from settling and breaking on the branches.

Tropical rainforest trees grow up to 50 metres tall. The tree's massive lower trunk holds it steady as its top sways in the wind.

BROAD LEAVES

In cool climates with long dark winters, broad-leaved trees lose their leaves in autumn. Tropical broad-leaved trees keep their leaves all year because the day length does not change much between seasons.

Losing leaves in autumn is an energy-saving trick because there is not enough sunlight in winter for the leaves to make food for the tree. The food and water pipes in the stem are sealed off, starving and drying out the unwanted leaves.

DECIDUOUS TREES

Deciduous trees are broad-leaved, with spreading crowns and roots that reach deep into the soil to find water. These trees have flowers in spring. The flowers develop into fruits. The horse chestnut (shown here) has pinkish-white flowers. Its seed, the conker, is cased inside a spiky shell. In autumn, the tree sheds its leaves and the bare branches are exposed, with the new buds ready to burst into leaf when spring comes. In autumn, the leaves fall to the ground because there is not enough light during winter for efficient photosynthesis.

Shedding its leaves means a deciduous tree needs less water in winter. The food supply is cut off near the bud, and the leaf dies.

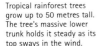

Horse chestnut (summer)

Horse chestnut (winter)

Blossom and leaf

Seed (conker)

Roots take in water and minerals and anchor the tree to the ground. Some trees have long roots, with as much growth below the ground as above it. Other trees, such as conifers, have massive trunks, but shallow roots that collect water from a wide area.

Fig tree

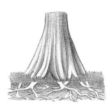

A giant sequoia in Sequoia National Park, California, USA. Sequoias are shorter than redwoods, but thicker. These conifers can live for over 4,000 years.

The banyan of India has hundreds of air roots which hang down from the branches.

The cycad looks like a palm tree with cones. Cycads first grew 225 million years ago.

Some bristlecone pines in the United States are an amazing 4,000 years old and are still growing.

The tree has already stored enough food to allow its next year's buds to grow, and without its food supply, the leaf dies. The chlorophyll that made the living leaf green is broken down and re-absorbed by the tree for use in next year's leaves. The coloured chemicals that remain in the leaf produce the brilliant red, yellow and brown colours of autumn.

INSIDE A TREE

When a tree is cut, you can see the annual growth rings in the slice of trunk. The outer bark, or cork, is a layer of dead tissue which is hard, to protect the softer living parts of the tree inside. Bark can stretch, to let the trunk and branches grow more thickly underneath.

Beneath the bark of most trees is the phloem carrying food through tiny tubes. The inner wood, or xylem, is stiffened with waxy substances for extra strength. Some of the younger xylem acts as a network of tubes to carry water, called sap, from the roots to the leaves. Other plants move water through xylem, but cannot stiffen it to make wood. It is the stiffness of the woody xylem that gives trees their rigidity, and allows them to grow so much taller than other plants.

TREES AND THE ENVIRONMENT

The leaves of a tree, like the leaves of any green plant, make food through a process called photosynthesis. The waste product of this process is oxygen, which is passed back into the air through tiny pores in the leaves called stomata. By absorbing carbon dioxide and releasing oxygen, trees keep the atmosphere healthy. Forests can be thought of as the Earth's 'lungs'.

Trees also protect the land from erosion. Their leaves and branches absorb heavy rains. Their roots bind the soil and stop it from being washed or blown away. Trees also suck up water and prevent flooding.

CONIFEROUS TREES

Coniferous trees, like this Chilean pine or monkey puzzle tree, rarely drop their leaves (needles). They have shallow roots and make their seeds inside cones. Cones may take three years to ripen and release their seeds, which then flutter to the ground. The monkey puzzle is unusual because its male and female cones grow on separate trees. Many conifers produce both male and female cones on the same tree. The male cones release pollen and fertilize the eggs in the female cones.

Bark

Female cone

Chilean pine (monkey puzzle tree)

SEE ALSO

40–1 Classification of livinng things, 44–5 Plant anatomy, 150 Wood and paper

BIOMES AND HABITATS

A biome is a region of Earth characterized by climate and containing distinctive plant and animal life. A biome is made up of different habitats.

Deserts are defined as areas in which more water evaporates than falls as precipitation. One fifth of the world's land is desert. Although deserts are harsh, dry habitats, they are far from lifeless. Animals and plants have developed many adaptations in order to survive.

Pleats on the stem of the saguaro cactus absorb water and unfold after heavy rains

If Earth is viewed from space, it appears to be covered with large, distinct areas – for example, deserts, oceans and forests. Each of these areas is called a biome. The boundaries between the different biomes are separated by variations in the climate, particularly in the average annual rainfall and temperature. The climate determines what type of vegetation is found in each particular region. In turn, the vegetation determines the type of animals that are found there. Together, all the biomes make up the biosphere – the region of Earth that supports life.

PLACES TO LIVE

There are many different biomes on Earth, and not all scientists agree on a definitive classification. However, the generally accepted biomes include tundra, taiga, temperate forest, scrub land, tropical rainforest, grassland, desert (which includes the polar regions), marine, fresh water and estuary.

Howler monkeys are commonly found in the Amazon rainforest of South America. Howlers feed mostly on the abundant plant matter surrounding them, but also on insects, small mammals and birds.

A biome is made up of smaller distinct regions called habitats. A habitat is defined as where an organism lives. For example, a pond skater's habitat is a pond or lake, an earthworm's habitat is the soil and a conifer tree's habitat is the soil, as well as the space above and around the tree. The organisms that live in a particular habitat are called a community.

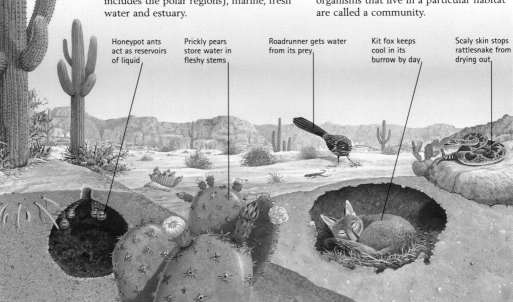

Honeypot ants act as reservoirs of liquid

Prickly pears store water in fleshy stems

Roadrunner gets water from its prey

Kit fox keeps cool in its burrow by day

Scaly skin stops rattlesnake from drying out

The ocean is a huge biome with many habitats, from the shallow coast to the deep sea. The ocean contains countless invertebrates, fish, reptiles and mammals.

Key to illustration: 1 Herring **2** Sperm whale **3** Prawn **4** Rat-tail **5** Angler fish **6** Grenadier **7** Cod **8** Gulper eel **9** Sea spider, tube worms, clam, white ghost crab **10** Tripod fish **11** Viper fish **12** Swallower **13** Lantern fish **14** Swordfish **15** Yellowfin **16** Giant squid **17** Hammerhead shark **18** Barracuda **19** Portuguese man-of-war **20** Plankton **21** Green turtle **22** Sea lion **23** Common dolphin

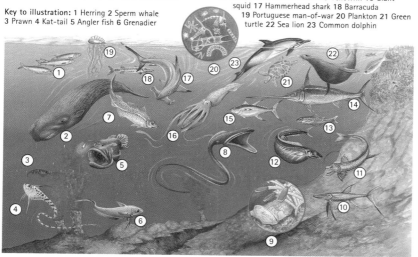

Penguins are flightless birds that live in the cold Antarctic. They have layers of fat to insulate their bodies. Adults and chicks huddle together in large groups to keep warm.

Arctic hares change the colour of their coat with the changing seasons. In summer, they have brown fur, but in winter, they grow a new white coat to blend in with the snow.

In any habitat, a number of species live and depend on one another in what are known as mutually beneficial relationships. Plants provide food or shelter for animals. In return, animals may help to pollinate the plants. In turn, plant-eating animals are eaten by other animals.

WHERE ANIMALS LIVE

Every habitat provides the right conditions for the plants and animals that live in it. Penguins and polar bears are adapted in different ways for life in the bitterly cold polar regions. The kit fox and fennec fox look similar, but are not related. The kit fox lives in the deserts of North America, while the fennec is a desert fox of North Africa. If the habitat in which an organism lives changes, some species may be able to change their way of life and adapt to the change in conditions. However, often other species cannot cope, and they have to move or they will die. This ability to adapt is the driving force behind the process known as evolution.

UNUSUAL HABITATS

Some animals have adapted to live in the most harsh habitats. For example, some polar fish have evolved a natural form of antifreeze in their blood and tissues. This substance prevents ice crystals forming, even below freezing-point, and the fish are able to survive. Japanese macaque monkeys keep warm in winter by sitting in hot-water thermal pools.

Coral reefs are formed by colonies of tiny coral invertebrates. The reef offers shelter to fish and other sea creatures.

SEE ALSO

16–17 The oceans, 52–3 Classification of living things

MARINE INVERTEBRATES

The first multi-celled animals were invertebrates
(creatures without backbones) and lived in sea.
The oceans are still full of invertebrate animals.

Sea anemones look more
like plants than animals.
They cling to rocks,
catching their prey in
gently waving tentacles.

The descendants of prehistoric
invertebrates still swim and crawl in
the oceans. They include worms, corals,
clams, snails, starfish, octopuses and squid.

NO BACKBONE
Being without a backbone is no handicap
to these very successful animals. In fact,
there are far more invertebrates than
vertebrates. In the sea, invertebrates can be
very big. The North Sea bootlace worm
grows to be up to 25 metres long. The
giant clam of the Pacific can weigh more
than 300 kilograms. Invertebrates can also
live to great ages, as long as 220 years in
the case of the Atlantic quahog clam.

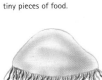

Sponges are a collection
of individual cells
organized into one body
on the sea floor. A sponge
cannot move. It sucks in
sea water and filters out
tiny pieces of food.

CORALS
Coral is made of the limestone
'skeletons' of tiny coral polyps.
These grow by 'budding', each
new polyp burying its parent. In this way,
the mass of coral grows ever larger, to
form plant-like shapes and huge reefs.
Some types of coral that live in warm
shallow water have single-celled plants
called algae inside them. The algae make
sugar using sunlight, which they share
with their coral hosts. The corals protect
the algae, and provide them with the
minerals they need to make proteins and
other important chemicals.

Medusa (jellyfish) can be
as small as a pea or as
big as a table. The mouth
and tentacles dangle
beneath the soft body.
The tentacles pull food up
into the animal's mouth.

A jellyfish catches its prey with stinging tentacles, and
swims by opening and closing its umbrella-shaped body
to squeeze out water.

TENTACLES AND SPINES
Jellyfish and anemones have tentacles
armed with stinging cells. The Portuguese
man-of-war is actually thousands of
individual animals acting as one. Its long
tentacles paralyse any fish they touch. The
outsides of the tentacles are covered in
stinging cells that contain tiny poisonous
needles, to paralyse the man-of-war's prey.
Echinoderms (starfish, sea urchins, sea
cucumbers, brittlestars and featherstars)
have spiny skins and tube feet. They
pump liquid into their feet to make them
expand. When a starfish loses a leg, it will
usually grow another.

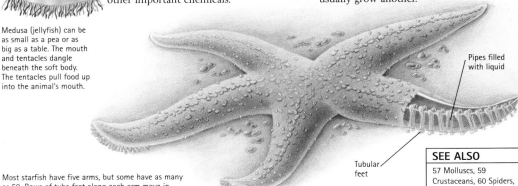

Pipes filled
with liquid

Tubular
feet

Most starfish have five arms, but some have as many
as 50. Rows of tube feet along each arm move in
rhythm as the starfish hauls itself over the sea floor.

SEE ALSO
57 Molluscs, 59
Crustaceans, 60 Spiders,
centipedes and scorpions,
61–3 Insects

MOLLUSCS

There are more than 100,000 kinds of molluscs, ranging from tiny snails to enormous giant squid. Some live on land, but most live in water.

A mollusc has a soft body covered by a protective mantle. Many molluscs have a chalky shell attached to their mantles, which may be outside the body (as in a limpet or snail), or inside (as in cuttlefish). The octopus is a mollusc which does not have a shell at all. Molluscs live on land, and in fresh and salt water.

Some molluscs, like the mussel, hardly ever move. They cling to one place and feed by opening their shells. Other molluscs have a single foot, and can move slowly. Limpets move around to feed, but always return to exactly the same spot.

ONE SHELL OR TWO

Snails have one shell, inside which the animal can coil itself to escape danger. Other molluscs have a shell in two halves, hinged so that the halves can be opened. The two-shelled molluscs are known as bivalves, and most are filter-feeders. Many live buried in the seabed and seldom move from one spot.

OCTOPUS AND SQUID

Octopus and squid are the largest and probably the most intelligent of all molluscs. They have large brains and excellent eyesight, and move rapidly when they need to escape danger. They can also change the colour of their mantle.

An octopus uses long sucker-arms to catch prey, and then bites with its beak-like mouth. Most octopuses are small, but giant squid can be 15 metres long.

These animals swim by jet-propulsion. They suck water under their soft mantles and force it out of a nozzle.

Garden snails move slowly on a trail of moist slime. The snail has sensitive tentacles on its head which it uses to feels its way around objects.

SNAILS AND SLUGS

Snails live on land and in water. Some eat plants, others are carnivorous. In dry weather, land snails seal up their shells to keep moist. Slugs are very similar to snails, except they either have tiny shells inside their mantle or no shells at all.

Snails and slugs have strong teeth made of a substance similar to iron ore. Some types can use their teeth to grind through the shells of other molluscs. Many scientists think snails use their iron teeth like a compass, so they do not get lost.

The squid hunts by stealth, but can make a quick getaway by contracting muscles to squirt a jet of water out through its body. Using this method, a squid can reach 30 km/h to escape a hungry fish or sperm whale.

The piddock is a bivalve mollusc. Its paired shells have fine teeth, which grind away like drill bits to make holes in solid rock. The piddock hides inside.

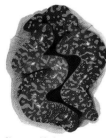

Clams are bivalve molluscs. They open and close their hinged shell with a powerful muscle. Clams filter food out of seawater using hairlike structures called cilia.

Many molluscs burrow into sand to hide from predators. They feed through siphon tubes that protrude from the sand.

SEE ALSO

16–17 The oceans, 56 Marine invertebrates, 58 Worms

WORMS

Worms are legless invertebrates – animals without backbones. Some live in the soil or in water, while others are parasites, living inside other animals.

The bristleworm is a good swimmer, propelling itself through the water with undulating movements of its body. It has feelers on its head and leg-like projections on each body segment.

There are more than 55,000 different species of worms. The four main groups are ribbon worms, flatworms, roundworms and segmented worms.

PARASITES

Most simple worms, such as flatworms, are microscopic organisms too small to be seen with the naked eye. However, some parasitic tapeworms grow to more than 20 metres in length. Parasitic worms live in or on plants or animals. The larvae (young) of one nematode worm can be ingested by humans, and migrate to the lungs. The human host coughs up and swallows the larvae, which are taken to the stomach. Here, they feed off the food eaten by the host and lay eggs that pass out in the host's faeces. These eggs may be picked up to infect another person.

EARTHWORMS

The earthworm is a segmented, or annelid, worm. Earthworms swallow soil and digest plant matter in it. Waste soil is then excreted as a worm cast.

The worm's body is made up of a series of ring-like segments, covered with tiny bristles that aid movement. Earthworms are hermaphrodite, which means they have both male and female sex organs.

MARINE WORMS

The body structure of marine worms, or polychaetes, is similar to that of annelids, but the bristles on their bodies are longer than on annelids. Some marine worms eat plants, but many are hunters, shooting out a long tube (proboscis) to grab prey.

The underside of a typical garden earthworm is covered with tiny bristles, which the worm uses to pull itself along. Earthworms breathe through their skin.

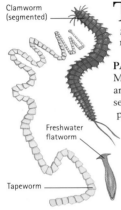

Clamworm (segmented)

Freshwater flatworm

Tapeworm

Leech

Not all worms resemble the familiar segmented earthworms found in the garden. Leeches look like slugs with suckers, while the clamworm, a marine worm, is covered with bristly projections.

Pharynx

Head

Nerve cord carries nerve signals

Muscle

Intestine

Nerve cord

Close-up of one segment

Casts of undigested soil pass out of anus

Clitellum produces cases for worm eggs

Body is made up of ring-like segments

SEE ALSO

40–1 Classification of living things, 56 Marine invertebrates

CRUSTACEANS

Crustaceans form part of a large group of organisms called arthropods. There are more than 38,000 different species, many of which live in the ocean.

A woodlouse has seven pairs of legs. Its body is covered with horny, overlapping plates, which protect it from predators.

Male fiddler crabs have one extra-large claw, which they wave to warn off other males.

Barnacles are small crustaceans that cling to rocks, ships and piers. They thrust out feathery feet to catch food.

Many crustaceans, including prawns, lobsters and crabs, are as familiar in the kitchen as they are in biology books. But there are many more species – woodlice, water fleas and barnacles all form part of the diverse group of animals called crustaceans. Crustaceans have hard outer shells, called exoskeletons, that protect their soft body tissues, and all have jointed bodies and legs.

A few crustaceans are terrestrial, or land-dwelling, organisms. For example, the woodlouse inhabits damp, dark places. When alarmed, it rolls itself into a ball to protect itself. Most other crustaceans live in water, however, although some crabs can live ashore – robber crabs even scramble up trees.

FROM EGG TO ADULT

Crustaceans start life as eggs. These hatch into tiny larvae. Millions of these larvae form the plankton that floats in the sea. Many crustacean larvae look nothing like their parents. Barnacle larvae are agile swimmers. But when the larvae become adult, however, they clamp themselves tightly onto a rock and never move.

An American lobster feasts on a herring. Lobsters are the heaviest crustaceans, weighing up to 20 kilograms. Lobsters are important in the fishing industry, since they are extremely popular as food items.

OUTGROWING THEIR SHELLS

As the larva grows, it sheds its shell and grows a new and larger one. This process is called moulting. Crustaceans continue to moult all their lives, which poses a problem for the hermit crab. Having no shell, it makes its home in the empty shells of whelks, winkles and other molluscs. When the crab outgrows its home, it looks for a larger shell to move into. During the time it is changing shells, the crab is vulnerable to enemies.

Crabs, crayfish, lobsters and shrimp all have ten legs. The front pair are adapted as pincers that are used for defence, for catching and tearing apart food, and even to signal to other individuals.

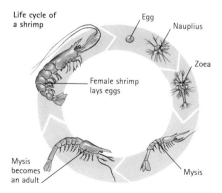

Life cycle of a shrimp

Egg

Nauplius

Zoea

Female shrimp lays eggs

Mysis becomes an adult

Mysis

Shrimps lay eggs that hatch into swimming larvae. The larvae go through several stages of development, eventually becoming mature adult shrimps.

Shrimps (foreground) and lobsters (top right) are among the best-known of the crustaceans. Along with crabs and crayfish, these creatures form a large order of crustaceans called decapods, all of which have five pairs of limbs.

SEE ALSO

16–17 The oceans, 56 Marine invertebrates, 57 Molluscs

SPIDERS, CENTIPEDES AND SCORPIONS

These animals are arthropods, like crabs and insects.
Spiders and scorpions are known as arachnids, and
centipedes and millipedes are myriapods.

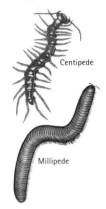

Centipede

Millipede

Arachnids are almost all hunting animals, and they are armed with poison fangs and an armoury of traps. Although centipedes are hunters, millipedes are vegetarians (herbivores). And some millipedes are poisonous, but use their venom only in self-defence against their predators.

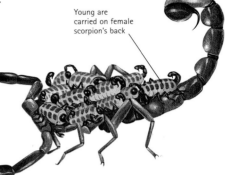

Young are
carried on female
scorpion's back

Few arthropods are caring parents. However, some female scorpions give birth to live young and carry them on their backs, to ensure their offspring are not eaten.

SPIDERS

Spiders are among the most successful hunters in the animal world. Some spin silk webs to trap flying insects. Others chase their prey at speed, or lurk in burrows waiting to pounce. Crab spiders live inside flowers and run sideways. Fisher and water spiders hunt close to – or actually in – water.

All spiders spin silk, even those that do not make webs. Spiders paralyse their prey with poison fangs. Some kinds, such as the black widow spider, can give a painful bite, though it is seldom fatal to humans. Spiders can eat only liquid food, so they squirt digestive juices onto their victims, which turn them into a liquid that is then sucked up by the spider.

Not all centipedes have 100 legs. Some have only 30, while others as many as 170 pairs. Millipedes have many more legs, up to 375 pairs. Centipedes are hunters. Millipedes eat decaying vegetation.

Trap-door

SCORPIONS

Scorpions are found mostly in warm, dry climates. They have pincer claws for seizing prey, and a poison sting in the tail. The tail curls over the scorpion's head when ready to strike. The sting of many types of scorpion can make a person ill, but very rarely kills them.

CENTIPEDES AND MILLIPEDES

Like spiders, centipedes have a poisonous bite. The first pair of a centipede's legs are modified as fangs. Centipedes hunt at night, scuttling at high speed to catch their prey – molluscs, worms and small insects. Millipedes have twice as many legs as centipedes. They eat the leaves of plants and rotting material.

Spider
in burrow

Many spiders hunt by lying in wait for unsuspecting passers-by. The trap-door spider hides in its burrow, opening the hinged door and pouncing when it senses the movements of a passing insect.

Bird-eating
spider

The bird-eating spider is a kind of giant tarantula. Its body is almost 9 cm long, and it is powerful enough to kill nestlings in birds' nests, though insects make up most of its day-to-day diet.

Orb-web
spider

Orb-web spiders make the most complex of all spiders' webs. The spider may wait in the middle of the web, or hide close by holding a signal thread to alert it when the sticky web catches an insect.

SEE ALSO

57 Molluscs, 58 Worms,
59 Crustaceans

INSECTS

Insects have only one limitation – size. They can live anywhere in the world, and eat any kind of food, but they cannot grow to more than a few centimetres long.

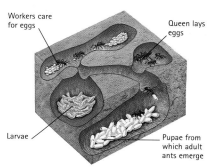

Ants are social insects. In the communal nest, the queen lays eggs. Workers look after the eggs and larvae. Soldiers guard the colony.

Workers care for eggs

Queen lays eggs

Larvae

Pupae from which adult ants emerge

A human louse, like all sucking lice, feeds on the blood of mammals. Some lice can bite. All lice have mouthparts adapted for their way of feeding.

The male Goliath beetle of Africa is the world's heaviest insect. It is 110 mm long and weighs 100 g. If a beetle were any heavier, it would collapse under its own weight.

Insects are by far the most numerous of all animal species. More than a million species are known – more than all the other animal species added together.

AN INSECT'S BODY

Insects range in size from tiny fleas, that can be seen only through a microscope, to beetles as large as your hand. Insects have no bones. Instead their bodies have hard outer coverings, or exoskeletons.

All insects have a similar body plan. The body is in three parts: a head, thorax and abdomen. The head has eyes, jaws and feelers. The thorax is the middle part, to which the legs and wings are attached. All insects have six legs, and many have one or two pairs of wings. The rear section, the abdomen, contains the stomach, reproductive organs and breathing tubes called spiracles. These tubes take the place of lungs, but because of them, insects cannot grow to a large size. If the insects were any larger, they would suffocate because air could not get into their bodies quickly enough through the spiracles.

Many insects have compound eyes, made up of hundreds of lenses, each of which gives a tiny image. The insect's brain combines these images into one picture.

SOCIAL INSECTS

Ants, termites and many wasps and bees are all social insects. They function only as members of a colony. The life of the colony centres on a single egg-laying queen. Males exist solely to fertilize new queens that fly off to found new colonies. After they mate with a queen, the males die. Workers carry out various tasks to keep the colony going. They are divided into castes for doing different jobs. Soldier ants, for exmaple, protect the colony.

Social insects build complex structures. Wasps make paper nests from chewed-up plants. Bees build geometric honeycombs of wax cells. And ants and termites build huge nests out of soil. ▶

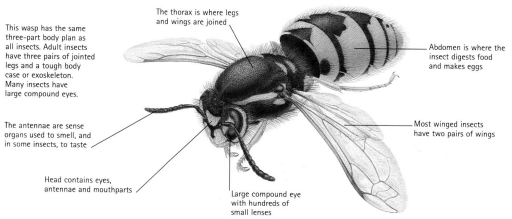

This wasp has the same three-part body plan as all insects. Adult insects have three pairs of jointed legs and a tough body case or exoskeleton. Many insects have large compound eyes.

The antennae are sense organs used to smell, and in some insects, to taste

Head contains eyes, antennae and mouthparts

The thorax is where legs and wings are joined

Large compound eye with hundreds of small lenses

Abdomen is where the insect digests food and makes eggs

Most winged insects have two pairs of wings

Flies are one of the largest insect groups, with more than 750,000 kinds. Many flies lay their eggs in dung and rotting food. Some spread diseases, such as cholera and malaria.

Dragonflies are the fastest fliers in the insect world. They can hover and even fly backwards as they hunt, catching prey with their long legs. The dragonfly nymphs develop underwater in ponds.

A grasshopper lays eggs that hatch into wingless nymphs. As the nymphs grow wings, they look like smaller versions of their parents. The nymphs can only hop, so they are found near to the ground. The adults can fly onto stems.

Adult grasshopper

Winged nymph

Eggs

Wingless nymph

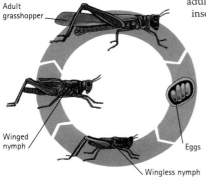

► Mating for some insects is a dangerous business. A female praying mantis eats anything within reach of her jaws – including the male with which she has just mated.

LIFE CYCLE OF A MONARCH BUTTERFLY

The butterfly lays its eggs on a food plant, milkweed. Each egg hatches into a larva, or caterpillar. The caterpillar eats and grows rapidly. Hungry birds are warned to keep away by the caterpillar's vivid markings. When full-grown, the caterpillar becomes a pupa or chrysalis. Inside this apparently lifeless case, many changes take place. The pupa splits open, and the adult crawls out. As soon as its wings are dry, it flies away.

1 Female lays eggs on leaves of milkweed

2 Caterpillars hatch and start to eat

3 Full-grown caterpillar spins silken thread

4 Skin is shed revealing a chrysalis (pupa)

FLYING
Most insects have two pairs of wings, though worker ants and fleas never have wings. A beetle does not appear to have wings. Its front wings form hard cases that cover the delicate back wings, with which the beetle flies. Faster fliers, such as wasps, have two pairs of wings. True flies have one pair of flying wings and their second pair have evolved into flexible rods that are used for balance.

LIFE CYCLES
All insects lay eggs. The young of most insects go through four stages of growth and development. In butterflies and moths, the stages are: egg, larva, chrysalis, adult. The immature insect is a caterpillar, which looks nothing like the adult. Caterpillars do little else but eat plants. The main role of adult butterflies and moths is to mate.

Other insects such as grasshoppers and cockroaches go through a three-stage growing process – from egg to nymph to adult. The nymph looks like a miniature version of the adult insect, although nymphs usually do not have wings, whereas the adult form normally does.

FOOD
Evolution has equipped insects to eat an astonishing variety of foods. Some are carnivorous, hunting by speed (like the dragonfly) or by stealth (like the preying mantis). Others are plant-eaters, chewing leaves, sucking sap or boring into wood.

Swallowtails are large butterflies of Africa, Europe and Asia. Most butterflies fly by day.

5 Inside, an adult butterfly forms and emerges as skin splits

6 Butterfly clings to twig as blood flows into new wings

7 When its wings are hard and dry the butterfly flies away. The cycle can now begin again.

Moths fly mainly at night. They tend to be less colourful than butterflies. The luna moth is about 11 cm across.

PESTS AND HELPERS

A few insects, such as the desert locust of Africa, are destructive pests. Swarms of locusts can eat an entire field of crops in a few hours and cause a famine in days. Many other insects, however, are helpful to people. Without bees and other flying insects, flowering plants would not be pollinated and fruit trees would not bear fruit. Honey, made from the sugary nectar of flowers by bees, has been collected by humans for thousands of years.

Scavenging insects, such as burying beetles, feed on dead matter and help to make the soil fertile. Ladybirds and some wasps are helpful because they eat other insects, such as aphids, that are pests. The silkworm (the larva of the silk moth) is reared by people for the silk that it spins when turning into a pupa.

MINIATURE MARVELS

There are some extraordinary record-breakers in the insect world. A cicada's song can be heard 500 metres away. A queen termite may live for 50 years. Many insects can lift or drag objects 20 times their own weight. Yet the largest insect, the Queen Alexandra birdwing butterfly, measures only 28 cm across.

An emperor moth. Moths hold their wings open when resting. When the wings are flapped, their spots can startle enemies.

Locusts, usually solitary, form vast migratory swarms, when thousands come together to lay eggs. Armies of wingless nymphs march overland, then take wing. Some swarms eat every single leaf in sight.

Large butterflies, like this Rajah Brooke birdwing, flutter in tropical forests. Butterflies fold their wings when at rest.

SEE ALSO

60 Spiders, centipedes and scorpions, 74–5
Animal reproduction

FISH

Fish were the first animals with backbones (vertebrates) and skeletons made of bone, and are the animals best adapted to life in water.

There are three groups of fish, all of which are found in sea water and fresh water. The cartilaginous fish have gristly, rather than bony, skeletons and leathery skins, not scales. They include the sharks and rays. The smallest group are the jawless fish, such as lampreys and hagfish. By far the largest group are the bony fish, which have bony scales covering their bodies and bony skeletons. Bony fish are found in both ancient and evolved forms.

LIVING IN WATER
The first fish appeared in the oceans about 540 million years ago. By breathing through gills, fish are fully adapted to life in water. Most fish cannot live out of water. Only the lungfish can breathe air because it has primitive lungs. Lungfish live in stagnant water that does not have much oxygen in it, so they gulp breaths of air at the surface.

SALT OR FRESH?
About 60 per cent of fish species live in salt water. A few kinds can live in either salt or fresh. Sea fish need salt and other chemicals to live. Because sea water is saltier than their own body fluids, they lose water through their skins by a process called osmosis. They drink sea water to prevent their bodies drying out. It is the opposite for freshwater fish. Their bodies are saltier than the water, so they

The deep-sea angler fish lives in darkness and uses a luminous lure to attract prey. Many deep-sea fish make their own light, sometimes to attract mates.

absorb it. Freshwater fish get rid of the extra water by passing a lot of urine.

MIGRATION
Some fish make long migrations to their breeding grounds. Salmon begin their lives in fresh water, swimming downriver to the sea. When ready to breed, adult salmon make an astonishing return journey, to the same river in which they spawned. There they mate and breed, and die – their life-cycle complete. American and European eels make similar migrations, from freshwater rivers to the salty water of deep oceans.

PARENTAL CARE
All female fish lay eggs, usually a great many, which are fertilized by males. When the babies hatch, they must fend for themselves. Only a few fish care for their young. Sticklebacks build nests, which the male defends aggressively. Some cichlids shelter their young in their mouths. Baby seahorses hatch and develop in a pouch on their father's body.

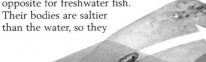

Lungfish
in water

Lungfish
in burrow

PARTS OF A FISH

The body of a typical bony fish is streamlined for swimming, although many fish have round or flat shapes. Fish use tails for swimming and fins for balance and steering. Fish senses include the lateral line system which detects subtle changes in water pressure.

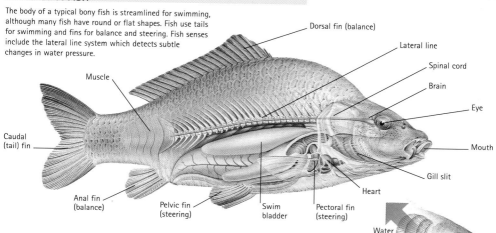

Dorsal fin (balance)

Lateral line

Spinal cord

Brain

Eye

Mouth

Gill slit

Heart

Water out

Muscle

Caudal (tail) fin

Anal fin (balance)

Pelvic fin (steering)

Swim bladder

Pectoral fin (steering)

SHAPES AND ADAPTATIONS

Fish come in a variety of body shapes. Eels look more like snakes. Flatfish such as plaice begin life the right way up and then lie on their sides. One eye travels across the head as it grows, so that the adult fish can lie hidden on the seabed with two eyes looking up. Porcupine fish have prickly skins and blow themselves up, like balloons, to frighten away or baffle a predator. Some fish, like the scorpion fish and sting rays, have poisonous spines, which can even kill humans.

Fish have many unusual adaptations. Deep-sea fish, living in a gloomy world that is without sunlight, make their own chemical lights. Angler fish wave a lure to entice prey within snapping reach. Freshwater archer fish catch prey by spitting jets of water at insects, knocking them out of the air.

Some fish can leave the water for a short while. Mudskippers use leg-like fins to crawl over mud. Flying fish use their long, stiffened fins as wings and glide through the air to escape enemies.

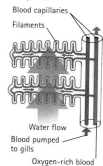

Water in

Gills

Fish breathe through gills on either side of the head. Water passes out through the gill slits.

Blood capillaries

Filaments

Water flow

Blood pumped to gills

Oxygen-rich blood

The gills contain tiny blood-filled filaments. These remove oxygen from water the fish gulps in through its mouth.

A shoal of black-striped salema. Fish swim in groups for protection. One fish is hard to catch among so many.

SEE ALSO

16–17 The oceans,
66–7 Amphibians,
68–9 Reptiles

AMPHIBIANS

Amphibians are the smallest class of vertebrates. Although they were the first animals to colonize the land, amphibians must return to the water to breed.

A long-tailed salamander spends most of its life on the land.

Like all newts, the smooth newt hunts for food in the water. It uses its long tail to whip through the water as it swims.

Like many frogs, the green tree frog inflates a throat sac to 'sing' during the mating season to attract potential female mates.

Caecilians live in tropical regions. They hunt at night, and live in burrows underground. There are only 160 species in this group of amphibians.

Amphibians are the smallest group of vertebrates (animals with a backbone), having some 3,000 different species. Like fish and reptiles, amphibians are cold-blooded animals. This means that amphibians cannot regulate their own temperature and rely on the Sun to warm their bodies. Most amphibians begin their lives in the water, breathing with gills. As they grow, they develop lungs and legs, and are able to move about on dry land.

WHERE AMPHIBIANS LIVE

Amphibians are found throughout the world except the polar regions. They live in a number of different habitats, including rainforests, ponds, woodlands and lakes, as well as high mountain grassland and even deserts.

Although adult amphibians can survive dry periods, most need to live in a damp environment, such as river or pond. In the humid tropical rainforests, many frogs can survive without the need for permanent water – they use tiny droplets of water that accumulate in the leaves of plants.

Since amphibians are cold-blooded creatures they become inactive in cold conditions. In extreme cold, they may hibernate, often in mud at the bottom of a pond or under a log.

THE THREE GROUPS

Frogs and toads make up 80 per cent of all amphibian species. Their back legs are long and powerful, enabling the creatures to leap, although many toads prefer to

1 Frogs lay eggs, or spawn, in water. The eggs are encased in jelly.

crawl along the ground. Newts and salamanders form the second group, which have short legs and long tails. The third and smallest group are the caecilians. They look like worms and live underground.

FOOD AND FINDING MATES

Many people find it difficult to tell frogs and toads apart. As a general rule, frogs have smoother skins than toads and spend more of their time in water. Most frogs and toads prey on insects and other small animals, staying perfectly still and waiting for their prey to pass. Some use their long, sticky tongues to catch prey.

Some toads and frogs have a long tongue with a sticky tip that they shoot out to catch unsuspecting insect prey.

▶ The axolotl of Mexico is a curious animal – a salamander like no other. Axolotls never grow up. Instead, they remain in the tadpole stage of development – complete with gills – all their lives, and even breed as tadpoles.

AMPHIBIAN LIFE CYCLE

The life-cycle of the common frog starts when a male and female mate. Male frogs compete vigorously to mate with females. Many males make loud cries to attract females and deter rivals. Females produce masses of eggs or spawn that are usually fertilized externally. After about ten days, the eggs hatch into tadpoles with gills, which live and feed in the water. Eventually, the tadpole grows back legs, then front legs, and its tail shrinks. Finally, after the adult frog has developed lungs, it clambers out to find food on dry land.

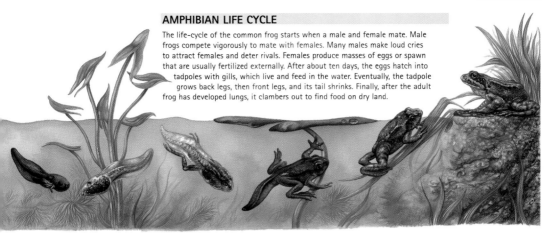

2 After around ten days, spawn hatches into tadpoles with gills.

3 The back legs of the tadpoles grow first.

4 The legs and lungs grow and the tail shrinks.

5 With fully developed lungs, the adult frog can leave the water.

Male frogs croak to call females during the mating season. They do this by forcing air over their vocal cords. The loudest frogs have an inflatable vocal sac that puffs up with air like a balloon.

FROM EGG TO ADULT

Amphibian courtship is often a frantic affair, with males and females gathering in large numbers. Once the eggs have been fertilized, most amphibians take no further interest in their young.

However, some take measures to protect the eggs. For example, caecilians coil themselves around their eggs inside their burrows. Baby Surinam toads hatch from eggs encased in tiny pockets on the surface of their mother's skin. The male midwife toad attaches its eggs to its back legs and carries them about for around three weeks until they hatch.

The largest amphibians, such as the 1.8-metre Chinese giant salamander and the cane toad, have few enemies. Most of the smaller species rely on camouflage and escape to evade a predator. Others are protected by their bright colours.

POISON FROGS

The startling colours of the little poison dart frogs of South America serve as danger signals to potential predators. Poison frogs are venomous. The golden poison dart frog of Colombia, for example, contains enough toxin in its body to kill around 1,000 people.

The bullfrog of North America is noted for its loud croaking mating call.

Gliding frogs stretch folds of skin beneath their toes. The folds act like parachutes when they leap.

A poison dart frog, from South America, is brightly coloured to warn off potential predators.

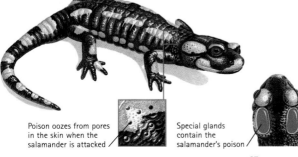

Poison oozes from pores in the skin when the salamander is attacked

Special glands contain the salamander's poison

Similar to all amphibians, the fire salamander has slime-making glands that lie just under the surface of the skin. The slime these glands produce helps to keep the skin moist and also provides a chemical defence against predators. The striking colours of the fire salamander warn other animals that it is poisonous. In ancient times, some people believed that salamanders could live in fire.

SEE ALSO

64–5 Fish, 68–9 Reptiles, 74–5 Animal reproduction

REPTILES

Reptiles are cold-blooded animals and they live in warm climates. They are characterized by their dry, scaly skin. Most reptiles lay leathery-shelled eggs.

Sea turtles spend most of the year in the ocean. Once a year, however, females leave their ocean habitat to lay their eggs.

1 The female makes her way to a sandy beach and hauls herself ashore, using her paddle-like flippers as legs.

2 She digs a shallow hole above the level of high tide, covers her eggs with sand and returns to the sea, until next time.

3 After a short period of development, the hatchlings break out of their shells and dig their way out of the sand.

4 Finally, the newly hatched turtles race for the water. Many are eaten by predators, but about 1% survive into adulthood.

For over 150 million years, reptiles were the dominant life forms on Earth. The best-known of these animals were the dinosaurs, but there were many others, including the flying pterosaurs and ocean-dwelling plesiosaurs and ichthyosaurs.

Today, there are four main groups of reptiles: alligators and crocodiles (about 25 species), tortoises and turtles (about 250 species), and the snakes (about 2,700 species) and lizards (over 3,700 species). The fourth group contains just one species, the tuatara of New Zealand.

LIVING ON LAND

Most reptiles are excellent swimmers and some, such as the turtles and terrapins, spend most of their lives in water. But reptiles can lay their eggs on land, unlike most amphibians. The leathery shells of the eggs stop the embryos drying out.

Reptiles have dry, scaly skins. Many are agile, fast-moving animals, the fastest being the snakes. All reptiles are considered to be cold-blooded animals, but the term is misleading. Although a reptile's

Crocodiles have huge jaws for seizing prey, but can pick up their young with surprising gentleness.

body warms in sunlight and is less active when cold, reptiles have a temperature regulation mechanism. Scientists have still to discover how this mechanism works.

CROCODILIANS

Crocodiles, and their relatives the alligator, the caiman and the gharial, are collectively called crocodilians. These reptiles are large carnivores, with strong jaws and powerful tails. They either lie on river banks, basking in the sun or stay almost submerged in the water, with just their eyes and nostrils showing. Crocodilians are

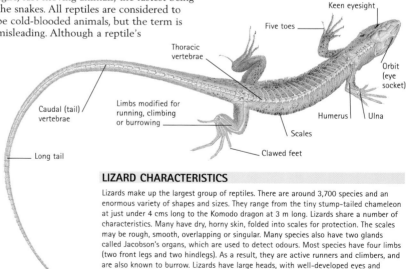

Keen eyesight

Five toes

Thoracic vertebrae

Orbit (eye socket)

Caudal (tail) vertebrae

Limbs modified for running, climbing or burrowing

Humerus

Ulna

Scales

Long tail

Clawed feet

LIZARD CHARACTERISTICS

Lizards make up the largest group of reptiles. There are around 3,700 species and an enormous variety of shapes and sizes. They range from the tiny stump-tailed chameleon at just under 4 cms long to the Komodo dragon at 3 m long. Lizards share a number of characteristics. Many have dry, horny skin, folded into scales for protection. The scales may be rough, smooth, overlapping or singular. Many species also have two glands called Jacobson's organs, which are used to detect odours. Most species have four limbs (two front legs and two hindlegs). As a result, they are active runners and climbers, and are also known to burrow. Lizards have large heads, with well-developed eyes and eyelids. Food, either plants or other animals, is seized in the jaws. Many lizards have long tails – a tempting target for a predator. Often, this is all the hunter ends up with, since the tail will snap off. The lizard makes its getaway and grows a new tail.

The largest snakes, such as the anaconda from South America, are constrictors. They grab their prey with their fangs and tightly coil themselves around it. The victim's major blood vessels rupture and it suffocates.

caring parents. Females lay their eggs in sand or in nests of vegetation and guard their newly hatched young fiercely.

TURTLES AND TORTOISES

Turtles and tortoises are covered in a hard shell – only the head, legs and tail are exposed. When alarmed, the tortoise hides its head inside the shell. Marine turtles are fast swimmers but are near-helpless when they come ashore to lay their eggs.

SNAKES AND LIZARDS

Snakes move by wriggling their bodies. Most species are found in deserts and tropical forests but some live in the ocean.

Snakes are carnivorous animals, which means they feed only on meat. They sense their prey by smell, using their flicking tongues to 'taste' the air, or with special heat-sensing organs. Poisonous snakes can bite as soon as they hatch. Vipers and rattlesnakes have long fangs – cobras and sea snakes have shorter fangs.

Many snakes are not poisonous. They kill their prey with a bite alone or, like boas and anacondas, by crushing prey to death in powerful coils.

Most lizards have legs, with the exception of slow worms. Some, like the Australian thorny devil, have spines and some, like the Gila monster of North America, are poisonous. Some lizards and snakes give birth to live young, but most lay eggs.

CATCHING FOOD

Many reptiles prey on insects and small mammals. Some snakes pursue mice into their burrows or climb up trees to take young birds from nests. A small lizard called the gecko has sucker pads on its feet, enabling the animal to run across flat ceilings of houses to catch insects. The chameleon moves slowly, whipping out a long sticky tongue to catch its next meal.

The largest lizard of all is the Komodo dragon from Komodo Island in Indonesia. Up to three metres long, it will eat small deer or wild pigs.

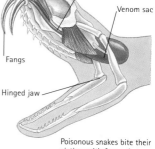

Venom canal

Venom sac

Fangs

Hinged jaw

Poisonous snakes bite their victims with fangs that fold down from the roof of the mouth. The venom is injected from a sac at the back of the head.

The giant tortoise of Aldabra is a slow-moving herbivore. Other giant tortoises live on the Galapagos Islands in the Pacific.

A Galapagos iguana can dive up to 15 metres underwater to feed on algae, kelp and other marine plants that grow near the rocky shore where it lives.

The chameleon changes colour to blend in with its surroundings. This is triggered by anger, fear or by variations in light intensity or temperature.

SEE ALSO
54–5 Biomes and habitats, 66–7 Amphibians

BIRDS

Birds are the largest group of warm-blooded vertebrates. All birds have feathers, beaks and a pair of front limbs that are modified into wings.

Birds of prey have hooked claws or talons that can grasp and crush prey.

Perching birds have one backward-pointing toe that provides a firm grip.

Swimming birds have webbed feet that they use like paddles.

▲ Most birds have four, clawed toes. The feet and claws of different species are adapted to suit their different ways of life.

A hummingbird hovers in front of a flower, sipping at the sugary nectar inside with its long tongue.

Birds are one of nature's finest examples of adaptation. They are found all over the world, from the coldest polar ice cap to the hottest desert. They live in the air, on land, and on and in water. The ability to fly is found in many insects and some vertebrates, but it is most developed in birds. It has allowed birds to occupy every land environment and to travel great distances in search of food.

Several unique features have enabled birds to become masters of the sky. They are the only animals with feathers, have hollow bones for lightness in the air and strong breast muscles to work their wings.

FEATHERED FINERY

Feathers are perfectly designed for flight. They provide not only a lightweight, aerodynamic surface, but also an excellent insulation for the bird's body. Feathers are made from a protein called keratin and there are several types. For example, the soft down feathers are closest to the skin and provide insulation. Flight feathers are larger and more stiff. Depending on the species, a bird has between 900 and 25,000 feathers on its body.

Male birds often have more colourful feathers, or plumage, than females. In the breeding season, the males use their distinctive plumage as a courtship display to attract females as mates.

Since a bird depends on its feathers to fly, it needs to keep them in good condition. It does this by preening – cleaning and combing its feathers.

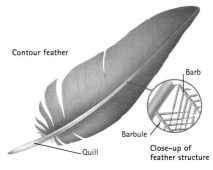

Contour feather

Barb

Barbule

Quill

Close-up of feather structure

▲ A flight feather has a central rod called a quill. In close-up, the thread-like barbs can be seen. The barbs are held in place by hooked structures called barbules.

BIRDSONG

Birds call or sing as part of their mating behaviour. Specific calls are used to attract a mate or to warn other males about territorial limits. The sounds are produced by the syrinx, a structure unique to birds that is located at the base of their trachea, or windpipe. The syrinx contains elastic membranes that vibrate when air is expelled from the lungs. Changes in the tension of the membranes alter the pitch of the call or song.

NESTS AND EGGS

All birds hatch from eggs. The egg has a hard shell, which protects the embryo inside. Most birds lay their eggs in nests, but some just lay them on the ground. The materials from which nests

At nearly 3 metres tall, the flightless ostrich is the largest of all the birds. The ostrich also lays the largest egg, weighing 1.7 kilograms.

The spotted harrier has a broad tail and wings so it can glide on air currents. Hawks rely on their keen eyesight to spot their prey.

1 Woodpecker (drill) 2 Crossbill (nutcracker) 3 Kestrel (tearing) 4 Spoonbill (detector/sieve) 5 Oystercatcher (probe)

◄ The size and shape of a bill determines the type of food a bird eats. A kestrel's hooked bill, for example, is used for tearing flesh, but the probing bill of the oystercatcher is used to find and open shellfish.

The plover lays its eggs in a scrape on the ground.

are made usually blend in with the surroundings so that the chicks that hatch in them remain safe. The eggs themselves are often coloured for camouflage to avoid the attention of predators.

Some birds lay one or two eggs, others as many as ten. They are incubated by one or both parents sitting on them until the hatchlings are ready to emerge.

TAKING FLIGHT
To fly, birds need to generate muscle power to beat their wings. The breast muscles are the source of propulsion for powered flight. A well-oxygenated blood supply is essential for the muscles to work efficiently. As a result, birds have evolved a well-developed pair of lungs and, like mammals, a four-chambered heart to

ensure that the muscles get the maximum oxygen from the blood. Large birds, such as albatrosses and vultures, can flap their wings slowly, or hover in the air on rising air currents. Smaller birds need to flap their wings faster to stay in the air.

Some birds have lost the ability to fly. The ostrich turns quickly to escape its enemies, and penguins have become masterful swimmers and divers.

FINDING FOOD
Birds eat high-energy foods, including fish, fruits, insects, seeds, worms and other birds. These provide the bird with sufficient energy to maintain their body temperature and fly. Birds avoid low-energy foods such as grasses.

Birds of prey are probably the most determined bird predators. The fastest is the peregrine falcon, which dives vertically through the air at speeds of up to 300 kilometres an hour. Vultures are scavengers, flying high to spot a dead animal and gliding down to pick the bones clean.

Water birds, such as grebes, nest on or beside lakes, streams and rivers.

Swallows make mud-nests stuck to the walls and rafters of buildings.

The American robin uses twigs and grasses to make a cup-shaped nest.

▲ Birds' nests can be little more than a scrape in the ground or a delicate structure of mud, plant material and saliva.

► The Australian lyrebird shows off its long tail feathers to attract a potential mate. Many male birds perform courtship displays during the breeding season.

◄ A North American bald eagle catches a fish, snatching it from the water with its feet. Eagles are one of the largest birds of prey. They seize prey in their talons, tearing the flesh into chunks with their hooked bills.

SEE ALSO
54–5 Biomes and habitats, 72–3 Mammals

71

MAMMALS

Mammals are warm-blooded animals that feed their young from milk-producing glands. Some mammal species are the most intelligent creatures on Earth.

All the animals belonging to the class Mammalia are warm-blooded creatures with hair. Among them are the most intelligent, fastest and largest animals on Earth.

Like humans, the adult male bonobo, or pygmy chimpanzee, of Zaire, is a primate. Primates are considered to be the most intelligent animals of all.

Mammals are an extremely diverse group of organisms, and they inhabit almost every type of habitat on Earth. They live on land, in hot or cold climates, in the sea, and they have even taken to the air.

Mammals are warm-blooded animals, often with leathery or furry skins. This means that all species control their body temperatures by sweating or panting when it is hot and shivering when it is cold.

GROUPS OF MAMMALS

Mammals appeared quite late in the evolutionary timescale. There were mammals during the age of dinosaurs, but they were small animals, rather like shrews, squirrels and badgers of today.

Mammals belong to the class Mammalia. This class consists of three main groups. The most primitive lay eggs, like reptiles and birds. These are the monotremes, and only two species survive today – the platypus and the echidna.

MARSUPIALS

Although marsupials give birth to live young, the offspring have only partially developed. As a result, the offspring stay in their mother's marsupium (pouch), feeding on her milk until they are fully developed. The kangaroo is probably the best-known marsupial. Most species are found in Australia and New Zealand.

THE HIGHER MAMMALS

The placental mammals are the most advanced group. The term 'placental' refers to the fact that this group gestates (develops) their young inside their bodies, providing nutrients through the placenta in the uterus. There are many different species. These include flying mammals (bats), sea mammals (seals, dolphins and whales), large herbivores or plant-eaters (elephants and giraffes) and powerful carnivores (dogs, cats and bears).

The primates include apes, monkeys and human beings.

THE LARGEST MAMMALS

Whales are the largest mammals that have ever lived on Earth. These humpback whales are 'lunge-feeding', scooping up mouthfuls of small krill from the water and straining the food through bristle-like growths in their mouths. The huge size of whales means that they need to consume vast quantities of krill every day. Since mammals need air to breathe, whales must come to the surface regularly to take in air. The blue whale is the largest animal on the planet. This huge marine mammal may reach up to 30 metres in length and weigh 135 tonnes.

A female kangaroo carries its offspring, or joey, in a pouch called a marsupium. The youngster stays in the pouch until it can look after itself.

Mammals that feed only on insects are known as insectivores. The giant anteater of South America has strong claws, a long snout, and a sticky tongue to help it dig into ant hills and devour the ants inside. A thick covering of hair protects it from the insects' bites.

LOOKING AFTER THE YOUNG

Mammals owe much of their success to parenting. They are generally the most caring parents in the natural world. The female feeds the young with milk from her body and looks after them until they can fend for themselves. During this time, the offspring learn essential survival techniques, such as social behaviour and methods of obtaining food. Some mammal young, such as mice, are born blind and helpless, and require a period of intensive parental care. Others, such as deer, stand and run within minutes of being born.

MAMMAL SENSES

Mammals have highly developed senses, which have also contributed to their success. Some have two eyes set either side of the head. Each eye provides a different view of the surroundings. Others have binocular vision – the eyes are at the front of the head and work together. This type of vision enables the animal to judge distances more accurately.

Some mammals have specialized senses, such as the bat's sonar (echolocation) and the mole's sensitive whiskers. For some mammals, the sense of smell is most important. For example, dogs use scent-messages to mark their territory.

In cold climates, some mammals, such as the European dormouse, hibernate. They do not eat during this period – rather they live on the fat stored in their bodies.

The duck-billed platypus of Australia is one of the few mammals that lays eggs. The female cares for the helpless young in a burrow, feeding them with milk from special glands on her body.

Instead of teats, the female platypus has milk pores

The newborn platypus laps milk oozing from its mother's milk pores

The eggs hatch after 10 days

Bats eat insects and feed from flowers. Bats are the only truly flying mammals. The wing of a bat consists of a thin membrane that stretches between the fingers, body and leg.

SEE ALSO

70–1 Birds, 74–5 Animal reproduction

ANIMAL REPRODUCTION

The longer an animal takes to reproduce, the larger it is and the longer it lives. Some animals are ready to breed within hours of being born; others take many years.

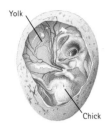

Yolk

Chick

Inside a bird's egg, the chick is fed from the yolk. The chick of a hen starts to hatch about 21 days after the egg is laid. By this stage, its feathers and claws are fully formed.

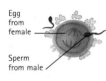

Egg from female

Sperm from male

In sexual reproduction, a sperm from the male fertilizes the larger egg cell of the female. In fertilization, chromosomes from the male and female sex cells come together.

Although there is a great variety of life on Earth, there are only a few basic methods of reproduction. An animal's life span is determined chiefly by the time it needs to reach adulthood, mate and reproduce. In animals, this procedure takes two forms: asexual, when only one parent produces the young, as in sponges and corals; and sexual, when male and female cells combine to form a new animal.

PLACENTAL MAMMALS

Mammals need most time to reproduce because most mammal babies take months or even years to develop. In the higher mammals (the placentals), the unborn baby develops inside its mother's body, to which it is joined by a two-way filter called the placenta. The placenta provides the baby with food and oxygen from the mother's blood, and takes away the growing infant's waste.

After birth, a mammal baby suckles on milk from the mother. The milk is produced in breasts, or mammary glands. These glands are unique to mammals and give them their name.

Mating involves pairing of males and females. Courtship rituals often involve elaborate behaviour, like the dancing of these egrets. Some mated pairs stay together for life.

REARING YOUNG

Birds that lay a clutch of eggs every year produce far more young in a lifetime than an elephant, which, every five years, gives birth to a baby that has taken nearly two years to develop inside its body.

Rearing young takes up most of a parent's (usually female) energies. Bears are by nature solitary animals. After giving birth, the female bear guards her young with care, teaching them to find food, so they will be able to look after themselves. Cubs will usually stay with their mother for one or two years.

▲ The offspring of hydra are created by budding off from their parent.

▶ The fertilized egg is the start of a new life. The first cell divides, and subsequent cells divide many times in a process called mitosis. Eventually, cells specialize to form organs.

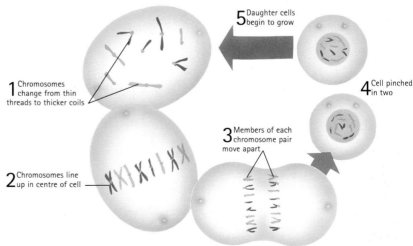

5 Daughter cells begin to grow

1 Chromosomes change from thin threads to thicker coils

4 Cell pinched in two

3 Members of each chromosome pair move apart

2 Chromosomes line up in centre of cell

GESTATION PERIODS

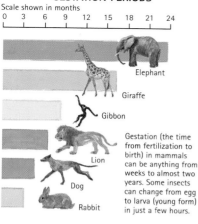

Scale shown in months

0 3 6 9 12 15 18 21 24

Elephant

Giraffe

Gibbon

Lion

Dog

Rabbit

Gestation (the time from fertilization to birth) in mammals can be anything from weeks to almost two years. Some insects can change from egg to larva (young form) in just a few hours.

Mammals form bonds between mates and within larger groups. Male lions father cubs, but do little else. Females rear the young and do most of the hunting for the group, or pride, working together as a team.

PARENTING AND LEARNING

Animal parents show various levels of concern for their young. A male king penguin keeps its egg, and chick, snug beneath a flap of skin on his feet all winter, until the comparatively warm weather of a polar spring arrives.

Parents teach by example, and the more complex an animal's way of life, the more there is for young animals to learn. After all, it takes humans many years to learn how to look after themselves without their parents' help.

Bird chicks can recognize their parents' voices. However, most of what a bird does is instinctive. Fox cubs must learn to hunt by imitating their parents, and through play. Monkeys and other animals which live in groups also learn by watching and copying. Even adults may pick up new behaviour in this way, by copying the food-gathering techniques of a more ambitious or daring individual.

LIFE SPANS

No animal lives as long as the oldest plants. A mayfly emerges from its larval stage, breeds and dies in a few hours. Over 20 years is old for most mammals. Some fish, such as carp and sturgeon, can live from 50 to 80 years. Elephants live to be over 60; tortoises and turtles to 100 or more. Humans in developed countries usually live to about the age of 70, slightly longer than chimpanzees (average 50–60).

CLONES

Clones are genetically identical organisms. This can happen naturally. The offspring of plants and animals that can reproduce asexually – that is without joining a male and female cell together – are clones of their parents. Most higher animals do not form clones naturally, except when identical twins are produced.

However, scientists have learned how to clone living things in the laboratory. The ability to clone animals and plants will be very useful in farming and medicine, but there are fears that the technology could be misused.

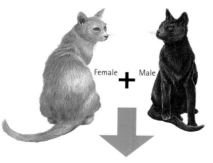

Female + Male

Kittens

Cloning techniques make it possible for scientists to make identical copies of an animal. Dolly the sheep, the first cloned adult mammal (1997), was an exact replica of her mother, and had no father.

◄ Pairs of chromosomes carry two genes for the same character. One gene is usually dominant. But sometimes two genes may be equal and mix their effects. So if a black male cat and a female ginger cat have kittens, some kittens will be black, some ginger, and some a mixture of the two.

SEE ALSO

50–1 Fruits and seeds, 107 Genes and chromosomes

FACTS AND FIGURES

BRANCHES OF LIFE SCIENCES

Anatomists study the structures of living organisms, often using microscopy.
Biologists study the structure, behaviour, and evolution of living things of all kinds.
Botanists are biologists who study plants.
Ecologists study the relationships between living organisms and the environments in which they live.
Embryologists study the formation and development of plants and animals from fertilization until they become independent organisms.
Entomologists study insects.
Ethologists study the inherited behaviour of animals in their natural environments.
Ichthyologists are zoologists who study fish.
Marine biologists study life in the oceans.
Mycologists study fungi.
Ornithologists study birds.
Naturalists are people with an interest in nature. They may specialize in particular species, or they may just enjoy watching and recording animals and plants.
Palaeontologists study fossils to gather information about forms of life that existed millions of years ago.
Taxonomists classify plants and animals in an ordered system.
Zoologists are biologists who study animals.

TYPES OF BIOMES

A biome is a plant and animal community that covers a large geographical area. The boundaries of a biome are determined mainly by climatic conditions.
Deserts are very dry regions where few plants grow. They may be cold or hot.
Grasslands are most common in temperate regions. In tropical regions with a long dry season, the typical grassland is savanna, grassland with scattered clumps of trees.
Oceans form by far the largest biome in terms of extent. The species that live in a given ocean habitat are determined by the depth, sunlight penetration, temperature, water conditions and nutrient availability in that particular location.
Scrublands are areas where bushy forms of vegetation dominate. Summers are hot and dry and fires are frequent.
Taigas, also called **boreal forests**, are regions of subarctic coniferous forests. Winters are cold and long.
Temperate forests are found between the tropical and polar regions. The climate is mild with moderate rainfall. Temperate forest may be coniferous or deciduous.
Tropical rainforests grow where the weather is hot and humid all year. They form the richest biome in terms of its variety of plant and animal species.
Tundras are cold and dry regions where the subsoil is permanently frozen.

CLASSES OF LIVING THINGS

Plants
So far, around 300,000 species of plants have been identified and classified. Plants range in size and complexity from simple algae to massive trees. Scientists predict there could be at least as many species still to be discovered, many of them growing in forests and on mountains where they are difficult to reach. There are far fewer plants than animals.

Fungi
There are approximately 100,000 known species of fungi. A fungus is a single-celled or multi-cellular organism that absorbs nutrients directly through its cell walls. Many fungi are parasites, taking their nutrients from other organisms.

Animals
Taxonomists group animals into around 30 major classifications, which they call phyla. Some phyla include many thousands of species. The phylum Nematoda, for example, consists of at least 12,000 species of roundworms. Among the so-called higher animals, the main groups are:
Molluscs: Soft-bodied, boneless marine animals that typically have a protective shell. Snails, bivalves such as cockles and cephalopods such as squids, are all examples of molluscs. There are around 100,000 species in this group.
Arthropods: Animals with jointed legs. Around one million arthropods have been identified, most of which are insects. There may be as many as 10 million insects still waiting to be named and described.
Fish: Aquatic animals that fit into three types. The Osteichthyes, or bony fish, are a class of around 22,000 known species, including cod. Sharks and rays are members of a class of around 5,000 known as Chondrichthyes, or cartilaginous fish. Lampreys and hagfish are members of the superclass Agnatha, or jawless fish.
Amphibians: There are about 3,000 known species, including frogs, toads and newts.
Reptiles: There are about 6,500 species, including snakes, lizards and crocodiles.
Birds: There are some 9,000 species in the class Aves. All species of Aves lay hard-shelled eggs and have feathers. Of the Aves, more than 5,700 species are perching birds of the order Passeriformes.
Mammals: The mammals are classified into 18 orders, in two sub-classes, placental mammals and the marsupials. There are 4,500 species of mammals, which include the primates – monkeys, apes and humans. These animals have body hair and have mechanisms that regulate their body temperature.

KEY DATES

77	Roman naturalist Pliny the Elder completes *Historia Naturalis*, the first encyclopedia about nature.
1665	British scientist Robert Hooke pioneers the use of microscopes to study cells and organisms.
1758	Swedish naturalist Carolus Linnaeus develops a system that is still used for naming animals.
1830s	German scientists Matthias Schleiden and Theodor Schwann show that the cell is the basic unit of all plants and animals.
1865	Austrian monk Gregor Mendel demonstrates the principles of heredity using pea plants.
1872	Yellowstone National Park is created in the United States. It is the first case of a park being created for the preservation of its natural environment.
1879	What is now the Royal National Park is set up in Australia.
1898	Kruger National Park is created in South Africa.
1909	Sweden opens the first national parks in Europe.
1910	US biologist Thomas Morgan shows that chromosomes carry genetic information.
1953	British biophysicist Francis Crick, US biochemist James Watson and British chemist Rosalind Franklin discover the structure of DNA.
1982	First cloning of mouse cells, and the creation of 'giant mice' in the United States through genetic engineering.
1988	First egg hatched by Californian condors in captivity. This species, of which there were only 27 birds remaining, had been taken into captivity to try to prevent it becoming extinct.
1989	The United States and member countries of the European Union ban ivory imports in an attempt to protect African elephants.
1996	A new squirrel-like mammal, the Panay cloudrunner, found in the Philippines.
1997	First clone of an adult mammal, Dolly the sheep, at the Roslin Institute in Scotland.
1998	A new species of deer, the Truong Son muntjac, found in Vietnam.
1998	A report from the International Union for the Conservation of Nature and Natural Resources states that 34,000 plant species are in danger of extinction – around 12 per cent, or one in eight, of Earth's plant species.

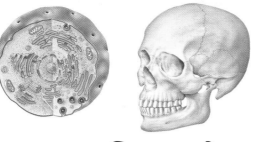

CHAPTER 3
HUMAN BIOLOGY

Humans are the most intelligent and advanced of the millions of living organisms on Earth. Our intelligence is such that we are self-aware and also capable of studying human biology – how our bodies are built and how they work.

There are now more than six billion people on Earth. Each person is unique in terms of personality, appearance, body shape and movement, and skin, hair and eye colour. Even identical twins – who share the same genetic material – have unique fingerprints and other differences. Despite these variations, all human bodies work in largely the same way.

For thousands of years, theories about the structure, workings and diseases of the human body relied more on myth and magic than they did on scientific observation. For example, it was not until the 16th century that the first accurate studies of anatomy were performed.

Since the 17th century, biologists and doctors have used ever more scientific methods to investigate the human body and its diseases. Inventions such as microscopes and X-rays quickened the pace of discovery, so that by the end of the 20th century, the workings of the body were well understood, and medicine could treat most diseases. Current research seeks to identify the human genome – the blueprint for human life – and to devise ways to treat diseases encoded in human genetic material.

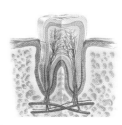

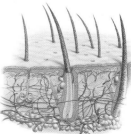

BODY ORGANIZATION

The human body consists of more than 50 trillion microscopic living units, called cells. These cells perform specific tasks to ensure that the body works.

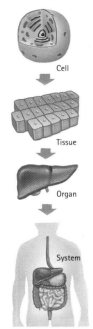

Cell

Tissue

Organ

System

Groups of liver cells form a tissue. This tissue with others make up the organ called the liver. Together, the liver and other linked organs make up the digestive system.

The human body is organized according to a hierarchy, or sequence, of different levels of complexity, starting from simple molecules right up to the body itself. Molecules such as carbohydrates, lipids, nucleic acids and proteins form the building blocks from which cells are made. They also take part in the chemical reactions collectively called metabolism. The body's metabolism interacts with the body's building blocks to form tiny living units called cells. To stay alive and give the body energy, each cell needs a constant supply of food and oxygen.

Individual cells that are similar in structure and function join together to form tissues. These perform different roles in the body. Several different types of tissues form structures called organs, each of which have a specific task or tasks. Organs include the eyes, kidneys, liver, lungs and stomach. For example, the role of the stomach is to store and break down food during digestion. The stomach works with other linked organs to form the digestive system. This not only digests food, but also absorbs useful nutrients from food into the bloodstream and eliminates any waste. The digestive system is one of twelve systems, all of which work together to ensure that the body carries out the functions it needs to survive.

Each young person in this photograph looks completely different. Apart from the differences between males and females, however, they all share the same basic body structure, which works in exactly the same way.

TISSUES

The body is made up of four basic types of tissues. Epithelial tissues are tightly packed cells that form leakproof linings to surfaces such as the skin and the lining of the digestive system. Connective tissues hold the body together and provide a framework. They include cartilage and bone. Muscular tissues consist of cells that contract (tighten) to move the body. Nervous tissue, in the brain and nerves, consists of a network of cells that carry electrical signals. Most organs contain all four types of tissue. Within tissues, cells are surrounded by tissue fluid. The fluid provides cells with a stable environment, delivers oxygen and food to the cells and removes waste products.

BODY SYSTEMS

There are 12 major systems in the human body. Seven of these are shown here. The systems not shown here include the respiratory system, integumentary system (skin and nails), male and female reproductive systems, urinary system and the immune system. Each system carries out one or more processes essential for life. For example, the circulatory system – the heart, blood vessels and blood – delivers food and oxygen to all body cells and removes their waste products.

▶ Individual bones make up the skeletal system, which supports the body.

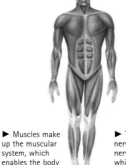

▶ Muscles make up the muscular system, which enables the body to move.

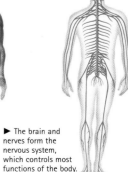

▶ The brain and nerves form the nervous system, which controls most functions of the body.

Cell membrane surrounds cell

Jelly-like cytoplasm supports the cell's organelles

Nucleus is the cell's control centre

Mitochondria supply energy to the cell

Ribosomes, the spheres on the surface of the endoplasmic reticulum, make proteins

Endoplasmic reticulum transports materials around the cell

Although different cells come in various shapes and sizes, they look very similar inside. Organelles ('tiny organs') inside the cell have specific functions. They all work together to produce a living cell.

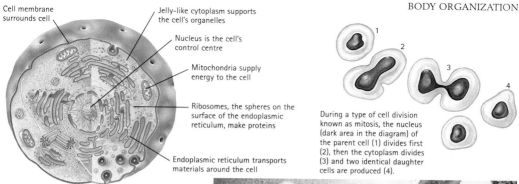

During a type of cell division known as mitosis, the nucleus (dark area in the diagram) of the parent cell (1) divides first (2), then the cytoplasm divides (3) and two identical daughter cells are produced (4).

CELLS

Although different cells perform different tasks, they all share the same structure. A plasma membrane separates each cell from its surroundings and allows material into and out of the cell. Inside the cell, tiny organelles – the microscopic equivalent of the body's organs – float in a watery jelly called cytoplasm. Organelles do different things, but they all cooperate to produce a living cell. The most important organelle is the nucleus, the cell's control centre. The nucleus contains genetic material in the form of deoxyribonucleic acid (DNA). This provides the blueprint for building and running the cell. Other organelles include mitochondria, ribosomes and endoplasmic reticulum.

Cells reproduce by dividing in one of two ways. Mitosis, which occurs throughout the body, enables the body to grow and repair itself by replacing worn-out cells. Meiosis, which occurs only in the testes and ovaries, produces sex cells – sperm and eggs – that take part in reproduction.

▲This magnified image shows a cell called a lymphocyte. These cells are found in blood. The nucleus of a lymphocyte takes up much of the space inside the cell. Lymphocytes play a vital part in defending the body against disease.

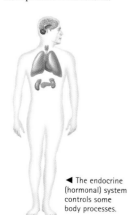

◀ The endocrine (hormonal) system controls some body processes.

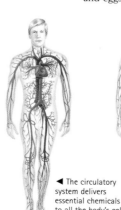

◀ The circulatory system delivers essential chemicals to all the body's cells.

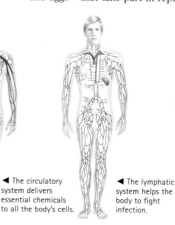

◀ The lymphatic system helps the body to fight infection.

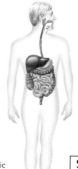

◀ The digestive system digests foods and absorbs nutrients into the body.

SEE ALSO

86–7 The brain and nervous system, 96–7 Heart and circulation, 99 Lymphatic system

THE SKELETON

The skeleton is a flexible framework that shapes and supports the body, protects vital organs such as the brain and anchors the muscles that move the body.

There are more than 20 bones in the human skull. Together, they provide a number of clues about the shape of the face and head. Scientists can use these clues to rebuild muscles and skin around the skull using clay. As a result, experts can recreate the faces of people who died long ago.

For centuries, bones were regarded as lifeless structures whose main aim was to support the active softer tissues around them. Gradually, scientists realized that bones are very much alive. Indeed, they have their own blood vessels and are constantly being rebuilt and reshaped.

The skeleton is not just a supportive framework for the body. Flexible joints between different bones enable the bones to move when pulled by muscles. The skeleton also protects vital organs such as the brain inside the skull. Bones themselves act as a store of calcium. This mineral is essential for muscles and nerves to work properly. Bones also make different types of blood cells. The skeleton contains cartilage, which covers the ends of bones in joints, and forms part of the skeletal system itself in the ear and nose, and between the sternum (breastbone) and ribs.

TYPES OF BONE

The four main kinds of bone are classified according to their shape and size. Long bones, such as the femur (thighbone), are adapted to withstand stress. Short bones include the wrist bones. Flat bones, such as the ribs, are often protective bones. Irregular bones include the vertebrae.

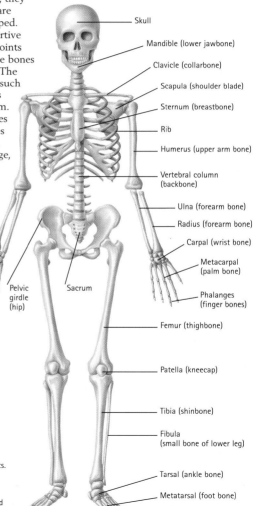

- Skull
- Mandible (lower jawbone)
- Clavicle (collarbone)
- Scapula (shoulder blade)
- Sternum (breastbone)
- Rib
- Humerus (upper arm bone)
- Vertebral column (backbone)
- Ulna (forearm bone)
- Radius (forearm bone)
- Carpal (wrist bone)
- Metacarpal (palm bone)
- Phalanges (finger bones)
- Femur (thighbone)
- Patella (kneecap)
- Tibia (shinbone)
- Fibula (small bone of lower leg)
- Tarsal (ankle bone)
- Metatarsal (foot bone)
- Phalanges (toe bones)
- Pelvic girdle (hip)
- Sacrum

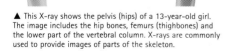

▲ This X-ray shows the pelvis (hips) of a 13-year-old girl. The image includes the hip bones, femurs (thighbones) and the lower part of the vertebral column. X-rays are commonly used to provide images of parts of the skeleton.

▶ An adult skeleton is made up of 206 bones. It can be divided into two parts. The axial skeleton forms the main axis of the body and consists of 80 bones that make up the skull, vertebral column (backbone) and ribs. This part of the skeleton protects the brain, spinal cord, heart and lungs. The appendicular skeleton consists of the upper and lower limbs and the pectoral (shoulder) and pelvic (hip) girdles that attach them to the axial skeleton. Of the 126 bones that make up the appendicular skeleton, all but 20 are found in the hands and feet.

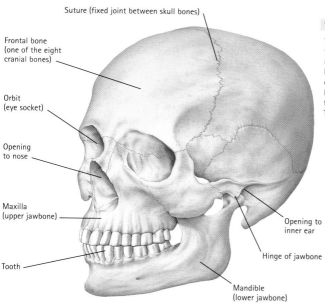

Suture (fixed joint between skull bones)

Frontal bone (one of the eight cranial bones)

Orbit (eye socket)

Opening to nose

Maxilla (upper jawbone)

Tooth

Opening to inner ear

Hinge of jawbone

Mandible (lower jawbone)

THE HUMAN SKULL

The skull forms the basic shape of the head and protects the brain. It consists of 22 bones. Eight cranial bones make up the cranium, which supports and protects the brain. There are 14 facial bones that form the structure of the face. All but the mandible (lower jawbone) are linked by fixed joints called sutures. The mandible is able to move freely, allowing the mouth to open and close. The skull also houses three pairs of ossicles or ear bones.

'Exploded' view of skull

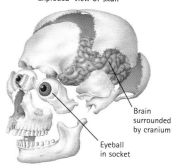

Brain surrounded by cranium

Eyeball in socket

AXIAL SKELETON

The axial skeleton consists of the skull, vertebral column (backbone), ribs and sternum. The skull houses the brain and the major sense organs – the eyes, ears, tongue and nose. It also contains openings to the digestive and respiratory systems. The flexible, S-shaped spinal column is made up of 26 irregular bones called vertebrae, which support the entire body. Muscles and ligaments attached to bony projections on the vertebrae help to keep the backbone upright. Seven cervical vertebrae support the neck and head, twelve thoracic vertebrae form joints with the ribs and five lumbar vertebrae carry most of the body's weight. The sacrum and coccyx are fused vertebrae that connect the spine to the pelvic girdle.

The ribcage protects the thoracic (chest) organs and also aids breathing. It is formed by the sternum (breastbone) and twelve pairs of curved, flattened ribs. The ribs form a joint with the thoracic vertebrae at one end. The upper seven, or true, ribs are connected to the sternum by flexible costal (rib) cartilages. The next three, called false ribs, are connected to the true ribs. The lowest two ribs (floating ribs) are attached only to the thoracic vertebrae.

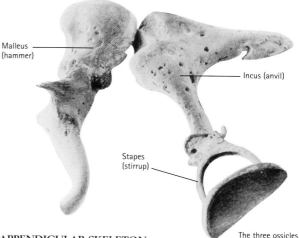

Malleus (hammer)

Incus (anvil)

Stapes (stirrup)

The three ossicles (ear bones) are the smallest bones in the body. The ossicles are located within the temporal bone on each side of the skull.

APPENDICULAR SKELETON

The appendicular skeleton is made up of the bones in the arms and legs and also girdles that link them to the body. The pectoral (shoulder) girdle consists of the scapula and clavicle. The pelvic (hip) girdle carries the weight of the upper body. The hands and feet contain many small bones. The hands can manipulate objects. The feet help to balance the body.

SEE ALSO

82–3 Bones and joints,
84–5 Muscles and movement, 92–3 Ears, hearing and balance

81

BONES AND JOINTS

Bones is living tissue that is both strong and light. The 206 bones that make up the skeletal system are linked together at joints, most of which move freely.

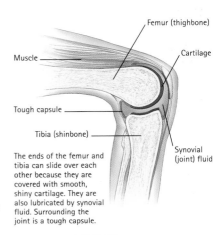

Femur (thighbone)

Muscle

Cartilage

Tough capsule

Tibia (shinbone)

Synovial (joint) fluid

The ends of the femur and tibia can slide over each other because they are covered with smooth, shiny cartilage. They are also lubricated by synovial fluid. Surrounding the joint is a tough capsule.

There are few structures that can rival bone in terms of its strength and lightness. All bones are made up of a hard material,called matrix, that contains widely spaced bone cells called osteocytes. Bone matrix consists of two main parts – a protein called collagen provides flexibility, and mineral salts, in particular calcium phosphate, provide strength. Together, these two components make bone as strong as steel but five times as light.

Matrix has two forms in bones: hard, compact bone forms the outer layer; lighter, spongy bone forms the inner layer. Long bones, such as the femur, contain a central cavity filled with bone marrow. This jelly-like material also fills the spaces within spongy bone. Red bone marrow, found in the skull, ribs and pelvis, is responsible for making red and white blood cells. Yellow bone marrow, found in the long bones of adults, stores fat.

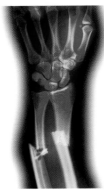

This X-ray clearly shows a fracture of the ulna (at left) and radius (at right). The fractured ends of the bones need to be put back in place by a doctor so that they join properly as the bones heal. The hand bones are shown at the top of the picture.

BREAK AND REPAIR

Bones fracture if they are put under stress. If this occurs, a blood clot forms between the broken ends, and bone cells secrete a new matrix. Fractures are compound (open) if the bones project through the skin or simple (closed) if they do not.

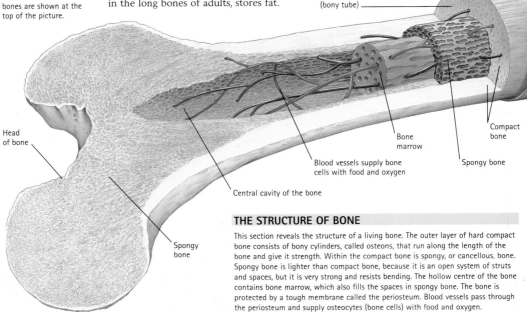

Osteon (bony tube)

Head of bone

Bone marrow

Compact bone

Blood vessels supply bone cells with food and oxygen

Spongy bone

Central cavity of the bone

Spongy bone

THE STRUCTURE OF BONE

This section reveals the structure of a living bone. The outer layer of hard compact bone consists of bony cylinders, called osteons, that run along the length of the bone and give it strength. Within the compact bone is spongy, or cancellous, bone. Spongy bone is lighter than compact bone, because it is an open system of struts and spaces, but it is very strong and resists bending. The hollow centre of the bone contains bone marrow, which also fills the spaces in spongy bone. The bone is protected by a tough membrane called the periosteum. Blood vessels pass through the periosteum and supply osteocytes (bone cells) with food and oxygen.

JOINTS

Joints, or articulations, are the points at which two bones meet. They are classified into three main groups – fixed, slightly movable and synovial – according to the amount of movement each permits. Fixed joints, as their name suggests, allow no movement. The sutures between the bones of the skull are examples of fixed joints. Their jagged edges are similar to the pieces of a jigsaw, locking the skull bones firmly together. Each tooth is another example of a fixed joint. Teeth are firmly locked in their sockets so that they do not move when food is chewed. Slightly movable joints allow limited movement between adjacent bones. These joints are found between vertebrae. Adjacent vertebra are separated by an intervertebral disc made of fibrocartilage. This allows partial movement between vertebrae. Collectively these joints give the backbone flexibility and allow it to bend backwards and forwards and from side to side.

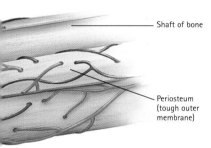

Shaft of bone

Periosteum (tough outer membrane)

SYNOVIAL JOINTS

Most joints – including the knee, elbow, knuckles, hip and shoulder – are freely movable, or synovial, joints. Synovial joints permit a wide range of movements. All synovial joints share the same basic structure. Bone ends are covered with glassy cartilage. Where the bones meet, they are separated by a synovial cavity filled with synovial fluid. Together, the cartilage and synovial fluid 'oil' the joint and reduce friction, producing a smooth movement. A joint capsule surrounds each synovial joint. Its inner membrane secretes synovial fluid. The outer part is continuous with tough straps, called ligaments, that hold the joint together.

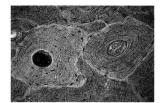

A magnified cross-section of compact bone from the femur reveals two osteons. In the middle of each osteon is a central channel that carries blood vessels. Dark spaces in the osteon contain bone cells.

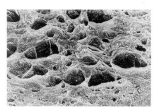

This magnified image of spongy bone from a bone in the foot looks very different to compact bone (at left). It consists of hard struts separated by linked spaces filled with bone marrow.

In a pivot joint, the end of one bone rotates within a space formed by another bone. At the top of the backbone, the atlas (first vertebra) rotates around the axis (second vertebra). This allows the head to turn from side to side.

A hinge joint works like a door hinge. The cylindrical end of one bone fits into the curved end of another bone. Hinge joints allow movement up and down but not from side to side. The knee (above) is an example of a hinge joint.

Ball-and-socket joints, such as in the shoulder (above), are the body's most flexible joints. The ball-shaped end of one bone fits into the cup-shaped socket of another bone, allowing movement in all directions.

In the saddle joint at the base of the thumb, the U-shaped ends of wrist and thumb bones fit together, permitting movement backwards and forwards and from side to side. This joint also allows the thumb to touch the tip of each finger.

Ellipsoidal, or condyloid, joints are found in the knuckles and between the lower arm and wrist bones. The egg-shaped end of one bone fits into the oval cup of another bone, allowing movement backwards and forwards and from side to side.

In a gliding, or plane, joint, the surfaces of each bone are level, allowing the bones to make short movements by sliding over each other. Gliding joints are found between carpals (wrist bones) in each hand and tarsals (heel bones) in each foot.

RANGE OF MOVEMENT

The shape of bone ends, the arrangement of muscles and the tightness of ligaments holding the joint together all determine the range of movement at a joint. The straightening and bending of the arm at the elbow is an example of extension (straightening) and flexion (bending). The raising and lowering of the lower jaw when chewing food is an example of elevation (lifting) and depression (lowering).

SEE ALSO

80–1 The skeleton, 84–5 Muscles and movement, 106 Growth and development

MUSCLES AND MOVEMENT

Every movement, from blinking an eye to running in a race, is produced by the body's muscles. Muscles are made of cells that have the unique ability to contract.

Three types of muscle are found in the body: skeletal, smooth and cardiac. Sprinting uses skeletal muscle, digestion requires smooth muscle, and a heartbeat involves cardiac muscle. As their name suggests, skeletal muscles move the bones of the skeleton and help support the body. The body has over 640 skeletal muscles that cover the skeleton and give the body overall shape. Skeletal muscles make up 40 per cent of the body's weight. They range in size from the powerful quadriceps femoris (thigh muscle) to the tiny stapedius in the ear. Tough cords called tendons attach the end of the skeletal muscle to the bone. Muscles extend across joints. When the muscles contract, bones move relative to one another.

All the actions involved in skipping, such as moving the arms and hands, bending the knees and lifting the feet, are produced by skeletal muscles. Instructed by the brain, these muscles pull the skeleton to produce coordinated movements.

Most muscles work in pairs, each with an action that is antagonistic to (opposes) the other. In the upper arm, for example, biceps brachii contracts (with triceps brachii relaxed) to bend the arm, while triceps brachii contracts (with biceps brachii relaxed) to straighten it.

Biceps brachii contracts

Triceps brachii relaxes

Biceps brachii relaxes

Triceps brachii contracts

▶ The body's skeletal muscles are arranged in overlapping layers. The muscles that are just below the skin are called superficial. Beneath these are the deep muscles. This anterior (front) view of the body shows some important superficial muscles and their actions. Muscles are given Latin names for different reasons. These include their action (flexor or extensor), their shape (deltoid means triangular), their relative size (maximus means largest) or their location (frontalis covers the frontal bone).

HOW MUSCLES WORK

Skeletal muscle cells, or fibres, are long, thin and packed with lots of parallel strands called myofibrils. Myofibrils contain two protein filaments – actin and myosin – which make skeletal muscle fibres look stripy. When a muscle receives a message from the brain along a nerve, the filaments slide past each other, making the fibre shorter, and the muscle contracts. Muscles can only pull, not push; they usually work in pairs, each pulling bones in the opposite direction.

Frontalis wrinkles the forehead

Orbicularis oculi closes the eye

Orbicularis oris closes the lips

Deltoid moves the upper arm in many directions

Pectoralis major pulls arm towards body and rotates it

Biceps brachii bends the arm

External oblique tightens the abdomen

Quadriceps femoris straightens the knee during walking and running

Gastrocnemius lifts the heel and bends the knee

Tibialis anterior straightens or lifts foot

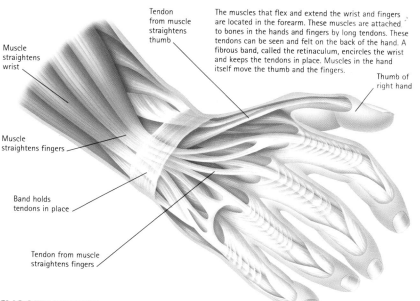

Muscle straightens wrist

Tendon from muscle straightens thumb

The muscles that flex and extend the wrist and fingers are located in the forearm. These muscles are attached to bones in the hands and fingers by long tendons. These tendons can be seen and felt on the back of the hand. A fibrous band, called the retinaculum, encircles the wrist and keeps the tendons in place. Muscles in the hand itself move the thumb and the fingers.

Thumb of right hand

Muscle straightens fingers

Band holds tendons in place

Tendon from muscle straightens fingers

Skeletal muscle fibres

Smooth muscle fibres

Cardiac muscle fibres

SMOOTH MUSCLE

Smooth, or involuntary, muscle is found mainly in the walls of hollow organs such as the oesophagus and bladder. Smooth muscle is vital to involuntary processes such as moving food along the alimentary canal during digestion (peristalsis). The short, tapering fibres of smooth muscle are packed into sheets and contract smoothly and rhythmically under the control of the autonomic nervous system – a person cannot consciously cause them to contract.

CARDIAC MUSCLE

Cardiac muscle is found only in the heart and makes up a large part of its structure. Its branched, striped fibres form an interconnected network. These fibres contract spontaneously without the need for an outside stimulus from the nervous system. Cardiac muscle contracts nonstop over 2.5 billion times in an average lifetime to pump blood around the body.

Muscle fibres differ in their appearance. Skeletal muscle fibres are long and striped. Smooth muscle fibres are short and tapered. Cardiac muscle fibres are striped and branched.

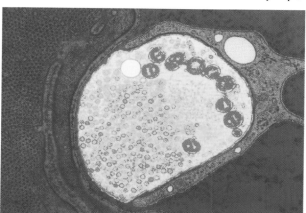

◄ A neuromuscular junction (here magnified 30,800 times) is where the end of a nerve fibre (yellow) meets a muscle fibre (red). When a nerve impulse arrives at a nerve ending, chemicals are released to make the muscle contract.

Sheath

Sheath

Muscle fibre (cell)

Skeletal muscle fibres are arranged in bundles that run along a muscle. Myofibrils inside each fibre consist of filaments that interact to make the muscle contract.

SEE ALSO

80–1 The skeleton, 86–7 The brain and nervous system

THE BRAIN AND NERVOUS SYSTEM

Billions of nerve cells, called neurons, link up to form the body's communication network, called the nervous system. The nervous system is controlled by the brain.

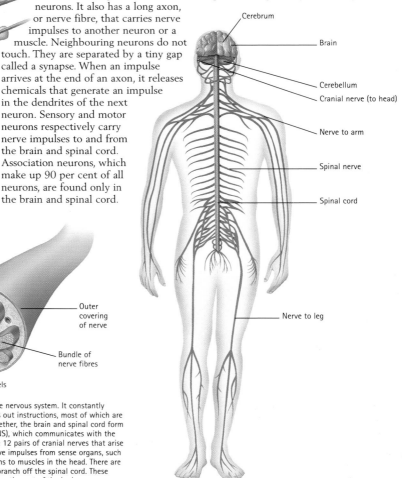

Spinal cord

Spinal nerve

Vertebra (bony segment of backbone)

The spinal cord extends from the base of the brain to the lower back. It is protected by the vertebrae. Spinal nerves branch off the spinal cord and carry nerve impulses to and from parts of the body.

Nerves consist of bundles of sensory neurons – which carry nerve impulses from sensors to the brain and spinal cord – and motor neurons – which carry nerve impulses from the brain and spinal cord to the muscles.

Nerve fibre

Outer covering of nerve

Bundle of nerve fibres

Blood vessels

The neuron is the basic unit of the nervous system. It is long, thin and transmits electrical signals, called nerve impulses, along its length. The cell body of the neuron is much like any other cell. It has many branched endings called dendrites that receive impulses from other neurons. It also has a long axon, or nerve fibre, that carries nerve impulses to another neuron or a muscle. Neighbouring neurons do not touch. They are separated by a tiny gap called a synapse. When an impulse arrives at the end of an axon, it releases chemicals that generate an impulse in the dendrites of the next neuron. Sensory and motor neurons respectively carry nerve impulses to and from the brain and spinal cord. Association neurons, which make up 90 per cent of all neurons, are found only in the brain and spinal cord.

SPINAL CORD

Essentially, the spinal cord is an extension of the brain. It is about 45 centimetres long and extends from the brain to the lower back. Spinal nerves along its length relay information between the brain and body. It also plays a vital role in reflexes. If a person touches a sharp object, for example, nerve impulses pass from the fingertip, through the spinal cord, directly to the upper arm muscles and instantly pull the finger away from danger.

Cerebrum

Brain

Cerebellum

Cranial nerve (to head)

Nerve to arm

Spinal nerve

Spinal cord

Nerve to leg

▶ The brain controls the entire nervous system. It constantly receives information and sends out instructions, most of which are relayed by the spinal cord. Together, the brain and spinal cord form the central nervous system (CNS), which communicates with the body through nerves. There are 12 pairs of cranial nerves that arise from the brain. Most relay nerve impulses from sense organs, such as the eyes, or carry instructions to muscles in the head. There are 31 pairs of spinal nerves that branch off the spinal cord. These relay nerve impulses to and from the rest of the body.

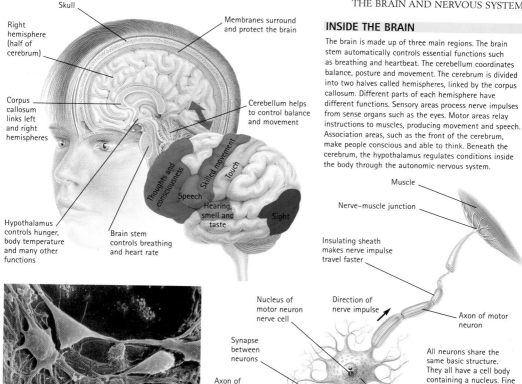

Skull

Right hemisphere (half of cerebrum)

Membranes surround and protect the brain

Corpus callosum links left and right hemispheres

Cerebellum helps to control balance and movement

Thoughts and consciousness

Skilled movement

Touch

Speech

Hearing, smell and taste

Sight

Hypothalamus controls hunger, body temperature and many other functions

Brain stem controls breathing and heart rate

INSIDE THE BRAIN

The brain is made up of three main regions. The brain stem automatically controls essential functions such as breathing and heartbeat. The cerebellum coordinates balance, posture and movement. The cerebrum is divided into two halves called hemispheres, linked by the corpus callosum. Different parts of each hemisphere have different functions. Sensory areas process nerve impulses from sense organs such as the eyes. Motor areas relay instructions to muscles, producing movement and speech. Association areas, such as the front of the cerebrum, make people conscious and able to think. Beneath the cerebrum, the hypothalamus regulates conditions inside the body through the autonomic nervous system.

Muscle

Nerve–muscle junction

Insulating sheath makes nerve impulse travel faster

Nucleus of motor neuron nerve cell

Direction of nerve impulse

Axon of motor neuron

Synapse between neurons

Axon of preceding neuron

Cell body

Dendrite

This image, which is magnified 494 times, shows association neurons from the cerebral cortex, the thin, outer part of the cerebrum (the 'thinking' part of the brain). Each neuron is linked to thousands of others.

BRAIN

The brain is made up of over 100 billion neurons. Each one communicates with thousands of other neurons to produce a complex communication and control network. The brain receives information about conditions both inside and outside the body, processes and stores this information and issues instructions based on what it has learned. The hypothalamus and brain stem control automatic processes such as breathing. The cerebellum regulates smooth body movements. The cerebral hemispheres control thought, imagination, memory, speech, emotion, sight, hearing, smell, taste and touch.

PARTS OF THE NERVOUS SYSTEM

The nervous system has two main parts: the brain and spinal cord form the central nervous system (CNS) and the nerves form the peripheral nervous system (PNS). Within the PNS, sensory neurons transmit nerve impulses from sense organs to the brain. Motor neurons transmit instructions from the brain and are of two types. Those of the somatic nervous system are under voluntary control and stimulate skeletal muscles to contract. Those of the autonomic nervous system (ANS) regulate processes inside the body such as breathing and digestion. The ANS has two divisions: sympathetic and parasympathetic. These have opposite effects and keep the body in a stable state.

All neurons share the same basic structure. They all have a cell body containing a nucleus. Fine processes, called dendrites, receive nerve impulses, via synapses, from other neurons. A long axon, or nerve fibre, carries nerve impulses away from the cell body. The cell body of this motor neuron is located in the central nervous system (CNS). It transmits nerve impulses to parts of the body instructing them to do something. For example, a nerve impulse to a muscle may cause it to contract. Similarly, a nerve impulse to a gland may cause it to release a secretion.

SEE ALSO

84–5 Muscles and movement

TOUCH

The sense of touch provides the brain with information about the body's surroundings. Touch sensors are scattered all over the surface of the body.

Sensors in the skin detect touch, pain, vibration, pressure, heat and cold. The softness of fur, the vibrations made by running the fingers over sandpaper, the pressure produced by holding a heavy weight, the pain of standing on a pin, the heat from a flame and the cold felt by plunging a hand into icy water – all of these are experienced when the skin's sensors are stimulated.

Sensors for light touch and pressure are in the upper part of the dermis of the skin. Those for heavy touch and pressure are larger and are found deeper in the dermis. Most of these sensors are enclosed in capsules. The sensors that detect heat, cold and pain are branched nerve endings near the junction between the epidermis and the dermis. These sensors are not enclosed in capsules. Information from all of the different sensors travels as electrical impulses along nerves that lead to the sensory area of the cerebrum (the main part of the brain). The brain interprets these impulses and provides a 'touch

This boy looks strange because the size of his body parts have been drawn according to how sensitive to touch they are. Some parts of the body, such as fingers and lips, are much more sensitive than other parts because they have many more touch sensors.

This girl is blind but is able to read by running her fingertips across the page. The words are 'written' » using the Braille system; its patterns of raised dots correspond to letters or numbers.

picture' of the person's surroundings, including information about pressure and warmth, for example. The feeling of pain warns the body about possible danger.

HABITUATION

When a person dresses in the morning, the clothes can be felt as they are pulled over the skin. After a short while, the clothes can no longer be felt. This loss of feeling is called habituation. The skin gets used to the stimulation of the clothes, and nerve impulses are no longer sent to the brain. Habituation is important because without it clothes would irritate the skin all day.

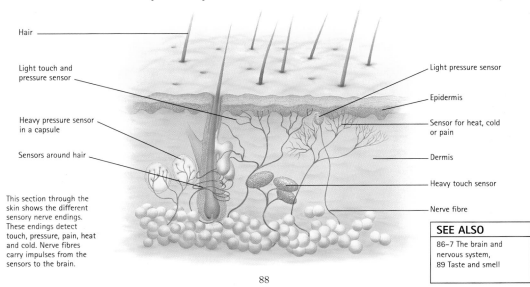

Hair

Light touch and pressure sensor

Heavy pressure sensor in a capsule

Sensors around hair

Light pressure sensor

Epidermis

Sensor for heat, cold or pain

Dermis

Heavy touch sensor

Nerve fibre

This section through the skin shows the different sensory nerve endings. These endings detect touch, pressure, pain, heat and cold. Nerve fibres carry impulses from the sensors to the brain.

SEE ALSO
86–7 The brain and nervous system,
89 Taste and smell

TASTE AND SMELL

Taste and smell are linked senses. Both detect chemicals, either in food or in the air. Together, they enable people to appreciate a wide range of flavours.

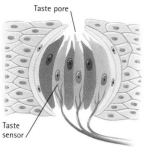

Taste pore

Taste sensor

A mixture of taste sensors and packing cells are clustered together in this taste bud, similar to the segments of an orange. The taste bud detects chemicals dissolved in saliva that enter through the taste pore, which is the opening onto the tongue's surface.

The organ of taste is the tongue. Scattered over its upper surface are about 10,000 taste buds. Taste buds detect four basic tastes – sweet, sour, salty and bitter. Bitter-tasting foods may be poisonous and can be spat out.

Taste buds are found on the sides of tiny projections, called papillae, that cover the tongue. Fungiform papillae resemble mushrooms; seven or eight large ridged papillae are at the back of the tongue; threadlike filiform papillae lack taste buds and help to grip food during chewing. When dissolved food chemicals reach a taste bud, sensory cells send nerve impulses to the brain.

Taste buds occur in zones on the tongue: sweet at the front, bitter at the back and salty and sour along the sides. The tongue also has receptors for heat and cold and pain receptors for spicy foods.

A view magnified 180 times of the tongue's upper surface shows pointed filiform papillae surrounding a fungiform papilla (yellow-orange) in the sides of which are taste buds.

SMELL

The sense of smell enables people to enjoy food and avoid dangerous substances in the air and in food. Humans can detect over 10,000 different odours. About 10 million olfactory (smell) receptors are located in the upper part of the nasal cavity in two patches of epithelium (lining), each the size of a postage stamp. Each receptor contains up to 20 hair-like cilia. When air is breathed in, molecules dissolve in a watery mucus and bind to the cilia. Smell dominates taste; a heavy cold makes food flavourless.

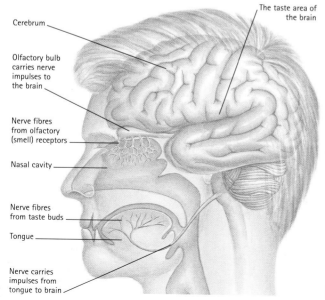

Cerebrum

Olfactory bulb carries nerve impulses to the brain

Nerve fibres from olfactory (smell) receptors

Nasal cavity

Nerve fibres from taste buds

Tongue

Nerve carries impulses from tongue to brain

The taste area of the brain

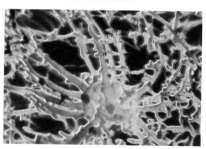

◀ Taste buds are found in the upper surface of the tongue. Impulses from taste buds travel along nerves to the taste area of the brain. Smell receptors are found in the upper part of each side of the nasal cavity. Nerve impulses from these receptors are sent to the part of the brain where smells are identified.

▲ Hair-like cilia (magnified 10,285 times) radiate from an olfactory (smell) receptor in the upper nasal cavity. When 'smelly' molecules touch these cilia, nerve impulses are sent to the brain.

SEE ALSO

86–7 The brain and nervous system, 90–1 Eyes and seeing, 92–3 Ears, hearing and balance

EYES AND SEEING

Vision is an extremely important sense. The eyes detect light from the body's surroundings and send messages to the brain, enabling a person to see.

The pupil is the hole that allows light into the eye. In dim light the coloured iris makes the pupil larger.

In bright light the iris makes the pupil smaller to prevent too much light entering the eye and damaging the retina.

Eyes are important because they provide the brain with much information about the body's surroundings. The retina, which lines the inside of the eye, contains photoreceptors, which are sensory cells that are stimulated by light. Photoreceptors make up 70 per cent of the sensory receptors in the human body, an indication of how important they are.

The two eyeballs, each about 2.5 cms in diameter, are found in orbits in the skull. Only a small part of the eye is visible from the front. Each eyeball moves using six extrinsic (external) muscles, which enable people to look from side to side and up and down. They cause tiny movements, saccades, of the eyeballs, that enable the eyes to constantly scan the surroundings.

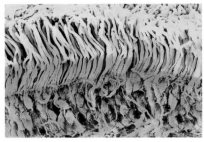

This is a section through the retina, the eye's light-sensitive layer. Rods and cones (yellow) respond to light and send messages to the brain along nerve fibres (pink).

BLIND SPOT

One region of the retina, known as the blind spot, does not contain any light sensors. This is where the optic nerve leaves the eyeball. The blind spot does not interfere with vision, however. Most of the time people do not notice any effect because the brain chooses to 'ignore' it.

THE EYE

The internal and external parts of the eye are revealed by this cutaway. Light enters through the clear cornea. The iris controls the amount of light that enters the eye so a person can see in both dim and bright light. The lens focuses light on the retina from both near and distant objects. The retina is packed with photoreceptors (light sensors).

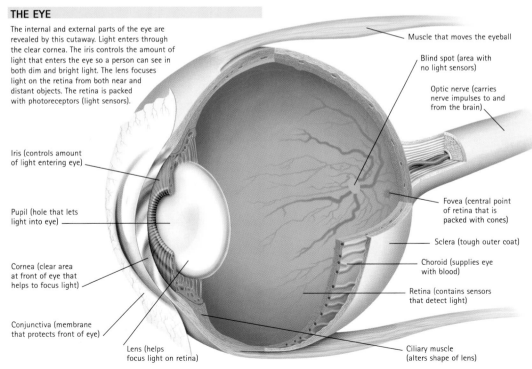

Muscle that moves the eyeball

Blind spot (area with no light sensors)

Optic nerve (carries nerve impulses to and from the brain)

Iris (controls amount of light entering eye)

Pupil (hole that lets light into eye)

Cornea (clear area at front of eye that helps to focus light)

Conjunctiva (membrane that protects front of eye)

Lens (helps focus light on retina)

Fovea (central point of retina that is packed with cones)

Sclera (tough outer coat)

Choroid (supplies eye with blood)

Retina (contains sensors that detect light)

Ciliary muscle (alters shape of lens)

HOW SIGHT WORKS

Light rays that enter the eye are refracted (bent) by the cornea and the lens to focus them on the retina. The ciliary muscle alters the thickness of the lens to focus light from near or distant objects. The iris controls the amount of light entering the eye. Its muscles continually alter the size of the pupil, making it large to admit more light or small to prevent excessive light from damaging the retina.

The retina is a thin layer of light sensors called rods and cones. The 120 million rods work best in dim light and are sensitive to black and white. About six million cones work best in bright light and detect colour. Cones are found mostly in the fovea, which generates the most detailed images. Three types of cone detect green, red and blue light, respectively. When they detect light, rods and cones generate nerve impulses that travel along the optic nerve to the visual areas at the back of the cerebrum, the main part of the brain. The brain reconstructs the images. Each eye detects a slightly different scene. The brain uses these differences to produce a three-dimensional picture of the world that enables people to judge distances.

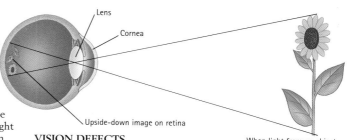

Lens
Cornea
Upside-down image on retina

VISION DEFECTS

Myopia, or short-sightedness, is an inability to see distant objects clearly because light from them is focused before it reaches the retina, producing a blurred image. This can be corrected by contact lenses or glasses. Hypermetropia, or long-sightedness, is an inability to see close objects clearly because light from them is focused 'behind' the retina, again producing a blurred image. This can be corrected by the use of glasses. Presbyopia may occur as part of ageing after the age of 45. Here, the ability to focus on near objects is lost, and glasses are needed for reading and other close-up work. Colour blindness, or colour deficiency, is the inability to distinguish between certain colours.

When light from an object enters the eye, the cornea and lens focus it to produce a clear, but upside-down, image on the retina. When hit by this light, sensors in the retina send nerve messages to the brain. There the image is 'seen' the right way up.

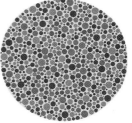

This pattern of dots is a test for colour blindness, an inability to tell certain colours apart. A colour-blind person lacks one of the types of cone (colour sensors) that detect red, green or blue light. Most common is red-green colour blindness, an inability to distinguish between those two colours. If you can see the number eight in this pattern, you are not colour blind. Colour blindness is more common in males than in females.

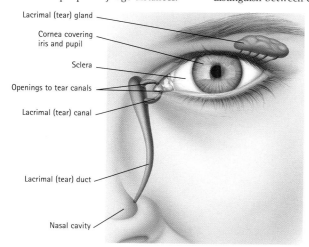

Lacrimal (tear) gland
Cornea covering iris and pupil
Sclera
Openings to tear canals
Lacrimal (tear) canal
Lacrimal (tear) duct
Nasal cavity

▲ Tears are produced by lacrimal (tear) glands. Tears spread over the eye's surface when a person blinks to wash away dirt and dust; they also contain the chemical lysozyme that kills bacteria. Tears then drain through two holes in the corner of the eye and empty into the nose.

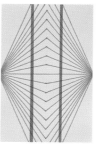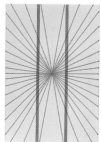

Optical illusions are images that trick the brain. Here both sets of red lines are straight, though they seem to curve inwards (left) or outwards (right).

SEE ALSO

86–7 The brain and nervous system, 88 Touch, 89 Taste and smell

EARS, HEARING AND BALANCE

Humans can detect over 400,000 different sounds. As well as detecting sound, the ears play an important role in balance and posture.

Sound is created by alternating waves of low and high pressure that pass through the air, similar to the ripples that spread across a pond when a stone is dropped into the water. These pressure waves pass into the ear where they are detected by sensors. These send messages to the brain, which interprets them as sounds.

Most of the ear lies hidden within the temporal bones of the skull. The part we can see, the pinna, channels sound waves into the auditory canal, a tube that secretes cleansing wax. The air-filled middle ear is bordered by the eardrum on one side and the oval window on the other. Its only opening is through the Eustachian, or auditory, tube that runs to the throat. It ensures that the air pressure on both sides of the eardrum is kept equal. If it is not, the eardrum cannot vibrate properly and hearing is impaired. Sudden pressure changes – such as when a train goes into a tunnel – can make the pressures unequal. Yawning or chewing forces air into, or out of, the Eustachian tube from the throat, and the ears 'pop' as the pressures equalize and normal hearing returns. The inner ear contains sound receptors and links to the brain. It is filled with fluid and sealed in a bony structure.

HOW HEARING WORKS

Sound waves arrive from the source that is making them, such as a radio, and enter the ear through the auditory canal. At the end of this canal, a taut piece of skin called the eardrum vibrates as the sound waves hit it. The eardrum passes on the vibrations to the three ossicles – the hammer, anvil and stirrup – in the middle ear. When these bones vibrate, the stirrup bone pushes and pulls the membrane that covers the oval window. This movement sets up vibrations in the fluid in the inner ear, which are detected by the sensors in the cochlea. The sensors send nerve impulses to the brain, which processes them. The person hears the sound. Loud sounds cause bigger vibrations in the fluid. The part of the cochlea near the oval window detects high-pitched sounds, while the tip of the cochlea's coil detects low-pitched sounds.

Sound waves usually arrive in one ear a split second before the other. The brain uses this tiny time difference to work out from which direction the sound came.

INSIDE THE EAR

The outer ear channels sounds into the ear. The middle ear is crossed by three tiny bones, called ossicles, that transmit sounds from the eardrum to the inner ear. The inner ear is filled with fluid and a series of channels. The cochlea contains sound sensors. The semicircular canals, and the saccule and the utricle, detect movement and position.

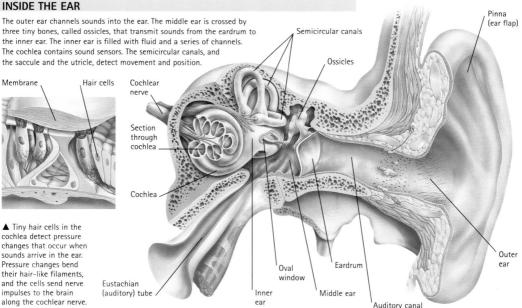

▲ Tiny hair cells in the cochlea detect pressure changes that occur when sounds arrive in the ear. Pressure changes bend their hair-like filaments, and the cells send nerve impulses to the brain along the cochlear nerve.

Membrane
Hair cells
Cochlear nerve
Section through cochlea
Cochlea
Eustachian (auditory) tube
Inner ear
Oval window
Eardrum
Middle ear
Auditory canal
Semicircular canals
Ossicles
Pinna (ear flap)
Outer ear

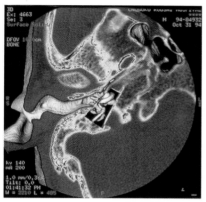

This CT scan shows a section through a living ear. The ear canal (white) runs from the left to the centre. The ear bones – ossicles – are highlighted in the centre. The larger grey, mottled areas are the bones of the skull.

This view (magnified 2,074 times) of the inner ear shows the region of the cochlea that detects sounds. When sounds arrive and cause vibrations in the fluid, the filaments of the hair cells (yellow and V-shaped) bend, and the hair cells send messages to the brain.

THE EARS AND BALANCE

Balance sensors are found within the inner ear inside two linked structures – the semicircular canals and the vestibule. These lie next to the cochlea and are also filled with fluid. The three semicircular canals are set at right angles to each other and detect movements of the head. At the base of each canal, sensory hair cells are embedded in a jelly-like cupule (cup-shaped structure). When the head moves, the fluid in one or more of the canals moves and bends both the cupule and its hairs. The hair cells then send nerve impulses to the brain. By analysing which semicircular canals sent nerve messages, the brain can tell which way the head and body are moving at any moment.

The vestibule contains two balance sensors, the utricle and the saccule. Both contain sensory hairs embedded in otoliths (ear parts made of calcium carbonate crystals). The utricle detects rapid acceleration and deceleration, while the saccule detects changes in the head's position. This information is combined with messages from the eyes, pressure sensors in the feet and receptors in muscles and joints to provide the brain with a complete picture of the body's position. The brain can then send out instructions to muscles to alter the body's position to maintain its posture and balance.

HEARING RANGES

Humans can hear a wide range of sounds, from low-pitched hums to high-pitched squeaks. Pitch is determined by a sound's frequency; that is, how rapidly one crest of the wave follows the previous one. Sound frequency is measured in hertz (Hz), or sound waves per second.

Young people can usually hear sounds between 20 Hz and 20,000 Hz. However, the range of sounds people can hear decreases with their age, so older people are unable to hear higher-pitched sounds. Some mammals hear high-pitched sounds that are inaudible to humans. Bats can hear sounds in the range 1,000 to 120,000 Hz, and cats from 60 to 65,000 Hz.

▲ The semicircular canals, and the utricle and the saccule play an important part in balance. They send information to the brain about the position and movement of the head. The brain instructs muscles to move and position the body so that it does not fall over.

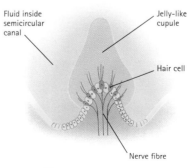

Fluid inside semicircular canal

Jelly-like cupule

Hair cell

Nerve fibre

◀ Inside each fluid-filled semicircular canal is a jelly-like knob called a cupule. Embedded in the cupule are hair cells. If the head moves, the fluid also moves and bends the cupule. Hair cells send messages to the brain so that a person is aware of their movement.

SEE ALSO

84–5 Muscles and movement, 86–7 The brain and nervous system

HORMONES

The endocrine system releases chemical messengers called hormones into the body. Different hormones control processes such as reproduction and growth.

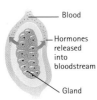

— Blood

Hormones released into bloodstream

— Gland

A gland is a group of cells that releases chemicals either into or onto the body. Endocrine, or ductless, glands (above) release hormones into the bloodstream. The blood carries them to the part of the body where they have their effect.

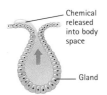

Chemical released into body space

— Gland

Exocrine glands include the salivary glands and sweat glands. Also called ducted glands, exocrine glands release their secretions, such as saliva or sweat, through a duct that opens into a space inside the body or onto the surface of the body.

The endocrine, or hormonal, system is made up of a number of endocrine (hormone-releasing) glands. Along with the nervous system, the endocrine system controls and coordinates the workings of the body. The endocrine system plays a key role in reproduction and growth and controls many other processes. The endocrine and nervous systems work in very different ways. In the nervous system, messages are carried in the form of electrical impulses. The endocrine system releases chemical messengers, called hormones, into the bloodstream. Carried by the blood to its target, the hormone alters the activities of cells by increasing or decreasing the speed of processes taking place inside them. Unlike the nervous system, hormones work slowly and have long-term effects. The pituitary gland controls most of the other endocrine glands. In turn, the pituitary gland is controlled by the hypothalamus in the brain. This provides a direct link between the endocrine and nervous systems.

THE ENDOCRINE SYSTEM

The glands that make up the endocrine system are scattered through the head, thorax and abdomen. The major endocrine glands are the pituitary, thyroid, parathyroid and adrenal glands. The pituitary gland releases more than nine hormones, controls the activities of most other endocrine glands and is itself controlled by part of the brain called the hypothalamus. The thyroid gland regulates the body's metabolic rate (the speed of chemical reactions inside body cells). Along with the parathyroid glands, it also controls calcium levels in the blood. The adrenal glands also affect metabolic rate and help the body withstand stress. Other organs also have endocrine sections. The pancreas controls glucose levels in the blood, but it also acts as an exocrine gland that releases digestive enzymes into the intestine. The testes in males and the ovaries in females produce sex hormones as well as making sperm and eggs.

REGULATING HORMONES

Hormone levels in the blood are regulated by negative feedback systems. These reverse unwanted changes, ensuring that hormones do not have too great or too little an effect. For example, thyroxine speeds up the body's metabolism. Too much thyroxine, and the body works too fast. Too little, and the body slows right down. Low thyroxine levels cause the pituitary gland to release thyroid stimulating hormone (TSH), and the thyroid gland produces thyroxine. High thyroxine levels have the opposite effect.

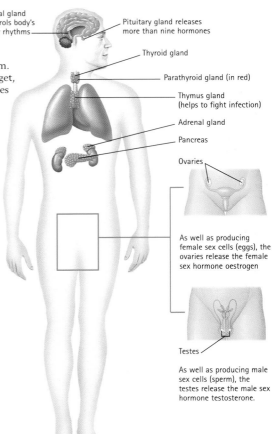

Pineal gland controls body's daily rhythms

Pituitary gland releases more than nine hormones

Thyroid gland

Parathyroid gland (in red)

Thymus gland (helps to fight infection)

Adrenal gland

Pancreas

Ovaries

As well as producing female sex cells (eggs), the ovaries release the female sex hormone oestrogen

Testes

As well as producing male sex cells (sperm), the testes release the male sex hormone testosterone.

The passengers on this roller coaster ride are experiencing the effects of the hormone adrenaline. Adrenaline is released by the adrenal glands. It helps the body to deal with dangerous situations. Adrenaline makes the heart beat and breathing faster. It also diverts blood to the muscles. Following the release of adrenaline the body is prepared to either confront dangerous situations or run away from them, a reaction called fight-or-flight.

PITUITARY GLAND

The pea-sized pituitary gland at the base of the brain helps to control the endocrine system. The pituitary gland releases more than nine hormones. Some control body functions directly, such as a growth hormone that stimulates growth. Others target other endocrine glands, such as follicle stimulating hormone that stimulates the ovaries to release the female sex hormone oestrogen.

The pituitary gland is made up of two parts called lobes. The larger anterior (front) lobe makes and releases most pituitary hormones. Their release is triggered by hormones secreted by the hypothalamus in the base of the brain. The smaller posterior (back) lobe stores and releases two hormones made by the hypothalamus.

ADRENAL GLANDS

The adrenal glands sit at the top of each kidney. The outer cortex of these glands releases hormones called corticosteroids, that help to control metabolism and regulate the concentration of substances in the blood. The inner medulla secretes adrenaline. If the brain perceives danger or stress, it sends nerve messages to the adrenal glands, causing them to secrete adrenaline. This prepares the body to confront the threat or run away from it.

PANCREAS

The pancreas is located below the stomach. It releases the hormones insulin and glucagon, which control blood glucose levels. Cells need a constant supply of glucose. If glucose levels are too high or too low, cells cannot take glucose in. Insulin and glucagon naturally balance each other to maintain a steady glucose level whether a person is hungry or has just eaten.

Every day this boy uses a special automatic syringe to inject insulin into his body. He suffers from diabetes, which means his pancreas does not produce insulin. Without the injections, the boy's body cannot control glucose levels in the blood and he would become very ill.

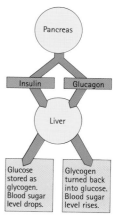

The pancreas releases insulin and glucagon. These hormones work in opposite ways to regulate levels of glucose in the bloodstream. If glucose levels rise, the insulin stimulates cells to take up glucose and the liver to store it as glycogen. If glucose levels fall, glucagon stimulates the liver to turn glycogen back into glucose.

SEE ALSO

86–7 The brain and nervous system, 98 Blood, 104–5 Reproduction

HEART AND CIRCULATION

The circulatory system supplies the body's cells with all their needs. It consists of the heart, the blood vessels and blood that flows through them.

English doctor William Harvey (1578-1657) was the first person to show that blood circulates in one direction around the body, pumped by the heart.

The magnified cross-section above shows the thick walls of an artery.

The heart pumps blood through a network of blood vessels, which extend 150,000 kilometres around the human body. There are three main types of blood vessels. Thick-walled arteries carry blood away from the heart. Thinner-walled veins carry blood back to the heart. Microscopic capillaries link veins and arteries. Formed as branches of arterioles (the smallest arteries), capillaries pass through tissues and supply groups of cells with essential substances. Waste products flow back into the capillaries. Capillaries link up to form venules, which unite to form veins. The human circulatory system is a double circulation with two 'loops'. One loop carries blood to the lungs. The other carries blood to the body. It takes blood about 60 seconds to complete a full circuit of the body.

ANATOMY OF THE HEART

The heart is a powerful, muscular pump that maintains a continuous flow of blood around the body. The heart is divided into two halves by the septum. Each half has a smaller upper chamber, called the atrium, and a larger lower chamber, called the ventricle. The right atrium receives oxygen-poor blood from the body through large veins called the superior (upper) vena cava and inferior (lower) vena cava. The pulmonary arteries carry blood pumped from the right ventricle to the lungs. The left atrium receives oxygen-rich blood from the lungs through the pulmonary veins. The left ventricle pumps the oxygen-rich blood to the body's cells along a large artery called the aorta.

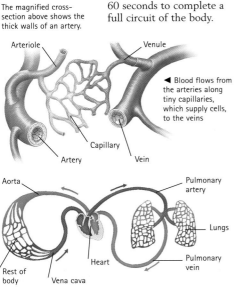

Arteriole

Venule

◄ Blood flows from the arteries along tiny capillaries, which supply cells, to the veins

Capillary

Artery

Vein

Aorta

Pulmonary artery

Lungs

Heart

Pulmonary vein

Rest of body

Vena cava

▲ The circulatory system is made up of two 'loops'. One carries oxygen-poor blood from the heart to the lungs (where it picks up oxygen) and back to the heart. The other carries oxygen-rich blood from the heart to the rest of the body (where it supplies oxygen to all body tissues) and then back to the heart.

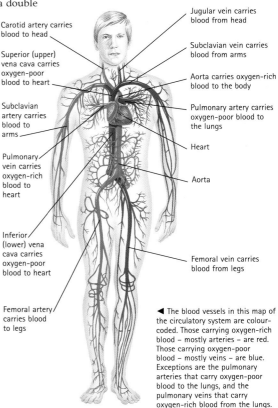

Carotid artery carries blood to head

Superior (upper) vena cava carries oxygen-poor blood to heart

Subclavian artery carries blood to arms

Pulmonary vein carries oxygen-rich blood to heart

Inferior (lower) vena cava carries oxygen-poor blood to heart

Femoral artery carries blood to legs

Jugular vein carries blood from head

Subclavian vein carries blood from arms

Aorta carries oxygen-rich blood to the body

Pulmonary artery carries oxygen-poor blood to the lungs

Heart

Aorta

Femoral vein carries blood from legs

◄ The blood vessels in this map of the circulatory system are colour-coded. Those carrying oxygen-rich blood – mostly arteries – are red. Those carrying oxygen-poor blood – mostly veins – are blue. Exceptions are the pulmonary arteries that carry oxygen-poor blood to the lungs, and the pulmonary veins that carry oxygen-rich blood from the lungs.

THE HEART

The front view of the heart (below) shows the main blood vessels carrying blood to and from the heart and the coronary artery that supplies the heart wall. A section through the heart (below right) shows the septum that divides the heart into left and right halves, the atria and the larger ventricles. Heart valves prevent blood flowing backwards.

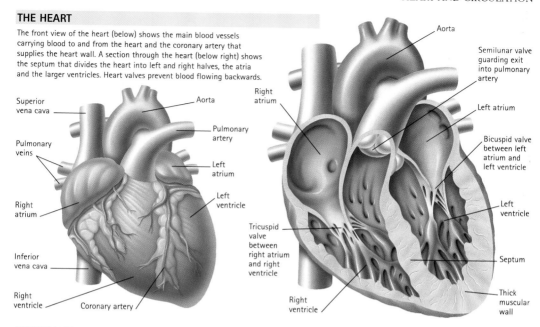

Superior vena cava

Pulmonary veins

Right atrium

Inferior vena cava

Right ventricle

Coronary artery

Aorta

Pulmonary artery

Left atrium

Left ventricle

Right atrium

Tricuspid valve between right atrium and right ventricle

Right ventricle

Aorta

Semilunar valve guarding exit into pulmonary artery

Left atrium

Bicuspid valve between left atrium and left ventricle

Left ventricle

Septum

Thick muscular wall

CORONARY CIRCULATION

Blood passes through the heart too quickly to supply the muscle cells in the heart wall with the oxygen and food they need. The heart has its own blood supply called the coronary circulation. Two coronary arteries branch from the aorta and supply all parts of the heart wall. Blood that has passed through heart muscle empties into the right atrium. If a coronary artery becomes blocked, the part of the heart it supplies may die and cause a heart attack.

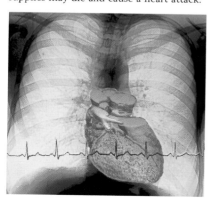

HEARTBEAT

During a single heartbeat, both ventricles fill with blood and then contract to pump blood out of the heart. As the heart fills with blood, semilunar valves close to prevent backflow into the heart from the aorta and pulmonary artery. As the heart empties, valves between the atria and ventricles close to prevent blood flowing back into the atria. As the valves between atria and ventricles close, they produce a long 'lubb' sound. As the semilunar valves close, they produce a shorter 'dupp' sound. Together, these sounds make up the heartbeat, which can be heard using an instrument called a stethoscope. The timing of each heartbeat is regulated by a 'pacemaker' in the wall of the right atrium.

▶ On average, the heart beats about 75 times a minute. Each heartbeat is a cycle of three stages – diastole, atrial systole and ventricular systole. These stages follow each other in a precisely timed sequence. Over the course of the three stages blood enters the atria, passes into the ventricles and is then pumped out of the heart.

◀ This X-ray shows the heart located inside the thorax (chest) between the two lungs (yellow). The ribs, also visible, surround and protect both heart and lungs. Also shown is an electrocardiograph (ECG), which is a record of the electrical changes taking place in the heart.

During diastole, the atria and ventricles are relaxed. Both atria fill with blood.

During atrial systole, the atria contract and squeeze blood into the ventricles.

During ventricular systole, the ventricles contract and push blood from the heart.

SEE ALSO

98 Blood, 99 Lymphatic system, 100–1 Lungs and breathing

BLOOD

Blood provides the body's trillions of cells with a delivery and removal system. It also helps to defend the body against infection and repairs damaged blood cells.

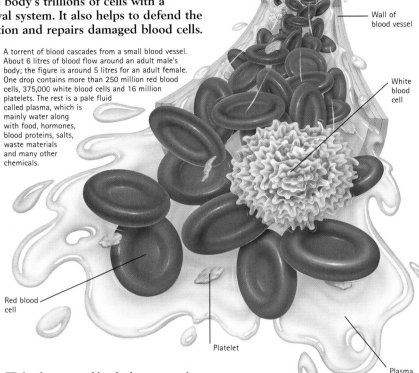

A torrent of blood cascades from a small blood vessel. About 6 litres of blood flow around an adult male's body; the figure is around 5 litres for an adult female. One drop contains more than 250 million red blood cells, 375,000 white blood cells and 16 million platelets. The rest is a pale fluid called plasma, which is mainly water along with food, hormones, blood proteins, salts, waste materials and many other chemicals.

Wall of blood vessel

White blood cell

Red blood cell

Platelet

Plasma

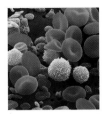

▲ Red blood cells, white blood cells (yellow) and platelets (pink) are made in bone marrow. About two million red blood cells are produced every second.

Red blood cells caught in platelet net

Damage to a blood vessel causes the platelets to form a 'net' of fibres that traps red blood cells.

Scab forms

Fibres and red blood cells form a clot to seal off the wound. The surface of the clot hardens into a scab.

Healed tissue under old scab

Underneath the scab, the blood vessel and skin repair themselves. Once this is done, the old, dried-up scab falls off.

Blood is pumped by the heart around the body through arteries, veins and capillaries. It supplies food and oxygen to the body's cells and removes wastes. Blood also maintains the body's temperature, fights disease and plays a role in repairing damaged blood vessels.

BLOOD AS A TRANSPORT SYSTEM
Oxygen is transported to cells by doughnut-shaped red blood cells. These are packed with haemoglobin, a substance that picks up oxygen as blood passes through the lungs and releases it into the body's cells. Plasma is a watery liquid that makes up about 55 per cent of blood. It is responsible for transporting food, waste products, chemical messengers called hormones and many other substances around the body. It also helps the body to maintain a temperature of around 37°C.

DEFENCE AND PROTECTION
Tiny disease-causing micro-organisms (pathogens) are constantly trying to infect the human body. White blood cells called phagocytes and lymphocytes destroy these invaders. Phagocytes hunt down and engulf any pathogens. Lymphocytes release killer chemicals called antibodies. These disable pathogens so that phagocytes can engulf them. Lymphocytes remember pathogens so that they can react even faster if the same pathogens invade again.

Platelets seal leaks in damaged blood vessels. Their action stops pathogens from getting into the body and also prevents the loss of blood from the damaged area.

SEE ALSO
96–7 Heart and circulation

LYMPHATIC SYSTEM

The lymphatic system is a transport system that drains fluid called lymph from the tissues into the blood. It also contains cells that defend the body against disease.

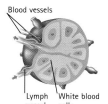

Blood vessels

Lymph vessel   White blood cells

A s blood flows along the circulatory system, a substance called tissue fluid passes out through the capillary walls. This fluid delivers oxygen and essential nutrients to tissue cells. Tissue fluid then removes wastes and returns back into the bloodstream through the capillary walls.

▲ A lymph node is a mass of white blood cells and a network of fibres. Debris and pathogens are removed from lymph as it passes through the node.

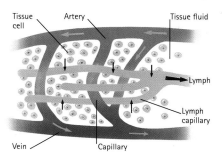

Tissue cell   Artery   Tissue fluid

Lymph

Lymph capillary

Vein   Capillary

Cells in the body's tissues are bathed in a fluid derived from nearby blood capillaries. Excess tissue fluid passes into blind-ending lymph capillaries and becomes lymph.

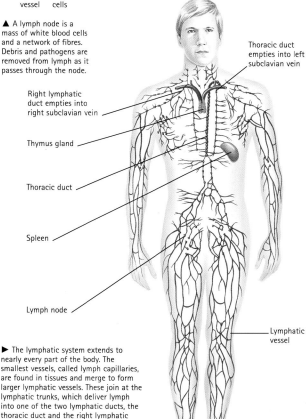

Thoracic duct empties into left subclavian vein

Right lymphatic duct empties into right subclavian vein

Thymus gland

Thoracic duct

Spleen

Lymph node

Lymphatic vessel

▶ The lymphatic system extends to nearly every part of the body. The smallest vessels, called lymph capillaries, are found in tissues and merge to form larger lymphatic vessels. These join at the lymphatic trunks, which deliver lymph into one of the two lymphatic ducts, the thoracic duct and the right lymphatic duct. The ducts empty lymph into the bloodstream through the left and right subclavian veins. Along the lymph vessels are swellings called lymph nodes. The spleen is an important part of the lymphatic system and is a primary filter for the blood. It also makes antibodies.

DRAINAGE

Every day, about 24 litres of fluid leaves the capillaries that pass through the tissues. Most returns directly back to the blood, but about 4 litres remains. The lymphatic system drains the excess fluid, now called lymph, and empties it back into the blood vessels in the upper chest.

Lymph is a colourless fluid that contains dissolved substances, debris and pathogens such as bacteria and viruses. It only flows in one direction – away from the tissues. Unlike blood, which has a heart to pump it, lymph moves through the lymphatic system with the help of skeletal muscles that push the fluid along when they contract. Valves in the lymph vessels stop the fluid flowing backwards.

DEFENCE

As lymph flows through lymph vessels, it passes through lymph nodes. Here, white blood cells called macrophages trap and engulf cell debris and pathogens. Other white blood cells, called lymphocytes, produce antibodies which are chemicals that mark pathogens for destruction. Other lymphatic organs have a similar role. The tonsils intercept pathogens that enter the mouth. Together, lymphocytes and macrophages form the immune system, the body's most powerful defence against disease.

SEE ALSO
78–9 Body organization, 96–7 Heart and circulation

LUNGS AND BREATHING

Humans need oxygen to live. It is obtained from air breathed into the lungs. In the lungs, oxygen passes into the bloodstream and is then carried to the body's cells.

Cells need a constant supply of energy to power their activities. Cells use oxygen to release energy from foods in a process called respiration. Respiration releases carbon dioxide, a poisonous waste product which must be removed from the body. Delivering oxygen and removing carbon dioxide is done by the respiratory system. This consists of a system of tubes that carry air in and out of the body, and a pair of lungs through which oxygen enters, and carbon dioxide leaves, the blood. Blood carries oxygen from the lungs to the cells, and transports waste carbon dioxide from the cells back to the lungs.

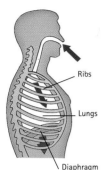

During inhalation, or the process of breathing in, the diaphragm contracts and flattens. The intercostal muscles also contract and pull the ribs upwards and outwards. This makes the lungs bigger and decreases the pressure inside them so that air is sucked into them through the mouth and trachea.

A SYSTEM OF TUBES

Air first passes through the nose. Hairs in the nostrils and sticky mucus lining the nasal cavity trap particles that would damage the lungs. Air then passes into the pharynx, through the larynx (voice box) into the trachea, which is reinforced by C-shaped pieces of cartilage. Mucus in the trachea also traps dirt. Fine hair-like cilia carry dirt up to the throat. The trachea branches into two bronchi, which themselves branch inside the lungs.

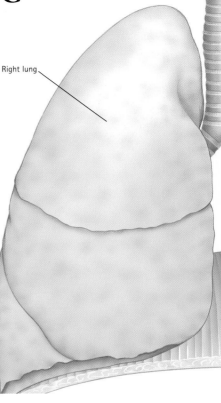

Right lung

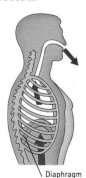

During exhalation, or the process of breathing out, the diaphragm relaxes and is pushed up by the abdominal organs beneath it. The intercostal muscles relax, so the ribs move downwards and inwards. This decreases volume inside the thorax. The lungs get smaller and pressure inside increases so that air is pushed out.

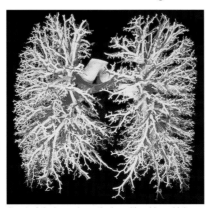

A cast of the lungs shows the bronchi and bronchioles (white) and the pulmonary artery (red). This network is called the bronchial tree – it resembles an upside-down tree with the trachea as trunk and bronchi as branches.

THE LUNGS

The lungs lie in the thorax (chest) and sit either side of the heart, protected by the backbone and ribcage. They rest on the diaphragm, a muscular sheet that separates the thorax from the abdomen. Healthy lungs are pink because they are full of blood. They also feel spongy, because they consist of a branching network of airways that end in millions of microscopic air sacs called alveoli, through which oxygen enters the bloodstream. Squeezed into the chest, all the alveoli provide a surface area for absorbing oxygen equivalent to two-thirds that of a tennis court. A thin pleural membrane covers the lungs; another lines the inside of the chest wall. Fluid between the membranes decreases friction and prevents pain during breathing.

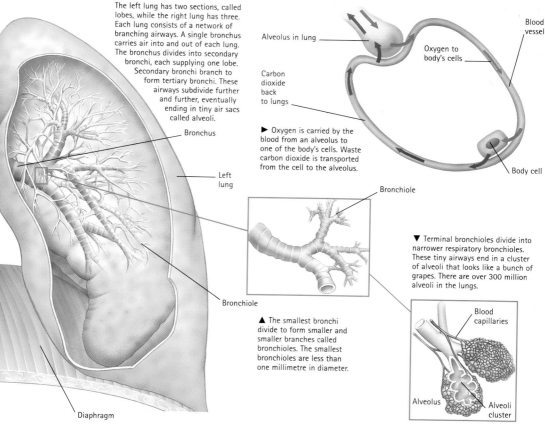

The left lung has two sections, called lobes, while the right lung has three. Each lung consists of a network of branching airways. A single bronchus carries air into and out of each lung. The bronchus divides into secondary bronchi, each supplying one lobe. Secondary bronchi branch to form tertiary bronchi. These airways subdivide further and further, eventually ending in tiny air sacs called alveoli.

Bronchus

Left lung

Diaphragm

Alveolus in lung

Carbon dioxide back to lungs

Blood vessel

Oxygen to body's cells

Body cell

▶ Oxygen is carried by the blood from an alveolus to one of the body's cells. Waste carbon dioxide is transported from the cell to the alveolus.

Bronchiole

Bronchiole

▼ Terminal bronchioles divide into narrower respiratory bronchioles. These tiny airways end in a cluster of alveoli that looks like a bunch of grapes. There are over 300 million alveoli in the lungs.

▲ The smallest bronchi divide to form smaller and smaller branches called bronchioles. The smallest bronchioles are less than one millimetre in diameter.

Blood capillaries

Alveolus

Alveoli cluster

GAS EXCHANGE

Gas exchange takes place continuously in the alveoli. This process ensures that the body's cells receive a constant supply of oxygen and are not poisoned by the accumulation of carbon dioxide. Oxygen dissolves in a thin layer of liquid that lines each alveolus, then moves by diffusion – the movement of molecules from a high concentration to a low concentration – across the thin wall of the alveolus into the blood capillary and into red blood cells. Carbon dioxide diffuses in the reverse direction from the blood into the air inside the alveolus and is exhaled. Inhaled air contains about 21 per cent oxygen and 0.04 per cent carbon dioxide. Exhaled air contains 16 per cent oxygen and about 4 per cent carbon dioxide.

BREATHING

Breathing, or ventilation, moves fresh air into the lungs to replenish supplies of oxygen. It also forces stale air out of the lungs to remove carbon dioxide. The elastic lungs depend on the diaphragm and rib muscles to change the shape of the thorax. Pleural membranes cover the lungs' surfaces and line the chest wall. They act like an 'adhesive pad', ensuring that the lungs follow the movement of the thorax. During inhalation, the diaphragm and intercostal muscles between the ribs contract. This makes the thorax and the lungs larger – and the pressure inside them lower – so that air is sucked in. During exhalation, the reverse occurs. The lungs never completely fill or empty – a reservoir of air remains and is refreshed.

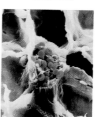

A view inside the lung magnified 410 times reveals a blood capillary filled with red blood cells. These cells pick up oxygen from the surrounding network of alveoli.

SEE ALSO

96–7 Heart and circulation, 98 Blood

FOOD AND NUTRITION

Food provides the body with vital substances called nutrients. Good nutrition means that the body has an adequate, balanced supply of nutrients.

A person who regularly eats fast food is not likely to be receiving a balanced diet. Although the meal shown above contains some carbohydrate in the bun and french fries, it is also very rich in protein and animal fats. It contains no fruit or fresh vegetables to provide vitamins or minerals.

The process by which humans obtain a regular supply of food to survive is called nutrition. Most food contains a variety of nutrients, which are released during digestion. Macronutrients – carbohydrates, proteins and fats – are needed in large amounts each day. Carbohydrates provide energy. They include complex starches found in potatoes and pasta and simple sugars in fruits and sweets. Proteins provide simple building blocks called amino acids for growth and repair. Fats provide energy and help insulate the body. Micronutrients – vitamins and minerals – are only needed in tiny amounts daily but are essential for cells to function. They include vitamins such as vitamins A and C, and minerals such as calcium. Also essential are water to maintain the body's fluid balance, and fibre, undigested plant material that keeps the intestinal muscles working properly.

Meals should be well balanced. Pasta and bread provide carbohydrate; beans and fish supply protein and vitamins with some fat; salad contains vitamins, minerals and fibre.

BALANCED DIET

The word *diet* refers to the type and amount of food a person eats each day. To maintain good health and avoid weight gain, a person's diet should be balanced, containing a range of nutrients in the right amounts. A balanced diet consists of about 55 per cent carbohydrates (mostly complex starches), about 15 per cent protein and 30 per cent or less fat (unsaturated fats from plant oils and oily fish are healthier than saturated fats from meat or dairy products). It should also include plenty of fresh fruit and vegetables.

THE FOOD PYRAMID

The food pyramid provides an easy way to plan a balanced diet. The bulk of a balanced diet should be made up of starchy, carbohydrate-rich foods, along with smaller amounts of proteins and fats (preferably not animal fats). It should also provide plenty of vitamins, minerals and fibre. The food pyramid provides a simple way of getting the balance right by showing the proportions in which the main types of foods should be eaten. Starchy foods, such as rice and bread, and vitamin-, mineral- and fibre-rich foods, such as vegetables and fruits, are found towards the base of the pyramid. Those that should be eaten in smaller amounts, such as meat and dairy products, are further up the pyramid. Those that should be eaten sparingly or not at all, such as sugary cakes and sweets, are located at the narrow top of the food pyramid.

Foods rich in starchy carbohydrates such as rice, bread, potatoes and pasta. Such foods release energy slowly throughout the day.

Foods rich in fat and sugar should only be eaten in small amounts.

Foods rich in protein such as beans, fish, chicken, meat and cheese. Protein is essential for growth and tissue repair. Meat and cheese also contain a lot of fat.

Fresh fruit and vegetables provide vitamins and minerals that are essential for good health and fibre (roughage) that keeps the digestive system working properly.

SEE ALSO

78–9 Body organisation, 103 Digestion

DIGESTION

The digestion of food releases simple nutrients in a form that can be used by the body's cells. This process takes place in the digestive system.

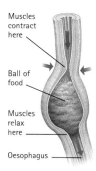

Muscles contract here

Ball of food

Muscles relax here

Oesophagus

Swallowed food is pushed down the oesophagus by wave-like muscular contractions called peristalsis. The muscles contract behind the food to push it downwards.

The nutrients essential for life are 'locked' inside the large molecules that make up food. The job of the digestive system is to break down these large molecules – such as carbohydrates, proteins and fats – to release simple nutrients such as sugars, amino acids and fatty acids.

The digestive process has four stages: ingestion, digestion, absorption and egestion. During ingestion, the food is taken into the mouth, chewed and swallowed. During digestion, food is broken down either by muscular crushing or by chemicals called enzymes. Absorption involves moving nutrients from the alimentary canal into the bloodstream. Finally, egestion ejects waste through the anus.

THE INTESTINES

The small intestine runs from the pyloric sphincter to the caecum of the large intestine. It is the most important part of the digestive system because most digestion and absorption takes place there.

The large intestine is 1.5 metres long and consists of the caecum, colon and rectum. Water is absorbed from the waste products of digestion as they pass through the colon. The waste forms semi-solid faeces, consisting of dead cells, fibre and bacteria. Faeces are stored in the rectum and then released through the anus.

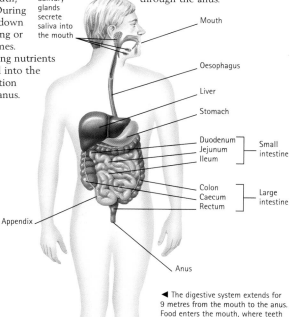

Salivary glands secrete saliva into the mouth

Mouth

Oesophagus

Liver

Stomach

Duodenum
Jejunum — Small intestine
Ileum

Colon
Caecum — Large intestine
Rectum

Appendix

Anus

◄ The digestive system extends for 9 metres from the mouth to the anus. Food enters the mouth, where teeth crush it into pieces and the salivary glands lubricate it with saliva. Peristalsis takes the processed food from the oesophagus to the stomach, which partly digests the food. The liver and pancreas release secretions into the small intestine, which completes digestion and absorption.

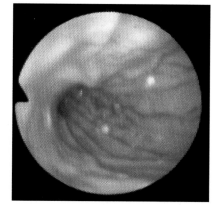

▲ An image of the stomach, seen through a device called an endoscope, clearly shows the slippery mucus that lines and protects the wall of the digestive system.

THE STOMACH

The stomach plays three roles in digestion. Firstly, its walls contract to crush food. Secondly, glands in the stomach wall release acidic gastric (stomach) juice that digests proteins in food. Thirdly, it expands to store food for up to four hours.

SEE ALSO

89 Taste and smell, 102 Food and nutrition

REPRODUCTION

Reproduction ensures that the human species does not become extinct. The male and female reproductive systems enable men and women to have children.

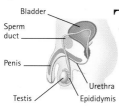

- Bladder
- Sperm duct
- Penis
- Urethra
- Testis
- Epididymis

Testes make millions of sperm a day. During sexual intercourse, the penis gets erect and is put into the female's vagina. Sperm travel along the sperm duct and out of the penis.

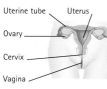

- Uterine tube
- Uterus
- Ovary
- Cervix
- Vagina

The ovaries contain a store of eggs, one of which is released each month. If it is fertilized by a sperm cell, it develops into a baby in the uterus. At birth, the baby passes out through the vagina.

The reproductive system does not become active until puberty during the early teens. Male and female reproductive systems differ, although both produce sex cells. Sex cells are made by a type of cell division called meiosis. Sex cells contain only 23 chromosomes (genetic material) in their nucleus, half the number found in other cells. The male sex cells, called sperm, are made in the two testes. More than 250 million sperm are made each day. The female sex cells, called eggs or ova, are produced in the two ovaries before birth. After puberty, one egg is released each month – ovulation – and the body prepares for possible pregnancy. Sperm and egg are brought together by an intimate act called sexual intercourse. The man inserts his penis into his partner's vagina and releases millions of sperm that swim towards the fallopian tubes. If sexual intercourse happens within 24 hours of ovulation, a sperm may penetrate the egg. This is called fertilization. The sperm's nucleus (23 chromosomes) fuses with that of the egg (23 chromosomes); the combined genetic material (46 chromosomes) gives the blueprint for a new human being.

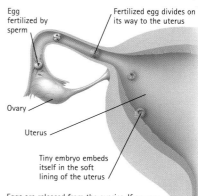

- Egg fertilized by sperm
- Fertilized egg divides on its way to the uterus
- Ovary
- Uterus
- Tiny embryo embeds itself in the soft lining of the uterus

Eggs are released from the ovaries. If an egg meets a sperm, it is fertilized. As it travels along the fallopian tube, the fertilized egg divides repeatedly. In seven days, the egg arrives in the uterus. It is now a hollow ball of cells.

CONCEPTION

Conception is the time between fertilization and implantation. As the fertilized egg passes along the fallopian tube it divides repeatedly to form a ball of cells called a conceptus. After seven days, the conceptus sinks into the soft lining of the uterus and becomes an embryo. If the two daughter cells separate when the fertilized egg first divides, the two cells will develop independently and result in identical twins. If two eggs are released during ovulation and both are fertilized by sperm, they will produce non-identical, or fraternal, twins.

HOW A BABY DEVELOPS IN THE WOMB

After fertilization, the fertilized egg travels to the uterus. What started as a single cell becomes a foetus made up of billions of cells. Development occurs within a fluid-filled sac, protected within the uterus. Food and oxygen pass through the umbilical cord from the placenta, where blood from the foetus and mother come into close contact.

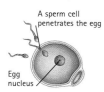

- A sperm cell penetrates the egg
- Egg nucleus

- Two cells

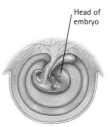

- Head of embryo
- Placenta

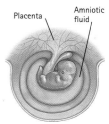

- Amniotic fluid

1 During fertilization, the nucleus of a sperm cell fuses with the nucleus of the egg to produce a fertilized egg.

2 Approximately 36 hours after the egg is fertilized, the egg has divided once, resulting in two cells.

3 About 72 hours after fertilization there are 16 cells. In a few days, the ball of cells will settle in the uterus.

4 After four weeks, the embryo is floating in a fluid-filled sac. The heart is beating and the brain has started to develop.

5 After five weeks, the embryo is the size of an apple pip. It has buds that will become arms and legs. The tail is shrinking.

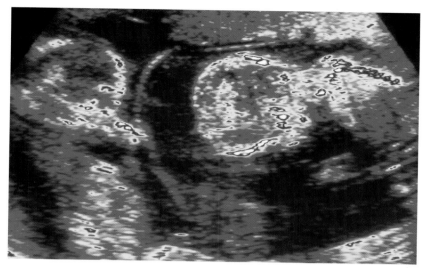

An ultrasound scan of the uterus of a pregnant woman shows that she has twins. The left foetus is seen in side view, with its body directly below its head. The right foetus has its head facing down and its body horizontal. Both their heads are at the top of the picture. Ultrasound is a safe, painless way of checking that the foetus is healthy. Sound waves are directed into the uterus, and the echoes that bounce back are processed to produce an image.

The picture above shows a healthy baby boy. The human body never grows as fast as it did in the womb. If growth continued at the same rate, a baby would be two kilometres tall by its first birthday.

PREGNANCY

Pregnancy is the time between conception and birth. For the first two months the developing baby is called an embryo. After this, when the organs are working, it is called a foetus. Amniotic fluid surrounds and protects the foetus. The growing baby is kept alive by the placenta, which is attached to the uterus. Inside the placenta, food and oxygen pass from the mother's blood to that of the foetus, and wastes pass in the opposite direction. The umbilical cord carries blood between placenta and foetus.

BIRTH

About 38 weeks after fertilization, the uterus starts to contract. This process, called labour, usually begins about 12 hours before the birth. Powerful contractions push the baby out through the vagina, and the baby takes its first breath of air.

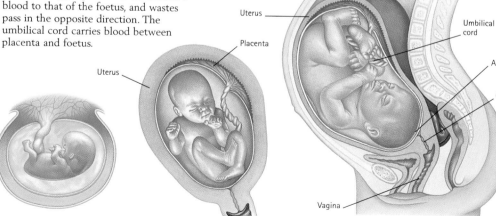

Uterus

Placenta

Uterus

Umbilical cord

Amnion

Cervix

Vagina

6 After eight weeks, the embryo – now called a foetus – is about the same size as a strawberry and has developed tiny fingers and toes.

7 After 28 weeks, the foetus is fully developed in the expanded uterus. It will further increase in weight before birth.

8 At full term, about 38 weeks after fertilization, the foetus has moved its head downward in preparation for birth.

SEE ALSO

78–9 Body organization, 106 Growth and development, 107 Genes and chromosomes

GROWTH AND DEVELOPMENT

Growth and development – from birth to adulthood – follow a fixed pattern in the first 20 years of life. By the age of 40, the first signs of ageing begin to appear.

Growth and development occur simultaneously. Growth is an increase in size. Development is where cells specialize to perform specific functions. During its first year, an infant is totally dependent on his or her parents for food and protection. However, the infant is already starting to develop skills such as talking, walking and interacting with others. These skills become more obvious and develop further as an infant gets older.

▲ Mother and baby make eye contact. This bonding process starts from the very moment a baby is born. Being held makes babies feel secure; they respond by smiling and making noises. Bonding reinforces the natural feelings parents have for their children.

▼ Every person follows the same pattern of growth and development, with slight differences between the development of the male and female reproductive systems. After rapid growth in the first year of life, children grow steadily until their early teens. Then, during puberty, the body grows rapidly and takes on an adult appearance. By the age of 20, the body has completed its growth.

Face and skull aged 6

Face and skull aged 16

Two photos of the same person taken at different ages show how the face changes shape between the ages of 6 and 16. The bones of the face – as shown by the shape of the skulls – grow rapidly during late childhood.

PUBERTY AND ADOLESCENCE
Puberty is a time of rapid growth that leads to sexual maturity. It starts around the age of 11 in girls and about 13 in boys. In both sexes, armpit and pubic hair grows. A girl's body becomes more rounded. Breasts develop and the hips become wider. Her ovaries start to release eggs and menstruation begins. A boy's body becomes more muscular and hairy. His shoulders widen, his voice deepens and the testes start to produce sperm. Puberty is part of adolescence, which also involves mental changes. These changes make a young adult become more independent and have sexual feelings.

AGEING
The body ages fairly rapidly after the age of 40. As cells become less efficient, the skin becomes more wrinkled, muscles less powerful, bones more brittle, senses less acute and the hair thins and turns grey. Eventually, one or more body systems stop working and the person dies.

Age 30–34

Age 2

Age 6

Age 10–12

Age 20–22

SEE ALSO

94–95 Hormones, 104–5 Reproduction, 107 Genes and chromosomes

GENES AND CHROMOSOMES

Chromosomes are found in the nucleus of nearly every cell. Each chromosome contains sets of instructions called genes.

DNA

Deoxyribonucleic acid, or DNA, stores the information needed to build a cell. Together, cells make up a functioning human body. DNA molecules are coiled up and packaged into thread-like chromosomes. There are 46 chromosomes in the nucleus of most cells. DNA molecules are organized into two linked strands that spiral around one another, forming a structure called a double helix. The strands are held together by four different chemicals called bases. The precise sequence of bases along a DNA molecule provides coded instructions for cell construction and operation.

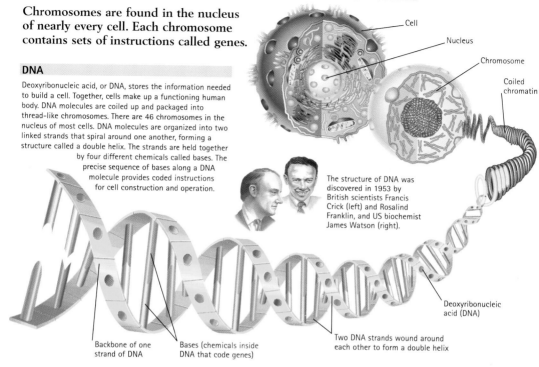

Cell

Nucleus

Chromosome

Coiled chromatin

The structure of DNA was discovered in 1953 by British scientists Francis Crick (left) and Rosalind Franklin, and US biochemist James Watson (right).

Deoxyribonucleic acid (DNA)

Backbone of one strand of DNA

Bases (chemicals inside DNA that code genes)

Two DNA strands wound around each other to form a double helix

There are about 100,000 genes in the human body. A single gene consists of a small section of a DNA molecule. Each gene instructs a cell to make a specific protein. Because proteins control cell metabolism, genes shape and operate our bodies. Apart from identical twins, gene combinations vary slightly for each person. Genes are arranged on a pair of matching chromosomes, one maternal (from the mother) and one paternal (from the father). There are two versions of the same gene on each pair of matching chromosomes. For example, a maternal chromosome may carry a gene for brown eyes, and a paternal chromosome a gene for blue eyes. In this case, only the brown gene is expressed and the child has brown eyes. The Human Genome Project, currently being carried out by scientists worldwide, aims to identify every human gene to find out what it controls.

CHROMOSOMES

Chromosomes contain thousands of genes. Genes are passed on from parents to their offspring. In the ovaries and testes, a process of cell division called meiosis makes sex cells (eggs and sperm) that contain 23 chromosomes. At fertilization, a sperm cell joins the egg to produce the full complement of 46 chromosomes. One pair of chromosomes, the sex chromosomes, differ from the other 22 pairs of chromosomes. While they carry genes, they are not the same in both sexes. Males have a longer (X) chromosome paired with a shorter (Y) chromosome. Females have two X chromosomes. The presence of XY chromosomes in the embryo ensures that male reproductive organs are formed.

▼ An electron micrograph reveals 8 of the 46 chromosomes found inside the nucleus of a human cell. This image was taken during mitosis. In this type of cell division, the chromosomes become much shorter and thicker.

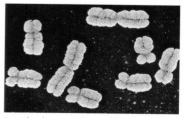

SEE ALSO

104 Reproduction, 106 Growth and development

FACTS AND FIGURES

BRANCHES OF HUMAN BIOLOGY AND MEDICINE

Anatomists study the structure of the body and how its parts fit together.
Biochemists study the chemical processes that occur in and around the body's cells.
Cardiologists study the heart and blood vessels and diseases that affect them.
Cytologists study cells.
Dermatologists study the skin and its disorders.
Endocrinologists study the hormonal system and its disorders.
Epidemiologists study causes of diseases and their spread through populations.
Geneticists study DNA, chromosomes, genes and the mechanism of inheritance.
Gynaecologists study the female reproductive system and its disorders.
Haematologists study the properties and disorders of blood and bone marrow.
Histologists study body tissues.
Immunologists study the immune system and its disorders.
Neurologists study the nervous system and its disorders.
Oncologists study the causes, mechanisms and treatment of cancers.
Ophthalmologists study the eye and its disorders.
Pathologists study and determine the causes of disease and death.
Physiologists study how the body's component parts function.
Psychiatrists study mental illness and its prevention and treatment.

BODY SYSTEMS

Circulatory system of heart, blood and blood vessels, transports materials to and from cells throughout the body.
Digestive system breaks down food so its nutrients can be used by the body.
Endocrine system releases hormones that control many body processes.
Immune system defends the body against micro-organisms that cause diseases.
Integumentary system of skin, hair and nails covers and protects the body.
Lymphatic system drains fluid from the tissues and destroys pathogens.
Muscular system moves the body and helps to support it.
Nervous system of brain, nerves and sense organs, controls the body and enables a person to think and feel.
Skeletal system of bones, cartilage and ligaments, supports the body, protects internal organs and permits movement.
Reproductive system enables humans to produce offspring.
Respiratory system carries oxygen into the bloodstream for transport to cells.
Urinary system removes waste materials.

KEY DATES

BC

c.500 Greek physician and philosopher Alcmaeon proposes that the brain, rather than the heart, is the organ of thinking and feeling.

c.420 Greek physician Hippocrates teaches a diagnostic approach to medicine based on observation.

AD

129 Birth of Greek anatomist and doctor Galen who wrote about the workings of the human body. His ideas, many of them false, held back the understanding of human biology and medicine for the next 1,000 years.

1037 Islamic physician and philosopher Avicenna dies. His medical texts continue to dominate European and Middle-Eastern medicine for more than 500 years.

1268 British philosopher and scientist Roger Bacon records the use of spectacles to correct eye defects.

1288 Syrian physician Ibn An-Nafis dies. He had demonstrated that blood flows through the lungs.

1543 Belgian anatomist Andreas Vesalius publishes an accurate description of human anatomy, correcting many of Galen's errors.

1628 British physician William Harvey publishes a description of how blood circulates around the body.

1661 Italian physiologist Marcello Malpighi discovers capillaries, the links between arteries and veins.

1674 Dutch microscopist Antoni van Leeuwenhoek observes and describes red blood cells.

1796 British physician Edward Jenner performs first vaccination and shows that cowpox fluid protects against smallpox.

1839 German physiologist Theodor Schwann proposes that animals consist of tiny living cells.

1846 First use of an anaesthetic – ether – for a tooth extraction – during surgery in a Boston, USA, hospital.

1848 French physiologist Claude Bernard describes liver function, and establishes concept of homeostasis.

1858 German biologist Rudolf Virchow shows that diseases occur when normal cells become defective, so establishing cell pathology.

1860s French chemist and biologist Louis Pasteur establishes the link between germs and diseases.

1865 British surgeon Joseph Lister uses carbolic acid as an antiseptic during surgery and dramatically reduces deaths from infection.

1880 French surgeon Paul Broca dies – he discovered certain parts of the brain control body functions.

1882 German physician Robert Koch discovers *Mycobacterium tuberculosis*, the bacterium that causes tuberculosis.

1895 German physicist Wilhelm Roentgen discovers X-rays.

1899 British physician Ronald Ross proves that mosquitoes carry malaria from human to human.

1900 Austrian neurologist Sigmund Freud publishes *The Interpretation of Dreams*, which contains the basic concepts of psychoanalysis.

1900 Austrian-born US pathologist Karl Landsteiner discovers blood groups A, O, B and AB.

1907 British biochemist Frederick Hopkins discovers vitamins.

1910 US biologist Thomas Morgan discovers how chromosomes carry genetic information.

1922 Canadian physiologists Frederick Banting and Charles Best discover insulin, providing a means to control diabetes.

1928 British microbiologist Alexander Fleming discovers penicillin, the first antibiotic.

1951 British physicist Francis Crick and US biologist James Watson discover the structure of DNA, helped by the X-ray evidence provided by British biophysicist Rosalind Franklin.

1952 US physician Jonas Salk develops the first polio vaccine.

1958 First ultrasound scan of a foetus in its mother's uterus.

1967 South African surgeon Christiaan Barnard performs the first successful heart transplant.

1972 Computerized tomography (CT) scanning introduced as a means of producing images of the internal organs of the body.

1979 World declared free of the killer disease smallpox as a result of a worldwide vaccination campaign.

1981 The first cases of Acquired Immune Deficiency Syndrome (AIDS) reported. AIDS is recognized as a new disease.

1990 Human Genome Project starts in the USA. Its aim is to analyse human DNA to find the genes in all 46 human chromosomes.

1995 Work starts to develop genetic engineering techniques as a possible means for treating inherited diseases.

1999 Chromosome 22 becomes the first human chromosome to have all its genes identified.

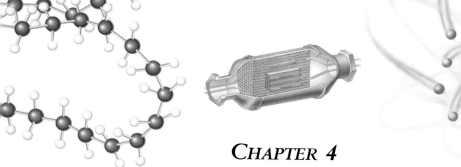

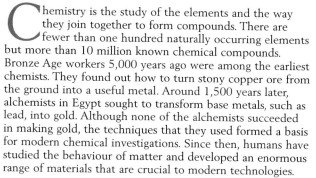

CHAPTER 4

CHEMISTRY AND THE ELEMENTS

Chemistry is the study of the elements and the way they join together to form compounds. There are fewer than one hundred naturally occurring elements but more than 10 million known chemical compounds. Bronze Age workers 5,000 years ago were among the earliest chemists. They found out how to turn stony copper ore from the ground into a useful metal. Around 1,500 years later, alchemists in Egypt sought to transform base metals, such as lead, into gold. Although none of the alchemists succeeded in making gold, the techniques that they used formed a basis for modern chemical investigations. Since then, humans have studied the behaviour of matter and developed an enormous range of materials that are crucial to modern technologies.

The chemical industry provides the concrete, metals and plastics to create buildings, machinery and vehicles; fuels for transport and heating; synthetic fibres for clothes; and fertilizers and pesticides to improve the yield of food crops. The knowledge of chemistry makes it possible to produce high-purity silicon for microprocessors. Biochemistry helps us to understand the processes that occur in living organisms, and pharmacological chemistry provides drugs and medicines to treat diseases, many of which were once impossible to cure.

THE ELEMENTS

Elements are substances that cannot be broken down by chemical methods. There are 92 naturally occurring elements and 20 synthetic elements.

Realgar

Pyrite

Malachite

Fluorite

All these substances are minerals found in the ground. Realgar contains the elements arsenic and sulphur, while pyrite contains iron and sulphur. Malachite is a mixture of copper carbonate and copper hydroxide which together consist of copper, carbon, oxygen and hydrogen. Fluorite contains calcium and fluorine.

Elements can be classified as metals and non-metals. Metals are usually shiny solids that conduct electricity. Most metals only melt at high temperatures. Metals are malleable, which means they can be hammered into different shapes. Many are also ductile, which means they can be stretched without breaking. Iron, copper, zinc and uranium are examples of metals.

With the exception of graphite – a form of carbon – non-metals do not conduct electricity. Solid non-metals, such as sulphur and phosphorus, are brittle (they break into pieces when hit). Most non-metals melt at much lower temperatures than metals; many are gases at room temperature. Chlorine, hydrogen and oxygen are non-metals.

There are 92 natural elements. With the exceptions of helium and neon, all can combine with other elements to form compounds. Chemical reactions can be used to break down compounds and free the elements they contain.

Element	Symbol	Details
Iron	Fe	The Latin name *ferrum* was used for over 4,000 years.
Lead	Pb	The Latin name *plumbum* was used for more than 3,000 years.
Copper	Cu	The Latin name for Cyprus, *cuprum*, was used for more than 5,000 years.
Sodium	Na	First isolated pure in 1807. Latin name *natrium*.
Aluminium	Al	First isolated pure in 1825.
Uranium	U	Discovered 1789; isolated pure 1841. Named after the planet Uranus.
Plutonium	Pu	First made in 1940; named after the planet Pluto.
Phosphorus	P	Discovered in 1669.

About one-fifth of the elements are non-metals; the rest are metals. Most elements exist in nature as compounds that contain two or more elements joined together.

SYMBOLS AND NAMES

Chemists use symbols of one or two letters to represent the elements. The first letter is always a capital letter and the second letter is always a lower-case letter. For example, the symbols for hydrogen and zinc are H and Zn.

Elements discovered before about 1800 were often given Latin names. The Romans called lead *plumbum*. Since lead is easily bent into shape, the Romans used lead to make pipes for carrying water. The Latin name gives the symbol Pb and also shows the origin of the words plumber and plumbing.

Metallic elements discovered more recently usually have names that end in -ium. Plutonium, for example, was discovered and named in 1940.

▲▶ Phosphorus is a non-metallic element. It must be stored underwater (above), since it catches fire when exposed to air, forming a compound called phosphorus oxide. Phosphorus also reacts violently in a stream of chlorine gas (right).

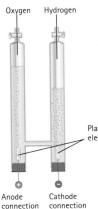

Oxygen   Hydrogen

Platinum electrodes

Anode connection   Cathode connection

An electrical current can break down some liquids and solutions. This process, called electrolysis, can be used to decompose water into its elements.

The airship *Hindenburgh* contained 190,000 cubic metres of hydrogen. In 1937, 35 of the 97 people aboard died when the hydrogen exploded.

Hydrogen 10%
Carbon 18%
Nitrogen 3%
Calcium 2%
Other elements 2%
Oxygen 65%

The human body consists mostly of water (hydrogen and oxygen), and compounds that contain carbon. Bones and teeth contain calcium.

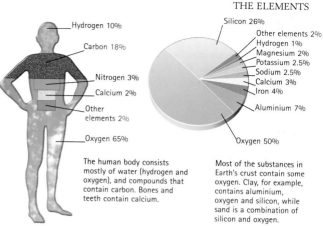

Silicon 26%
Other elements 2%
Hydrogen 1%
Magnesium 2%
Potassium 2.5%
Sodium 2.5%
Calcium 3%
Iron 4%
Aluminium 7%
Oxygen 50%

Most of the substances in Earth's crust contain some oxygen. Clay, for example, contains aluminium, oxygen and silicon, while sand is a combination of silicon and oxygen.

EARLY DISCOVERIES: METALS

A few elements are found in Earth's crust as pure substances. Gold is an element that occurs in some rocks as tiny flakes or small lumps of pure metal. This is because it does not combine easily with other elements. People first extracted and used gold about 5,500 years ago.

Some rocks contain metals combined with oxygen. Iron ore, for example is a compound of iron and oxygen. 3,500 years ago, people discovered how to make iron by heating its ore with charcoal. Copper, lead and zinc were made in similar ways.

EARLY DISCOVERIES: NON-METALS

Carbon and sulphur are the only non-metals that occur as pure substances in nature. Carbon is found as diamonds and graphite; charcoal (an impure form of carbon) was made for centuries by partly burning wood. It was used to manufacture iron. Sulphur is found as solid yellow lumps or powder around the craters of some volcanoes. From AD1200, it was used to make gunpowder and antiseptics.

LATER DISCOVERIES

Lavoisier founded modern chemistry in 1783 when he fixed the idea of elements at the centre of the subject. Just 26 pure elements were known at that time. As their apparatus and techniques improved, chemists discovered new elements with increasing speed. By 1900, all the naturally occurring elements had been identified, purified and given names.

SYNTHETIC ELEMENTS

The Universe consists mainly of hydrogen (90%) and helium (9%). The immense pressures and temperatures inside stars like the Sun cause nuclear reactions that turn hydrogen into helium. Further nuclear reactions squeeze hydrogen and helium together to make heavier elements. Earth formed from these elements when parts of the Sun broke away. Scientists use nuclear reactions to make heavy, artificial elements from natural elements. These synthetic elements are so unstable that they decay and fall apart, often in minutes or even seconds.

Nuclear reactions convert atoms of one element into atoms of other elements. An enormous amount of energy is released as this happens: the explosive force of an atom bomb can be equivalent to that of thousands of tonnes of ordinary explosives.

SEE ALSO

112 Atoms,
128 Carbon

ATOMS

The Universe is made up of tiny particles called atoms. Atoms are so small that billions of them would fit on the full stop at the end of this sentence.

British physicist and chemist John Dalton (1766–1844) produced an atomic theory of matter. Dalton believed that atoms were shaped like tiny spheres.

About 2,500 years ago, Greek philosophers argued about the make-up of matter. One group of thinkers, the atomists, believed that if it were possible to cut matter into smaller and smaller pieces, there would eventually be a piece so small that it could not be divided any more. The word *atom* comes from the Greek *atomos*, meaning 'uncuttable'.

Between 1803 and 1807, British chemist John Dalton worked on these ideas in his atomic theory of matter. He believed that atoms could not be created or destroyed. A pure sample of an element contains atoms that are all the same.

MODERN ATOMIC THEORY

Dalton's theory said nothing about the structure inside atoms. Then, in 1897, the first subatomic particle, called the electron, was discovered. In 1911, British physicist Ernest Rutherford (1871–1937) discovered that each atom has a dense, positively charged nucleus. In 1932, the neutron was discovered.

Modern atomic theory states that an atom consists of a nucleus of protons and neutrons surrounded by orbiting electrons. Neutrons and protons are more than 1,800 times as heavy as electrons. Protons have a positive charge, electrons have a negative charge, and neutrons are uncharged.

▲ Dalton invented symbols for different atoms. He suggested that atoms joined together to make compounds. This example of Dalton's diagrams shows two oxygen atoms joined to one carbon atom in carbon dioxide.

▶ The nucleus of an atom consists of protons and neutrons. The nucleus makes up most of an atom's mass. Electrons move in fixed orbits around the nucleus. This is because there is an electrical attraction between the negative charge of the electron and the positive charge of the protons in the nucleus. Neutrons help hold the nucleus together. Without them, the positively charged protons would repel one another.

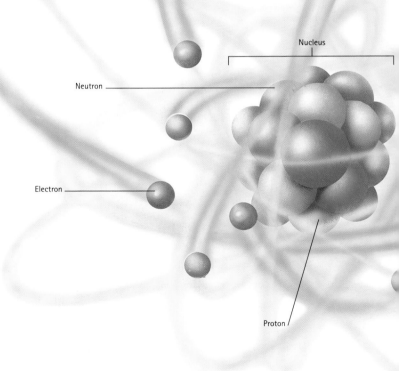

Nucleus

Neutron

Electron

Proton

ATOMS AND ELEMENTS

Atoms have equal numbers of electrons and protons. The negative charges of the electrons cancel out the positive charges of the protons. As a result, atoms have no overall electric charge.

The simplest element is hydrogen. It has one proton and one electron; it is the only element that does not contain a neutron. Other elements contain more electrons, protons and neutrons, and are heavier than hydrogen. For example, aluminium atoms have 13 protons, 13 electrons and 14 neutrons. Uranium is the heaviest naturally occurring element: it has 92 protons, 92 electrons, and 146 neutrons. Elements that are heavier than uranium are unstable. Their nuclei burst apart because the forces that draw the protons together are not strong enough to overcome the repulsion between their positive charges.

◀ The atomic-force microscope (AFM) was developed during the 1980s. A computer monitors the forces on a tiny diamond point as it scans across the sample.

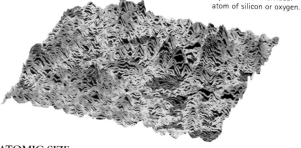

▼ This is a computer-generated AFM image of the surface of a piece of glass. Each raised area represents an individual atom of silicon or oxygen.

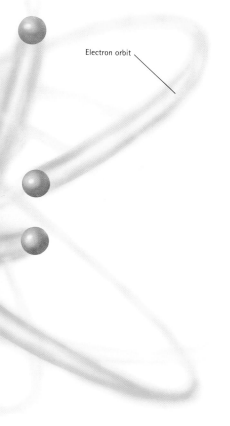

Electron orbit

ATOMIC SIZE

Atomic theories are models that scientists use to explain the results of their experiments. Because atoms are so small, nobody has actually seen one – an atom would have to be magnified 100 million times to produce an image one centimetre across. Another problem is that atoms are mostly empty space. A scale model with a tennis ball as the nucleus would use pin-heads for the electrons, and the whole model would be about 700 metres in diameter. Modern atomic theory also states that electrons move too fast to be able to estimate their location in an atom with certainty. They behave like a cloud of negative charge surrounding the nucleus.

SEEING IS BELIEVING?

Light waves are very large compared to the size of atoms. For this reason, ordinary optical microscopes cannot detect single atoms. They simply blur the images of millions of atoms together.

Atomic-force microscopes (AFMs) do not use light. Instead, a sharp probe moves backwards and forwards across the surface of a sample, sensing the electron cloud around each atom. A computer builds up a picture of the atoms in the surface.

In 1922, Danish physicist Niels Bohr (1885–1962) won the Nobel Prize for physics for his theory of atomic structure.

In Bohr's theory, electrons move around the nucleus of an atom in spherical shells called orbits.

SEE ALSO

110–1 The elements,
116–7 States of matter,
118–9 Atomic structure

THE PERIODIC TABLE

The periodic table lists the chemical elements in order of increasing atomic number. Elements with similar properties are grouped together.

Russian chemist Dmitry Mendeleyev (1834–1907) drew up the first periodic table in 1869. He left gaps in the table for elements not yet discovered, and predicted their properties by comparison with neighbouring elements.

The periodic table lists all the elements in eighteen vertical columns, or groups, and seven horizontal rows, or periods. The elements are arranged so that their atomic numbers increase from left to right through a period. An element's atomic number is equal to the number of protons in its nucleus and the number of electrons orbiting the nucleus.

The groups of the periodic table are labelled 1–18 from left to right, although other numbering systems are sometimes used. Elements in the same group have similar properties. The chemical properties of an element depend largely on the number of outermost electrons.

Sodium is in Group I. Although it is a metal, it is so soft that a knife can cut easily through a piece. Sodium is stored in oil to stop air or moisture reacting with it.

THE STRUCTURE OF THE TABLE

The periodic table consists of four main areas, or blocks, named with the letters s, p, d and f. Groups 1 and 2 form the s block on the left. The p block on the right contains groups 13 to 18. The d and f blocks form groups 3 to 12.

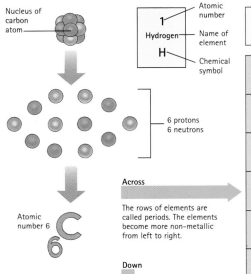

Nucleus of carbon atom

Atomic number

1
Hydrogen
H

Name of element

Chemical symbol

6 protons
6 neutrons

Atomic number 6

C

6

Across

The rows of elements are called periods. The elements become more non-metallic from left to right.

Down

The columns of elements are the groups. Elements in the same group have similar properties. As atomic number increases down a group, the elements show increasing metallic character.

▲ The atomic number of an element is the number of protons in an atom of that element. For carbon, this number is six. The chemical symbol for an element can be written with the atomic number written below and to the left of the main symbol.

1 Hydrogen H								

Periodic table key to colour code

s block f block

d block p block

1	2
3 Lithium Li	4 Beryllium Be
11 Sodium Na	12 Magnesium Mg

3	4	5	6	7	8	9
19 Potassium K	20 Calcium Ca	21 Scandium Sc	22 Titanium Ti	23 Vanadium V	24 Chromium Cr	25 Manganese Mn	26 Iron Fe	27 Cobalt Co
37 Rubidium Rb	38 Strontium Sr	39 Yttrium Y	40 Zirconium Zr	41 Niobium Nb	42 Molybdenum Mo	43 Technetium Tc	44 Ruthenium Ru	45 Rhodium Rh
55 Caesium Cs	56 Barium Ba	57–71 Lanthanide series	72 Hafnium Hf	73 Tantalum Ta	74 Tungsten W	75 Rhenium Re	76 Osmium Os	77 Iridium Ir
87 Francium Fr	88 Radium Ra	89–103 Actinide series	104	105 Element 105	106 Element 106	107 Element 107	108 Element 108	109 Element 109

57 Lanthanum La	58 Cerium Ce	59 Praseodymium Pr	60 Neodymium Nd	61 Prometheum Pm	62 Samarium Sm	63 Europium Eu	64 Gadolinium Gd	65 Terbium Tb
89 Actinium Ac	90 Thorium Th	91 Protactinium Pa	92 Uranium U	93 Neptunium Np	94 Plutonium Pu	95 Americium Am	96 Curium Cm	97 Berkelium Bk

S-BLOCK ELEMENTS

With the exception of hydrogen, the elements of the s block are all very reactive, soft metals with low densities. Most group 1 metals melt below 100°C, while most group 2 metals melt below 900°C. Compounds of s-block elements are used to colour fireworks. Sodium and potassium salts are needed for the nervous system to work properly, while magnesium compounds are vital for making green chlorophyll in plants.

D-BLOCK ELEMENTS

These elements are all hard, dense metals; most of them melt well above 1,000°C. The d-block metals include elements such as iron, copper and titanium. They are much less reactive than s-block metals.

All the d-block metals have just one or two outermost electrons, so they have similar chemical properties. Although atomic number increases across each of the three periods, the extra electrons orbit inside the atom, but closer to the nucleus.

P-BLOCK ELEMENTS

This block is a mixture of metals and non-metals. The elements above a diagonal line that runs from aluminium to polonium are non-metals. Elements in the line and below it are metals. Tin and lead are typical p-block metals. They are softer than d-block metals and less reactive.

The gases nitrogen, oxygen, fluorine and chlorine are at the top right of the p-block. The members of group 18 are called the noble, or inert, gases because they are almost completely unreactive.

F-BLOCK ELEMENTS

These elements are all rare metals. The members of the first row are all very reactive. The second row elements are all radioactive; many of them are synthetic elements made in laboratories or in the cores of nuclear reactors.

This motorcycle engine is made mostly from alloys of aluminium and iron. Both metals are mixed with small amounts of other elements to make alloys. These substances are designed to withstand mechanical wear and tear.

								18
								2 Helium **He**
			13	14	15	16	17	
			5 Boron **B**	6 Carbon **C**	7 Nitrogen **N**	8 Oxygen **O**	9 Fluorine **F**	10 Neon **Ne**
10	11	12	13 Aluminium **Al**	14 Silicon **Si**	15 Phosphorus **P**	16 Sulphur **S**	17 Chlorine **Cl**	18 Argon **Ar**
28 Nickel **Ni**	29 Copper **Cu**	30 Zinc **Zn**	31 Gallium **Ga**	32 Germanium **Ge**	33 Arsenic **As**	34 Selenium **Se**	35 Bromine **Br**	36 Krypton **Kr**
46 Palladium **Pd**	47 Silver **Ag**	48 Cadmium **Cd**	49 Indium **In**	50 Tin **Sn**	51 Antimony **Sb**	52 Tellurium **Te**	53 Iodine **I**	54 Xenon **Xe**
78 Platinum **Pt**	79 Gold **Au**	80 Mercury **Hg**	81 Thallium **Tl**	82 Lead **Pb**	83 Bismuth **Bi**	84 Polonium **Po**	85 Astatine **At**	86 Radon **Rn**

66 Dysprosium **Dy**	67 Holmium **Ho**	68 Erbium **Er**	69 Thulium **Tm**	70 Ytterbium **Yb**	71 Lutetium **Lu**
98 Californium **Cf**	99 Einsteinium **Es**	100 Fermium **Fm**	101 Mendelevium **Md**	101 Nobelium **No**	103 Lawrencium **Lr**

▲ One of the fuels used in nuclear power stations is uranium, which is element 92. It has the highest atomic number of all the naturally occurring elements. Energy is released when the nucleus splits to form other elements.

▲ The elements in group 18 of the periodic table are sometimes called the noble gases. The lightest, helium, has just two electrons. All the other elements in the group have eight electrons in their outermost shell. These arrangements are very stable which explains the lack of reactivity.

Many of the noble gases are used in lighting. The lamp in this lighthouse uses xenon to produce an intense bluish-white light.

SEE ALSO

110–1 The elements,
124–5 Bonding and valency

STATES OF MATTER

Solid, liquid and gas are the three common states of matter. They have distinctly different structures. Pure substances melt and boil at fixed temperatures.

Solid

The particles in a solid are packed tightly. Often, they form a regular pattern called a lattice. Particles in a solid do not move freely; they vibrate around fixed points in the lattice.

Gas

The particles in a gas are spread out much more than in solids. They move at great speeds – about 300 kilometres per hour – and collide with each other and the walls of their container.

Liquid

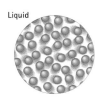

The particles in a liquid can move around and slide past each other like the particles in a gas. However, they are packed much closer, more like the particles in a solid.

▶ This graph shows the spread of the energies of particles in a solid, liquid or gas. The higher the energy, the faster the particles move or vibrate. Few particles have very high or very low energies.

Matter consists of particles. These can be separate atoms, molecules or ions (see page 166). Although particles often have complex shapes, chemists usually use spheres to make models of solids, liquids or gases.

In any substance, forces that attract particles towards each other oppose the energy of the particles, and this makes them move. This energy, called kinetic energy, increases with temperature. Whether a substance is solid, liquid or gas depends on the balance between kinetic energy and forces of attraction.

SOLIDS AND LIQUIDS

Substances are solids when the forces of attraction between their particles are strong enough to prevent the particles from moving freely. Solids have fixed shapes because the particles are held firmly together, often in a regular pattern called a lattice. Crystals are examples of highly regular lattices..

Liquids are fluid – in other words, they can change their shape. In a gravitational field such as on Earth, liquids collect at the bottom of a container and have a flat upper surface. In a liquid, the forces of attraction between the particles are too weak to hold them in a rigid formation. Instead, the particles can glide easily past each other.

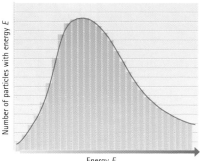

Energy *E*

White-hot molten iron is poured into moulds in a process called casting. When the temperature falls below 1,535°C, iron solidifies in the shape of the mould.

GASES

Substances exist as gases when the kinetic energy of their particles is large enough to completely overcome the forces that attract them. Similar to liquids, gases are fluid – they change their shape to fit their containers. Unlike liquids, however, gases have enough kinetic energy to spread out and completely fill their containers.

MELTING POINTS

Matter can change from solid to liquid, liquid to gas and so on, if the kinetic energy of the substance's particles alters. Kinetic energy is increased or decreased by changing the temperature, and the melting point of a substance is the temperature at which the kinetic energy of the substance's particles is just great enough to free the particles from the rigid lattice structure. The amount of energy needed to melt a solid depends on the strength of the attractive forces in the solid. The forces in iron, which melts at 1,535°C, are greater than oxygen, which freezes at –219°C.

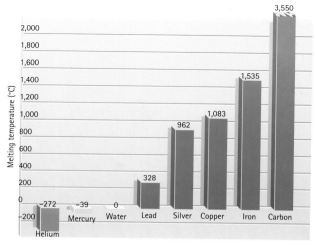

▲ The melting point of a substance depends on the attraction between its particles. This force is so weak for helium that it only solidifies at more than 25 times atmospheric pressure.

▼ Boiling points also depend on pressure and the force of attraction between particles. At the summit of Mount Everest, the boiling point of water is 28°C lower than at sea level.

BOILING POINTS

A liquid boils when bubbles of vapour grow in the liquid, rise to the surface and burst, forming a gas. The boiling point of a substance is the temperature at which the kinetic energy of the particles of that substance are great enough for them to escape the forces that pull the particles together. Just as with melting points, each pure substance has its own particular boiling point. For example, water boils at 100°C to form steam, liquid hydrogen boils at –260°C, and ethanol boils at 79°C.

Not all substances melt before they boil. Some solids turn into gas without passing through a liquid stage. This process is called sublimation. Solid carbon dioxide (dry ice) is a substance that sublimes – it becomes carbon dioxide gas at –78.5°C.

IMPURITIES AND PRESSURE

Impurities (small quantities of other substances) and pressure both affect boiling and melting temperatures. High pressure forces particles together, so they need more kinetic energy to melt or boil. This means that boiling point and melting point increase at high pressure.

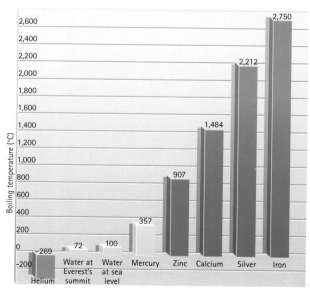

Impurities change boiling and melting points by interfering with the forces between particles. This is why ice melts when salt is sprinkled over it. Salt also increases the boiling point of water.

SEE ALSO

124–5 Bonding and valency, 144–5 Properties of solids, 198–9 Potential and kinetic energy

117

SOLUTIONS

Solutions consist of one or more substances dissolved in another substance. The most common solutions are solids or gases dissolved in liquids.

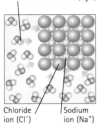

Water molecules (H₂O)

Chloride ion (Cl⁻) / Sodium ion (Na⁺)

If you stir salt into a glass of water, the solid salt crystals start to dissolve in the water to form a solution. In all solutions, the substance that dissolves is called the solute. The substance that dissolves the solute is called the solvent. Different solvents dissolve different solutes. For example, salt dissolves in water but not in pure alcohol or in petrol. Sugar behaves differently and dissolves in all three – water, pure alcohol and petrol.

Water molecules pull ions away from the crystal.

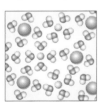

Water molecules surround the ions in solution.

▲ Granules of salt consist of sodium and chloride ions bound together in a formation called a crystal lattice. Water dissolves salt by pulling ions away from the lattice and surrounding them.

▶ Each litre of seawater contains about 32 grams of salt. Cooking salt is produced by trapping seawater in shallow pools. Heat from the Sun evaporates the water. The salt crystallizes and is raked into mounds to dry.

DISSOLVING

Solids consist of particles that are tightly packed in a fixed pattern. There are strong forces of attraction between the particles. The particles in a liquid are in constant motion. When a solid comes into contact with a liquid, particles in the liquid strike the surface of the solid. In these collisions, some of the particles in the solid become dislodged. A solution forms if the solid particles are more strongly attracted to the liquid particles than they are to each other. Solvent particles surround solute particles as the solid steadily dissolves. The result is a solution.

SOLUBILITY AND CRYSTALLIZATION

The mass of solute that can dissolve in one litre of solvent is called the solubility of the solute. A solution that contains the maximum possible amount of solute is called a saturated solution. The solubility of most solids increases with temperature.

If a solution is left in an open container, the volume of liquid decreases as the solvent evaporates. The solute does not evaporate. After some time, there is not enough solvent present to dissolve all the solute. The solution becomes saturated and crystals of solid solute start to form as the solvent continues to evaporate.

Frost crystals form when moist air cools below 0°C. Moisture that is dissolved in the air forms droplets of water that settle on cold surfaces and freeze. These elaborate patterns are crystals of ice.

SOLIDS

Solid solutions are made by allowing a liquid solution to solidify. Alloys form an important class of solid solution. Alloys are solid solutions of one or more metals or non-metals in another metal that form the major part of the solution. Alloys usually have very different properties compared to the original metal. Pure aluminium, for example, is very soft. By dissolving small amounts of copper and other elements a tough, light alloy called duralumin is produced. Duralumin is particularly light but very strong, so it is used to make the bodies and wings of aircraft.

As with other sorts of solutions, there is a limit to how much solute can dissolve in a solid solution. For example, pure iron is a soft, malleable metal. Dissolving tiny amounts of carbon in molten iron makes steel, which is much harder. The carbon atoms are scattered evenly throughout the solid solution. Iron can dissolve up to 0.4 per cent carbon. Adding more carbon results in tiny lumps of undissolved iron carbide, which make the steel brittle.

Bronze is an alloy, or solution, of up to 30 per cent tin in copper. It was first made more than 6,000 years ago and was used to produce armour, tools, weapons, helmets and ornaments.

GASES

Gases can also dissolve in liquids to form solutions. The solubility of gases decreases as the temperature increases. This is why small bubbles of air form in heated water long before it boils. Increasing the pressure of a gas makes more gas dissolve in a liquid. Gas solubility is measured at 0°C and one atmosphere pressure. Oxygen, for example, has a solubility of 49 cm^3 per litre of water under these conditions.

Liquids dissolve in gases, too. For example, water evaporates from the sea and the water vapour mixes with the air. When warm, moist air rises and cools, it can no longer dissolve all the water vapour it contains. Then tiny droplets of liquid water appear as clouds, mist and rain.

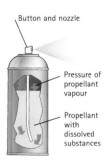

Button and nozzle

Pressure of propellant vapour

Propellant with dissolved substances

▲ Aerosol cans contain pressurized solutions of substances such as deodorants, insecticides and paints. When pressure is released by pressing the button, the solvent in the aerosol boils and forces the solution through the nozzle. A fine spray of the aerosol's contents comes out whenever the button is pressed.

Sealed container

Container opened

◄ Drinks are made fizzy by dissolving carbon dioxide gas at high pressure in the liquid. Opening the bottle allows the pressure to fall. The liquid can no longer dissolve all the gas.

No bubbles   Bubbles

► A solution is always transparent, even when it is coloured. This blue solution is cobalt chloride dissolved in water. A solid can be made to dissolve more quickly in a solvent by heating and mixing the solid with the solvent.

SEE ALSO

32–3 Clouds and fog, 116–7 States of matter, 140 Acids, 141 Bases and alkalis, 146 Iron, 148–9 Alloys

CHEMICAL REACTIONS

Chemical reactions change one set of substances into another set of substances. The speed of a reaction depends on the substances and temperature.

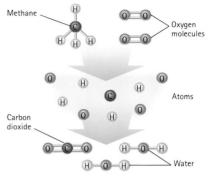

Methane

Oxygen molecules

Atoms

Carbon dioxide

Water

When natural gas burns, methane (CH_4) and oxygen (O_2) molecules break into atoms. The atoms recombine to form water (H_2O) and carbon dioxide (CO_2) molecules.

Iron filings and powdered sulphur

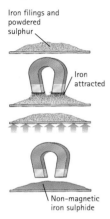

Iron attracted

Non-magnetic iron sulphide

A magnet separates iron from a mixture of iron filings and powdered sulphur. Heating the mixture makes a new substance, a compound called iron sulphide, which is not magnetic.

Chemical reactions change the chemical composition of substances. They can either break complicated substances down into smaller parts, or join together simple substances to make more complex ones. The substances present at the start of the reaction are called the reactants.

The substances present when the reaction has finished are called the products. Chemical reactions change the way the atoms of different elements in the reactants are grouped together.

Some chemical reactions, such as rusting, happen very slowly; others, such as explosions, happen very fast.

BUILDING UP
Iron and sulphur are elements. When iron filings and powdered sulphur are stirred together, they simply form a mixture. The mixture has all the properties of the two separate ingredients: the iron filings are attracted to a magnet, for example, and the individual substances can be seen with

a magnifying glass. Heating the mixture makes it glow as a chemical reaction occurs. Iron atoms join with sulphur atoms to form iron sulphide. Heat starts the reaction, just as a match lights a gas burner. This reaction is an example of simple substances joining together to make a more complex one. The product has different properties from the mixture: for example, it is not magnetic.

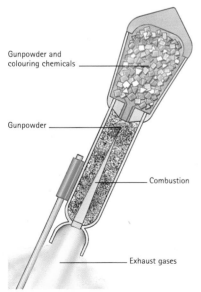

Gunpowder and colouring chemicals

Gunpowder

Combustion

Exhaust gases

◄ Rockets contain gunpowder, which is a mixture of potassium nitrate, sulphur and charcoal (carbon). When heated, potassium nitrate decomposes to form oxygen, which causes the sulphur and carbon to burn rapidly producing gases. These gases propel the rocket into the sky and scatter the colouring chemicals that provide the rocket's display.

► Chemical explosives decompose in a few thousandths of a second to form large quantities of hot, high-pressure gases. The expansion of these gases produces a blast that is strong enough to demolish a building.

BREAKING DOWN

When baking soda (sodium hydrogen-carbonate, $NaHCO_3$) mixes with vinegar, acid in the vinegar makes the baking soda break down into smaller parts. One of these products is carbon dioxide (CO_2), which makes the mixture bubble.

Gas stoves and heaters use oxygen from the air to burn methane gas. During the reaction, heat breaks down each molecule of methane (CH_4) into one carbon atom and four hydrogen atoms. These atoms then join up with oxygen to form carbon dioxide and water (H_2O).

REACTION RATES

The chemical reaction between iron and sulphur needs heat to make it happen. Heat is not necessary for the reaction between baking soda and vinegar to happen, but raising the temperature will make it happen faster: increasing the temperature increases the reaction rate.

The rate of a reaction is how quickly it changes reactants into products. Chemical reactions happen when molecules and atoms collide. The reaction rate increases when there are more collisions per second. Raising the temperature increases the kinetic energy of particles, so they move faster and collide more often. This is why reactions get faster at high temperatures.

The rate of a reaction also increases if more concentrated reagents are used. The higher concentration means that the

molecules that could react are crowded closer together; so they collide more often and the reaction happens faster. With gases, high pressure is the equivalent of high concentration. A diesel engine uses a combination of high pressure and high temperature to start an explosive reaction.

Rusting is a slow chemical reaction between iron, water and oxygen. In time, the iron in this old car will change completely to crumbly brown iron oxide.

1 Dilute acid and marble chips
2 Dilute acid and marble powder
3 Marble chips and concentrated acid
4 Marble chips and high temperature

1 Marble chips react with acid to form carbon dioxide gas. Reaction rates increase when:
2 powdering the marble increases contact between the reactants;
3 concentrated acid and
4 higher temperature increase the number of collisions per second.

◀ These stalactites (hanging down) and stalagmites (growing up) have taken thousands of years to form in this cave. Carbon dioxide in rainwater seeps down and reacts with limestone in the rock above the cave to form salts. These salts become insoluble and form solid deposits as the dripping water evaporates in the cave.

SEE ALSO
134 Catalysts, 136 Oxidation and reduction

CHEMICAL COMPOUNDS

Chemical compounds are substances made from atoms of two or more different elements in fixed proportions. Compounds are held together by chemical bonds.

Carbon

Oxygen

Bubbles decrease in size from the bottom to top

Limewater

Some compounds can be made by simply heating elements together. Passing oxygen gas over heated carbon forms carbon dioxide. During the reaction, each carbon atom bonds to two oxygen atoms. The limewater becomes milky and the bubbles become smaller as limewater dissolves the carbon dioxide.

There are more than 110 known chemical elements. Atoms of these elements join together in different combinations to make countless millions of different compounds. Strong forces of attraction called chemical bonds hold atoms together in these compounds.

Some compounds are very simple. For example, table salt contains just two elements, sodium and chlorine, bonded together. Its chemical name is sodium chloride. Other compounds are extremely complex, particularly substances found in living things, such as DNA and proteins.

CHEMICAL BONDS

Substances such as sodium chloride are made up of particles that have negative and positive charges. These particles are called ions. The positive ions are strongly attracted to the charge of the negative ions and vice versa. This attraction, called ionic bonding, holds the ions together like mortar holds bricks in a wall.

Other substances have bonds that join atoms together in groups called molecules. For example, water is a compound of the elements hydrogen and oxygen. Each water molecule consists of one oxygen atom and two hydrogen atoms. The bonds between atoms within a molecule are strong; the forces of attraction between molecules are much weaker.

▼ Carbon dioxide molecules contain just three atoms each. The attraction between the molecules is weak, so carbon dioxide is a gas at room temperature.

Temporary very weak bonds between moving molecules

▼ There are many different alcohols. All contain an oxygen atom joined at one end to a carbon atom and at the other end to a hydrogen atom. The chemical name for the alcohol with formula C_2H_5OH is ethanol.

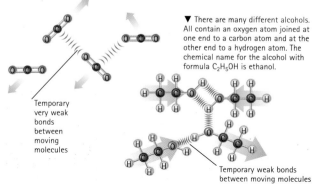

Temporary weak bonds between moving molecules

Polythene is a compound that contains carbon and hydrogen only. The carbon atoms join together to make long chains. Each carbon atom is also joined to two hydrogen atoms. A single polythene molecule contains between 3,500 and 200,000 carbon atoms. The chains tangle together to give a soft, waxy-feeling solid.

▼ This refuse worker bags polythene items together for recycling as dustbin liners and other products.

122

Carbon atoms form 'backbone'

METAL SALTS

Metal salts contain one or more metals bonded to one or more non-metals. Common salt, or sodium chloride, is a familiar example. The chemical formula of salt is NaCl, which shows that it contains equal numbers of sodium (Na) and chlorine (Cl) atoms. Another example of a salt is calcium carbonate ($CaCO_3$), which is the main component of limestone and chalk. Calcium carbonate has equal numbers of calcium (Ca) and carbon (C) atoms, and three times as many oxygen (O) atoms.

Salts are usually solids that melt at high temperatures. Sodium chloride, for example, melts at 804°C.

Hydrogen atoms are joined to the carbon 'backbone'

NON-METALS

Compounds that contain only non-metals mostly exist as molecules; many of these compounds are liquids or gases. Water, for example, consists of hydrogen and oxygen, which are both non-metals. The chemical formula of water is H_2O, which shows that each molecule is made up of two hydrogen atoms and one oxygen atom.

Most of the non-metal compounds that are solids melt at low temperatures. Candle wax, which is a mixture of compounds of carbon, hydrogen and oxygen, melts at around 70°C.

▲ The chemical name for polythene is poly(ethene). Chemists use the prefix 'poly' to indicate that a compound is made from the same simple part repeated many times. Polythene is made from ethene, a molecule that contains two carbon and four hydrogen atoms.

Ammonium dichromate, $(NH_4)_2Cr_2O_7$, is an unstable compound of chromium, hydrogen, nitrogen and oxygen atoms. Heating the orange crystals breaks the bonds between the atoms to form simpler substances – steam, nitrogen gas, and green chromium oxide.

THE NOBLE GASES

Helium, neon, argon, krypton, xenon and radon are the noble gases. They form group 18 of the periodic table. Helium, neon and argon are completely inert: they never take part in chemical reactions. Krypton, xenon and radon can react but do so only under extreme conditions. The noble gases are so unreactive because their electronic structures are extremely stable.

STABILITY

When elements take part in chemical reactions, they exchange or share electrons so as to achieve the same number of electrons as the closest noble gas. As a result, many compounds are more stable than the elements that make them up.

Some compounds, such as sodium chloride, are very stable. Often, highly reactive elements combine to form highly stable compounds. They seldom react with other substances and do not break down when heated because the bonds between their atoms are strong.

▶ Gold is found mixed with quartz rock deep underground. Most metals occur in their ores as compounds. Gold is so unreactive that it occurs naturally as pure metal.

SEE ALSO

110–1 The elements, 120–1 Chemical reactions, 124–5 Bonding and valency

BONDING AND VALENCY

Ionic bonds mainly form between metal and non-metal atoms. Covalent bonds form between non-metal atoms. Valency is the number of bonds an atom can form.

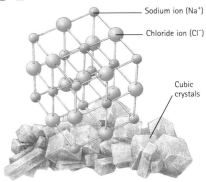

Sodium ion (Na^+)

Chloride ion (Cl^-)

Cubic crystals

Each ion in a sodium chloride crystal is surrounded by ions of opposite charge. The ions are arranged in a cubic lattice so that each salt crystal has a cubic shape.

E ach atom of an element contains a number of electrons that is exactly equal to the number of protons in its nucleus. The positive charges of the protons balance the negative charges of the electrons and the atom has no overall charge. The electrons orbit the nucleus in layers called shells. There is a limit to the number of electrons that each shell can hold. The first shell, which is closest to the nucleus, can hold up to two electrons. The second shell can hold eight electrons and the third can hold 18.

The rows of the periodic table list elements in order of increasing atomic number, which is the number of protons in an atom of an element. Each row starts with an element that has only one electron in its outermost shell. At the end of each row is a noble gas, which has a full set of electrons in its outermost shell. This arrangement of electrons, which is called a configuration, is unusually stable. This is why noble gases seldom react.

Sodium atom (Na)

Sodium ion (Na^+)

A sodium atom has 11 electrons in shells around a nucleus that contains 11 protons. A sodium ion has one negatively charged electron less, which gives it an overall charge of +1. Sodium ions are smaller than sodium atoms.

Chlorine atom (Cl)

Chloride ion (Cl^-)

A chlorine atom has 17 electrons and 17 protons. A chloride ion has one more electron and a charge of –1. Chloride ions are larger than chlorine atoms and, like sodium ions, chloride ions have a full outer shell of electrons.

▶ This picture shows sodium glowing brightly in chlorine gas. The chemical reaction that bonds the two elements together produces large quantities of heat and forms white crystals of sodium chloride, or common salt.

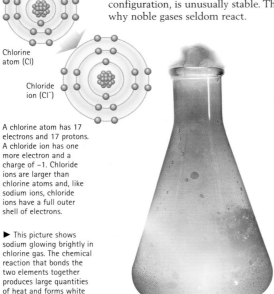

VALENCY

In chemical reactions, bonds form between atoms as they gain, lose or share electrons. As a result of these changes, each atom in a compound usually has a full outer shell of electrons. This is the stable electron configuration of the noble gas closest to each element in atomic number.

The valency of an element is the number of bonds it must make to have a noble gas configuration. Metals usually have only one or two electrons in their outer shells. They easily lose these electrons so that the next shell down becomes a complete outer shell. The non-metals at the far right of the periodic table are only one or two electrons short of a complete shell. For this reason, they easily accept electrons from atoms of other elements. The valencies of these elements are the numbers of electrons that they must gain or lose to form a complete shell.

Elements in the middle of the main block of the periodic table have outer shells that are three or four electrons short of a full shell. Carbon is an example: it has four electrons in an outer shell that can contain eight. It seldom accepts four electrons, however, because the negative charges would repel each other too much. Instead, carbon atoms overlap their outer shells (bond) with shells of other atoms and share four electrons to make up the full count. The valency of carbon is four.

IONS AND IONIC BONDING

Ionic compounds form when atoms of two or more elements trade electrons to form charged particles, or ions, of each element. The ions have noble gas configurations and their charges balance each other to give the compound no overall charge.

A sodium atom has 11 electrons. Two of these are in the first shell and eight are in the second shell. These two shells are complete. The last (11th) electron is alone in a shell that can hold up to 18 electrons. The outer shell of a chlorine atom is one electron short of being full. On their own, chlorine atoms link up in pairs to form Cl_2 molecules. In these molecules, a shared pair of electrons makes up the outer shell. If sodium is introduced, however, chlorine atoms give up this sharing arrangement to have an electron of their own. Each sodium atom loses its 11th electron and becomes a positively charged sodium ion (Na^+). At the same time, each chlorine atom gains an electron and becomes a chloride ion (Cl^-). The opposite charges of these two types of ion attract each other strongly. The ions bond together in a regular pattern called a crystal lattice.

Magnesium also forms an ionic chloride. Unlike sodium, magnesium has two electrons in its outer shell. Its atoms lose both of these to twice as many chlorine atoms. Magnesium ions (Mg^{2+}) form a salt with the formula $MgCl_2$. Magnesium has a valency of two.

In the five examples of covalently-bonded molecules below, each line represents a single covalent bond, which is a pair of electrons shared between two atoms. The valencies of the elements are:
carbon (C) = 4; chlorine (Cl) = 1; hydrogen (H) = 1; oxygen (O) = 2; phosphorus (P) = 3.

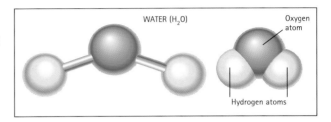

Oxygen nucleus

Carbon nucleus

4 shared electrons

4 shared electrons

Oxygen nucleus

In a carbon dioxide (CO_2) molecule, each oxygen atom is joined to the central carbon atom by a double covalent bond that consists of two shared pairs of electrons.

WATER (H_2O)

Oxygen atom

Hydrogen atoms

COVALENT BONDING

Ionic compounds mainly form between metals at the left of the periodic table and non-metals at the right. Compounds that contain only non-metals are usually held together in molecules by their atoms sharing pairs of electrons. This type of bonding is called covalent bonding.

Carbon dioxide (CO_2) is an example of a covalent compound. The valencies of carbon and oxygen are four and two, because each carbon atom needs four electrons to fill its outer shell and each oxygen atom needs two. In carbon dioxide, one carbon atom shares one pair of electrons with each of two oxygen atoms. In this way, all three atoms fill their outer shells.

▲ There are two ways to draw molecules. One is the ball-and-stick model (left), which shows the bonds between atoms. The second is the space-filling model (right), which shows the shape of the space occupied by the electrons in a molecule.

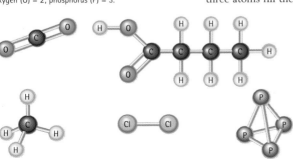

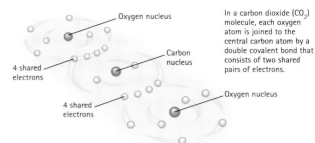

▲ Ethyne (C_2H_2), also called acetylene, burns to produce carbon dioxide (CO_2), water (H_2O) and enough heat to melt steel.

SEE ALSO

110–1 The elements, 112–3 Atoms, 114–5 The periodic table

SOLID STRUCTURES

Solids consist of atoms, molecules or ions bonded together. The properties of a solid depend on the strengths of the bonds that hold it together.

Iodine molecule I₂

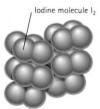

Solid iodine consists of iodine molecules arranged in a regular lattice. Each molecule contains two iodine atoms strongly bonded to each other. Iodine is soft because the bonds between the molecules are weak.

The structures of many solids are based on regularly repeating patterns of atoms, molecules or ions. These regular arrangements are called lattices.

There are four main types of lattice structure. Ionic solids, such a sodium chloride, consist of alternating positive and negative ions. Molecular solids, such as the element iodine, consist of simple molecules packed together in a lattice. Macro-molecular solids, such as diamond, graphite and glass contain huge molecules with millions of atoms each. Metallic solids consist of metal atoms held together by clouds of electrons that move freely from one atom to another.

IONIC SOLIDS
Ionic solids are hard and have high melting points. These properties result from strong forces of attraction between the oppositely charged ions in an ionic lattice. Ionic solids are also brittle. A force applied to the outside of a crystal can slide the layers of ions so that ions with similar charges are beside each other. The like charges repel strongly and the layers then break apart, cracking the lattice.

▼ Sodium chloride crystals consist of sodium and chloride ions. This cage model represents the ions as balls. The balls are spaced so that their 3-dimensional lattice structure is clear.

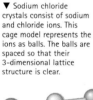

Each ion is surrounded by six ions of opposite charge. The attraction between charges holds the lattice together.

Chloride ion (Cl⁻)   Sodium ion (Na⁺)

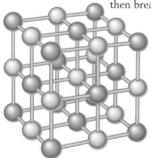

► This space-filling model is a scale model of the cubic sodium chloride lattice. It shows the relative sizes of the two types of ions and how they are arranged in space.

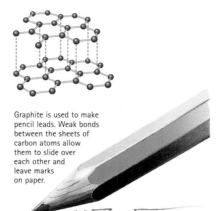

Graphite is used to make pencil leads. Weak bonds between the sheets of carbon atoms allow them to slide over each other and leave marks on paper.

MOLECULAR SOLIDS
Molecules consist of atoms held together by strong covalent bonds. For example, iodine molecules contain just two iodine atoms joined by a single covalent bond. In a molecular solid, the molecules are held together by weak forces of attraction. Molecular solids melt at low temperatures. Melting does not break the covalent bonds between the atoms; it breaks the weak attractive forces between molecules.

MACRO-MOLECULAR SOLIDS
The element carbon can exist in two forms: diamond and graphite. Both are macromolecular solids that contain only carbon atoms, but they have very different structures. In graphite, each carbon atom is joined to three others by covalent bonds that are short and strong. Hexagonal rings of six atoms join together to make flat sheets. The forces of attraction between the sheets are weak and the sheets can easily slide over each other. This is why graphite feels greasy and is used as a solid lubricant. In diamond, each carbon atom is joined to four others by strong covalent bonds. Billions of carbon atoms join together to make a 3-dimensional lattice of enormous strength. This makes diamond the hardest known solid. Glass has a structure similar to diamond. It consists of silicon and oxygen atoms instead of carbon atoms. Glass is less hard than diamond because its bonds are easier to break.

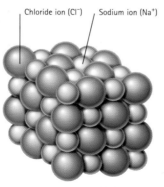

◀ Diamond is much rarer and more valuable than graphite. Like graphite, it consists entirely of carbon atoms. The structure of diamond (below) is entirely different from that of graphite, which is why the two solids are so different.

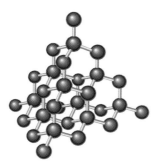

▲ At −78.5°C, solid carbon dioxide changes directly to a gas without melting. Dropped into water, it vaporizes and creates a cloud of ice crystals. This mixture is used to create a dense fog for stage effects.

METALLIC SOLIDS

Solid metals consist of separate atoms in a lattice structure. The atoms pack together in layers that stack on top of each other. Most metals have high melting points because of the strength of the bonds that hold metal atoms in lattices.

Metallic bonds are different from ionic and covalent bonds. Some electrons from each metal atom are free to move from one atom to the next. The metal atoms become positive ions when they give up their electrons. They are held together in a lattice by a 'sea' of free electrons.

If a voltage is applied to a sample of metal, the free electrons start to drift away from the negative terminal to the positive terminal. This is how electrical current flows through metal conductors.

Unlike ionic solids, metals can bend and stretch. This is because the layers of metal ions can slide over each other without the layers of the lattice breaking apart.

OTHER ELECTRICAL CONDUCTORS

An electrical current is a flow of electrical charges. The charge can be carried by either electrons or ions that are free to move. In some cases, the ability of a solid material to conduct electricity reveals information about its structure. Graphite, for example, is a rare example of a non-metallic conductor. This is because, in graphite, each carbon atom has only three of its outer electrons in bonds fixed between pairs of carbon atoms. The fourth bonding electron of each atom takes part in a huge bond that spreads through the whole sheet of carbon atoms. This type of bond, called a delocalized bond, is a pool of electrons that can move freely through the sheet and conduct electricity.

When an ionic salt melts or dissolves in water, its lattice breaks down and the ions become free to conduct electricity. This effect helped scientists to discover that ionic solids consist of charged particles.

Metal atoms are packed closely together in the solid state. The outermost electrons move freely and randomly between atoms. This 'sea' of electrons holds the metal together.

When an electrical current flows through a metal, the free electrons continue to move chaotically. But there is an average drift from the negative part of the metal to the positive part.

Metals and ionic solids have similar structures. Metal atoms or ions pack together to form layers of linked hexagons. These layers fit together to form a 3-dimensional lattice.

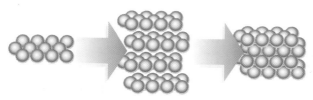

SEE ALSO

110–1 The elements, 112–3 Atoms, 144–5 Properties of solids

CARBON

The non-metal carbon is the basis for all life on Earth. It forms more compounds than any other element, but is not particularly plentiful in Earth's crust.

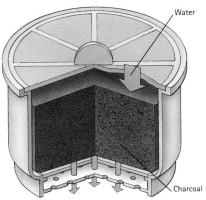
Water

Charcoal

Many household water filters contain charcoal. As water passes through the filter, the charcoal adsorbs dissolved substances, such as chlorine, from the water.

Carbon occurs in nature as graphite and diamond. Carbon forms many compounds with other elements and is found in many minerals. Limestone, chalk and marble are all different forms of calcium carbonate, $CaCO_3$, which formed from tiny marine organisms that died millions of years ago.

Most importantly, carbon is the only element whose atoms can bond with each other to form rings and chains with almost no size limit. Carbon forms more compounds than all the other elements combined. It is the basis of chemicals present in living things, fossil fuels, and petrochemicals. Carbon is constantly being exchanged between carbon dioxide in the atmosphere and compounds in plants and animals. Fossil fuels also produce carbon dioxide.

This baby, like all humans, is about 20 per cent carbon. About 4 per cent of grass and 40 per cent of an insect's shell is also carbon.

A NEW FORM OF CARBON
Buckminsterfullerene is a form of carbon made by heating graphite with an electric arc or a laser beam. It also occurs in soot. It consists of 60 carbon atoms in the shape of a ball. The carbon atoms form 12 pentagons and 20 hexagons on the surface of the ball. The compound was named after US architect Richard Buckminster Fuller (1895–1983), because its molecules resemble the dome-shaped buildings he designed.

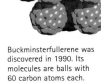
Buckminsterfullerene was discovered in 1990. Its molecules are balls with 60 carbon atoms each.

CHARCOAL
The carbon atoms in charcoal are randomly arranged. Charcoal is made by burning wood in an oven with little air. It can be used as a smokeless fuel, and glows red-hot as it reacts with oxygen in the air to form carbon dioxide.

Charcoal is very porous. It adsorbs many sorts of molecules by forming weak chemical bonds with them. Charcoal filters are used to purify gases and to decolourize liquids. For example, they are used in gas masks and in water filters. Charcoal is also used in the sugar-refining industry to remove the brown colour from sugar solution so that it crystallizes to produce pure white sugar.

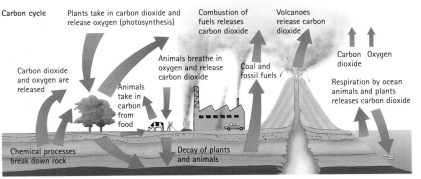

Carbon cycle

Plants take in carbon dioxide and release oxygen (photosynthesis)

Combustion of fuels releases carbon dioxide

Volcanoes release carbon dioxide

Carbon Oxygen dioxide

Carbon dioxide and oxygen are released

Animals breathe in oxygen and release carbon dioxide

Coal and fossil fuels

Animals take in carbon from food

Respiration by ocean animals and plants releases carbon dioxide

Chemical processes break down rock

Decay of plants and animals

Plants take in carbon dioxide and release oxygen during photosynthesis. Plants are eaten by animals, which take in oxygen and release carbon dioxide by respiration. The combustion of fossil fuels uses oxygen and releases carbon dioxide.

SEE ALSO
44–5 Plant anatomy, 112–3 Atoms, 120–1 Chemical reactions,

NITROGEN AND OXYGEN

Air is the mixture of gases that surrounds the Earth and makes up Earth's atmosphere. Ninety-nine per cent of air consists of the gases oxygen and nitrogen.

Oxygen (O_2) molecules have two oxygen atoms joined together by a double covalent bond.

Oxygen and nitrogen are both gaseous elements. Their molecules consist of two atoms joined by covalent bonds. In nitrogen molecules, N_2, three bonds hold the molecule together; in oxygen, O_2, there are only two. Oxygen is a very reactive element. It often releases great amounts of heat when it reacts. Nitrogen has much lower chemical reactivity.

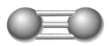

Nitrogen molecules have two nitrogen atoms joined together by a triple covalent bond.

CHEMICAL REACTIONS

Oxygen reacts with most other elements to form compounds called oxides. The most common example is hydrogen oxide, better known as water, H_2O. Iron reacts slowly with oxygen from the atmosphere to form rust, or iron oxide, Fe_2O_3.

When fuels burn and living things respire, they use oxygen to form carbon dioxide, CO_2. Plants use carbon dioxide and water to form oxygen and tissues. This process is called photosynthesis. Nitrogen is a key part of proteins in living things. The element constantly changes between nitrogen molecules in the atmosphere and compounds of nitrogen in the soil and in plant and animal proteins. This is called the nitrogen cycle.

◀ Steel wool burns brightly in pure oxygen. The reaction is faster than if it was a solid lump of steel because the fine strands of steel have better contact with the oxygen.

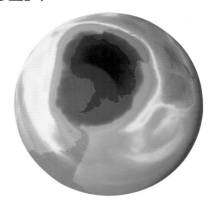

This computer-enhanced picture was taken by a satellite over Antarctica. It shows a hole in the ozone layer measuring 27 million square kilometres.

OXYGEN AND OZONE

A second form of oxygen, called ozone, has formula O_3. Its molecules have three oxygen atoms bonded in a triangle. Ozone forms a layer about 25 kilometres above Earth's surface. Ultraviolet (UV) radiation in sunlight splits oxygen molecules into separate atoms. These then combine with O_2 molecules to form ozone. Ozone acts as a sunscreen, preventing harmful UV from reaching Earth's surface. Aircraft exhausts, some aerosols and chemicals used in old refrigerators destroy ozone. Holes appear in the ozone layer and the increase in UV radiation damages plants and causes a rise in cases of skin cancer.

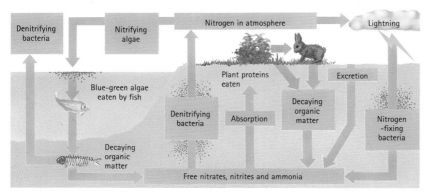

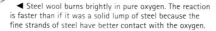

◀ The nitrogen cycle constantly exchanges nitrogen between the air and living things. Lightning and bacteria convert nitrogen gas into nitrates in the soil. Plants use these salts to form proteins which are eaten by animals. Decay and excretion return salts to the soil, where some bacteria release nitrogen.

Denitrifying bacteria
Nitrifying algae
Nitrogen in atmosphere
Lightning
Blue-green algae eaten by fish
Plant proteins eaten
Excretion
Denitrifying bacteria
Absorption
Decaying organic matter
Nitrogen-fixing bacteria
Decaying organic matter
Free nitrates, nitrites and ammonia

SEE ALSO

44–5 Plant anatomy, 120–1 Chemical reactions, 130 Air

AIR

The gases that make up the atmosphere, or air, are vital to survival of life on Earth. These gases interact continuously with living things and are affected by them.

A mason prepares to replace the face of a statue that has been badly damaged by acid rain. Acid rain is a dilute mixture of nitric and sulphuric acids. It forms when gases from burning fossil fuels mix with rain.

Almost 80 per cent of air is nitrogen; just over 20 per cent is oxygen and one per cent is argon. Air also contains small amounts of carbon dioxide and traces of noble gases other than argon. There is enough air to breathe at 10 kilometres above Earth's surface. If Earth were the size of a football, the breathable atmosphere would be less than one millimetre thick.

GASES FROM THE AIR

Many industrial processes use gases from the air. Oxygen is used in the production of steel and for welding. Nitrogen is used to make ammonia, which in turn is used to make dyes, fertilizers, explosives, medicines and plastics. Carbon dioxide gives fizzy drinks their bubbles. Argon is a noble gas with almost no chemical reactivity. It is used to fill lightbulbs. All these gases are obtained from air by a process called fractional distillation.

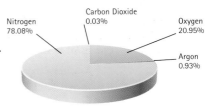

The main composition of the atmosphere remains fairly constant from day to day. The amount of pollution in the air varies with time and geographical location.

AIR POLLUTION

Almost all industries depend on fossil fuels for energy. Burning these fuels produces oxides of sulphur and nitrogen that cause lung diseases and acid rain. Acid rain gradually dissolves some types of stone and can kill trees and fish. Carbon dioxide in the atmosphere traps heat from the Sun and prevents Earth from being too cold to support life. This is known as the greenhouse effect. The carbon dioxide from burning fossil fuels is increasing the temperature of Earth's atmosphere. This effect, called global warming, could cause long-term environmental damage.

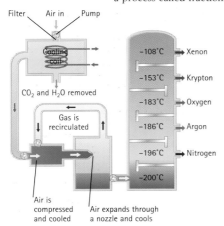

▲ At –200°C, liquid air can be separated into its component gases by distillation. Chilling removes carbon dioxide, CO_2, and water, H_2O, from the air. Repeated compression, chilling and expansion then cool the air until it becomes liquid.

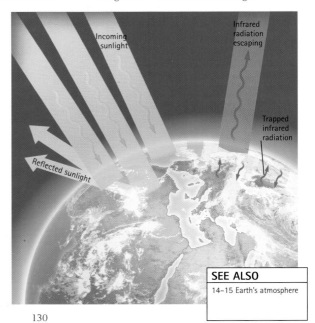

▶ Earth's surface is heated by the Sun. Earth loses heat in the form of infrared radiation. Carbon dioxide in the atmosphere traps some of this radiation, causing the greenhouse effect.

SEE ALSO

14–15 Earth's atmosphere

WATER

Water is a colourless liquid that has no odour. It has the chemical formula H_2O. At normal atmospheric pressure, water freezes at 0°C and boils at 100°C.

Most substances become more dense when they solidify. But ice is slightly less dense than liquid water. This is why icebergs float in the sea and ice cubes float in drinks.

A mature oak tree gives off about 250 litres of water vapour a day during the summer.

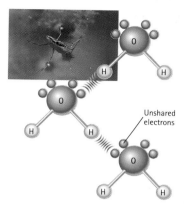

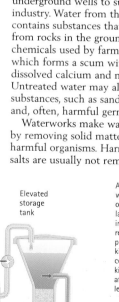

A lthough water is one of the most familiar chemical substances, it has some unusual properties. Ammonia (NH_3) and hydrogen sulphide (H_2S), similar to water, are both compounds of non-metals with hydrogen. Both are heavier than water molecules, which would normally make their boiling points higher. In fact, both are gases at room temperature, while water is a liquid. The unusually strong attraction between water molecules makes it a liquid at room temperature.

A water molecule consists of an oxygen atom bonded to two hydrogen atoms. The oxygen atom pulls electrons away from the hydrogen atoms. This gives the oxygen atom a small negative charge, while the hydrogen atoms have positive charges. These charges attract water molecules to each other. This makes water an excellent solvent for charged particles, such as the ions that make up salts.

WATER EVERYWHERE

Water covers 70 per cent of Earth's surface. Around 97 per cent of Earth's water is in the oceans. The remainder is mostly ice or snow. Less than one per cent is present in lakes and rivers.

Water is essential to life. It makes up about 70 per cent of the human body, while a lettuce is about 98 per cent water. Animals and plants use water to carry nutrients and waste products around inside them. For example, human blood is around 90 per cent water. Sap, which circulates in plants, is also mainly water.

The attractive forces between water molecules give water a surface skin. Because of this surface tension, some insects can walk on the skin without sinking.

WATER SUPPLIES

Water is taken from reservoirs, rivers or underground wells to supply homes and industry. Water from these sources contains substances that are dissolved from rocks in the ground, and can contain chemicals used by farmers. Hard water, which forms a scum with soap, contains dissolved calcium and magnesium salts. Untreated water may also contain solid substances, such as sand or soil particles and, often, harmful germs.

Waterworks make water fit for drinking by removing solid matter and by killing harmful organisms. Harmless dissolved salts are usually not removed.

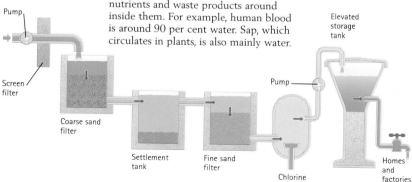

Pump

Screen filter

Coarse sand filter

Settlement tank

Fine sand filter

Chlorine

Pump

Elevated storage tank

Homes and factories

A waterworks takes in water from a lake, river or stream. Screens remove large objects, and increasingly fine filters remove all suspended solid particles. Chlorine, Cl_2, kills germs. In some countries, ozone is used to kill germs. Ozone, O_3, affects the water's taste less than chlorine does.

SEE ALSO
16—17 The oceans, 30—1 Rain and snow, 110–1 The elements

ORGANIC CHEMISTRY

Organic chemistry is the study of carbon compounds. Many of these substances are formed by living organisms. Others are made artificially.

There are around 3 million organic compounds. Most of these compounds contain carbon atoms linked together in rings and chains.

HYDROCARBONS

Hydrocarbons are organic compounds that contain only carbon and hydrogen atoms. Crude oil is a mixture of about 300 hydrocarbons, depending on the type of oil; and natural gas contains up to 99 per cent methane, CH_4. Hydrocarbons are used as fuels and include natural gas, bottled gas, petrol, diesel oil and kerosene.

ALKANES

Methane is the first member of a family of hydrocarbons called alkanes. The other members are ethane, C_2H_6, propane, C_3H_8, and butane, C_4H_{10}. Each member has one carbon atom and two hydrogen atoms more than the previous member. The -ane ending shows that a compound is a member of this group.

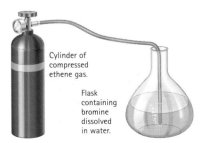

Propane, C_3H_8, is an alkane. It can be stored under pressure in steel cylinders. The liquefied gas vaporizes in the burner and burns with a hot, clean flame.

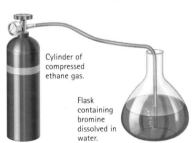

Cylinder of compressed ethane gas.

Flask containing bromine dissolved in water.

Bromine is a liquid halogen element. It is highly reactive and has a red-brown colour. The colour fades slowly as bromine replaces the hydrogen atoms in ethane one by one. This type of reaction is called a substitution reaction.

Cylinder of compressed ethene gas.

Flask containing bromine dissolved in water.

The colour of bromine disappears rapidly when mixed with an alkene. The carbon–carbon double bond opens up and the bromine atoms add on to the hydrocarbon. This type of reaction is called an addition reaction.

Methane has the molecular formula CH_4 which shows the types and numbers of atoms present. The structural formula shows how the atoms are grouped together.

The molecular formula of ethane is C_2H_6. The structural formula CH_3–CH_3 shows that each carbon atom is joined to three hydrogen atoms.

The formula of propane is C_3H_8. The prop- in the name shows that it contains three carbon atoms. Pent-, hex-, hept- and oct- indicate five, six, seven and eight carbons.

ALKENES

Alkene molecules have one or more double bonds between their carbon atoms. They are called unsaturated compounds because of these multiple bonds. Alkanes contain only single bonds and are saturated. The first members of the family are ethene, C_2H_4, and propene, C_3H_6. The structural formulae of ethene and propene show their double bonds: ethene is CH_2=CH_2; propene is CH_3-CH=CH_2.

Double bonds give alkenes increased reactivity compared to alkanes. One of the pair of bonds can break open and form bonds with other atoms. Polyunsaturated vegetable oils, which contain many double bonds, react with hydrogen to form solid, saturated fats. Their carbon–carbon double bonds open up to form carbon–hydrogen bonds, and leave single bonds where their double bonds were.

Simple alkenes can join together to make extremely long molecules called polymers. The polymerization reaction of ethene makes polyethene by a double bond in each molecule opening up and connecting it to two more ethene molecules as the polymer chain grows.

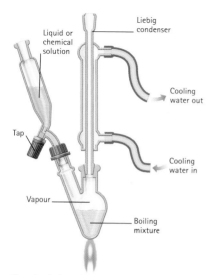

Slow chemical reactions can be made to go faster by heat. This apparatus has three main parts. The reaction takes place in the heated flask at the bottom. The tap controls the addition of liquids to the flask. The vertical condenser cools vapours that rise from the flask. This turns them into liquid that flows back into the flask.

DNA molecules are coiled inside cell nuclei. Based on the shape of a twisted ladder, the 'rungs' at the heart of the ladder carry a code that instructs each cell how to make proteins. An uncoiled DNA molecule would be about 1 metre long.

FUNCTIONAL GROUPS

The double bond in alkene molecules gives them their distinctive properties. It is an example of a functional group. New families of compounds result when different elements or groups of atoms replace hydrogen atoms in alkanes and other hydrocarbons.

The alcohols all contain a hydroxyl group, –OH, attached to a carbon atom. Ethanol, C_2H_5OH, is a sharp-smelling liquid present in alcoholic drinks, whereas ethane, C_2H_6, is a gas. A form of propanol, C_3H_7OH, is a useful cleaning fluid for video recorders and other appliances.

Vinegar contains an acid called ethanoic acid, CH_3COOH. The functional group in this case is carboxylic acid, –COOH. The smell of rancid butter and sweat is caused by butanoic acid, C_3H_7COOH.

There are only about 20 common functional groups. Millions of different organic molecules result from adding different combinations in different places on hydrocarbon molecules that are different lengths and shapes.

ORGANIC MOLECULES FOR LIFE

The chemistry of living beings is called biochemistry. We eat complex organic molecules in our food and break them down by digestion. Carbohydrates from starchy foods give glucose; proteins from meat and grains provide amino acids. Blood carries these small molecules to the cells in our bodies. Glucose breaks down further into water and carbon dioxide, and releases energy that we use to move about and to power other chemical reactions. Amino acids join together to make proteins for muscle and skin tissues, as well as other body structures. Some proteins are enzymes that help all these complex reactions to happen. Proteins control the shapes of our bodies and the way they work. The reactions that make our proteins are controlled by the DNA coiled inside each cell. We inherit our DNA structure from our parents.

Haemoglobin carries oxygen in red blood cells. It is a protein molecule that consists of four chains of 145 amino acids each, wrapped around an iron atom. The Austrian-born biochemist Max F. Perutz (1914–) worked out the structure of haemoglobin in 1959, after decades of research. He was joint winner of the 1962 Nobel prize in chemistry for his breakthrough.

SEE ALSO

103 Digestion, 107 Genes and chromosomes, 120–1 Chemical reactions, 124–5 Bonding and valency, 152–3 Oil and refining

CATALYSTS

Catalysts are substances that speed up chemical reactions without being used up themselves. Catalysts are used in many industrial processes.

Molecules can only react together when they collide with sufficient force. Some reactions are slow because too few particles collide with sufficient force to react. A catalyst works by breaking down one difficult reaction step into two or more easier steps.

In the first step, one of the molecules combines with the catalyst to produce a substance called an intermediate. The intermediate then reacts with a second molecule to form the product. In this second step, the catalyst is released from the intermediate. The catalyst is then free to react again. In a well-catalysed reaction, both of these steps require much less energy than the uncatalysed reaction.

Many catalysts are transition elements such as iron and platinum. This is because these metals form and break bonds easily with reacting atoms, molecules and ions. The solid metals are also easy to separate from liquids and gases after the reaction.

Honeycomb coated with metal catalysts

To exhaust

CO Carbon monoxide
CO_2 Carbon dioxide
NO_x Nitrous oxides
HC Hydrocarbons
N_2 Nitrogen
H_2O Water

Metal casing

Vehicle exhaust pollutants include unburnt hydrocarbons, oxides of nitrogen and carbon monoxide. A platinum catalyst converts them to harmless CO_2, H_2O and N_2.

In 1908, German chemist Fritz Haber (1868–1934) discovered that an iron catalyst helps hydrogen and nitrogen to react together. The product, ammonia, is an important raw material for making a wide range of chemicals, including plant fertilizers, dyes and explosives.

CATALYSTS IN ACTION

Catalysts help in the manufacture of a wide range of items. Nickel, for example, catalyses the reaction between hydrogen and vegetable oils to produce margarine.

Polyethene was first made more than 60 years ago by compressing ethene gas to dangerously high pressures. Using a catalyst mixture of titanium and aluminium compounds, the reaction happens at normal pressure and 60°C.

Platinum is an important catalyst for reactions between gases. Ammonia and oxygen react together as they pass through a platinum gauze. The product, nitrogen oxide, is used to make nitric acid.

Catalysts also help to 'crack' large molecules from crude oil to make fuels and raw materials for plastics.

Hydrogen   Nitrogen

▶ In the Haber process, a compressor pressurizes a mixture of hydrogen and nitrogen gas to 250 times normal atmospheric pressure. About 20 per cent of the mixture turns into ammonia as it passes through the iron catalyst at 450°C. The mixture then passes through a cooler, which liquefies the ammonia. Unchanged hydrogen and nitrogen return to the catalyst.

Pump

Catalyst beds

Cooler

Compressor

SEE ALSO

110 The elements, 120–1 Chemical reactions, 124–5 Bonding and valency

ENZYMES

Inside every living cell, there are biological catalysts that enable the biochemical reactions that support life to happen. They are called enzymes.

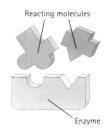

Reacting molecules

Enzyme

Product

▲ Reacting molecules lock into a part of an enzyme molecule called an active site. When the reaction is complete, the product detaches from the active site.

Enzymes are coiled protein molecules that catalyse biochemical reactions. These reactions can happen several billion times faster with enzymes than without them. Most enzymes catalyse one specific reaction only. Pepsin, for example, is an enzyme in digestive juices. It starts the digestion of proteins by breaking them into smaller pieces. Pepsin does not break down starch in food, however. That job is done by enzymes called amylases.

Enzymes work best over a narrow range of temperatures. They work only slowly below 30°C, and they break up above 40°C. Many inherited diseases are caused by the presence of faulty enzymes.

ENZYMES AT WORK

The part of an enzyme molecule that catalyses reactions is called its active site. When reacting molecules fit into the active site, they are held in the correct position for the reaction to happen. The active site in many enzymes contains a metal atom. Others contain a small molecule called a coenzyme, which is usually a vitamin. Traces of metal atoms and vitamins are present in a balanced diet. Lack of vitamins and trace metals can stop enzymes working properly and even lead to diseases such as scurvy.

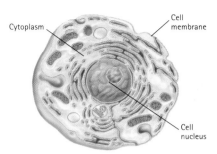

Cytoplasm

Cell membrane

Cell nucleus

A typical animal cell contains up to 100,000 different enzymes. These enzymes catalyse around 1,500 different life-supporting chemical reactions.

Enzymes are now made that can be used outside living cells. Biological washing powders contain enzymes that break down grease. These enzymes are extracted from plants. Enzymes that cause colour changes can be used to detect tiny amounts of substances. One of these enzymes is used in pregnancy testing kits.

▲ Exercise uses energy to drive muscles. Triose phosphate isomerase is one of the enzymes that catalyse the release of energy from glucose in the blood.

◀ A spider injects its prey with digestive enzymes. After a few hours, it will suck out the digested insides of the insect.

SEE ALSO

60 Spiders, centipedes and scorpions, 78–9 Body organization, 84–5 Muscles and movement

OXIDATION AND REDUCTION

Oxidation and reduction reactions involve the gain and loss of oxygen or the transfer of electrons between substances in a chemical reaction.

Oxygen is a common and reactive element that combines with most other elements. Heating copper in air, for example, forms a surface layer of black copper oxide. Hydrogen gas burns in air to form water. The products of both reactions are oxides. Chemists classify these types of reactions as oxidations.

REDOX REACTIONS

When hydrogen gas passes over hot copper oxide, the hydrogen removes oxygen from the copper oxide to form copper metal and water. The equation for the reaction between copper oxide and hydrogen shows what happens:

$$CuO + H_2 \rightarrow Cu + H_2O$$

Hydrogen combines with oxygen from copper oxide to form water, so hydrogen is oxidized in this reaction. At the same time, copper oxide is reduced to copper metal by removing its oxygen.

Reactions of this type are called redox reactions, because they are combined *red*uction and *ox*idation reactions. The substance that causes the oxidation – in this case copper oxide – is called the oxidant. The reductant is the substance that causes the reduction. In this reaction the reductant is hydrogen.

In the reaction between copper and oxygen, copper loses electrons and is oxidized. Oxygen gains electrons and is reduced. Copper oxide is a lattice of copper and oxide ions.

ELECTRON TRANSFER

Chemists describe redox reactions in terms of the transfer of electrons between substances. The reaction between copper and oxygen forms copper oxide, CuO, which contains copper ions, Cu^{2+}, and oxide ions, O^{2-}. In this reaction, each copper atom loses two electrons as it becomes a copper ion. These electrons are accepted by oxygen molecules, which form two oxide ions each. In general, oxidation is the removal of electrons from a substance, and reduction is the addition of electrons to a substance.

When redox reactions are described as transfers of electrons, many reactions that do not involve oxygen can be classified as oxidation and reduction reactions. For example, the reaction between sodium metal and chlorine gas is a redox reaction. Each sodium atom, Na, loses one electron to form a sodium ion, Na^+. At the same time, each chlorine molecule, Cl_2, gains two electrons to form two chloride ions. The result is the ionic compound sodium chloride, NaCl. In this reaction, chlorine is the oxidant and sodium is the reductant.

▲ The immense thrust of the main engines of a space shuttle comes from a roaring jet of steam. The steam forms in a redox reaction between hydrogen and oxygen.

▶ These iron rails are being welded together by the Thermite process. In this process, a mixture of aluminium and iron oxide is ignited to start the redox reaction between the two substances. The heat of the reaction is enough to melt the surface of the rails.

SEE ALSO

112–3 Atoms, 120–1
Chemical reactions, 292–3
Rockets and space shuttles

HYDROGEN

Hydrogen is the simplest element with the smallest mass of all. It is the most common element in the Universe and has important industrial uses.

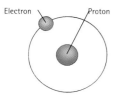

Electron Proton

The nucleus of ordinary hydrogen is a proton.

Neutron

The deuterium nucleus has one neutron and a proton.

Tritium has two neutrons and a proton in its nucleus.

▲ All three isotopes of hydrogen have a proton and an electron. Nuclei of deuterium, D, and tritium, T, also have neutrons.

Hydrogen is a colourless, odourless gas. It is the lightest element in the periodic table. Hydrogen gas consists of diatomic (two-atom) molecules, H_2. In hydrogen molecules, each atom contributes one electron to the covalent bond that holds the molecule together.

Hydrogen is highly flammable. It burns with oxygen to form water. Hydrogen is found in a range of compounds, including acids, hydroxides and hydrocarbons.

ISOTOPES

Isotopes are atoms of the same element that have different numbers of neutrons in their nuclei. Almost all hydrogen atoms consist of an electron orbiting a proton. Of every million hydrogen atoms, 150 have a nucleus that contains a neutron as well as a proton. This isotope of hydrogen is called deuterium, D.

Isotopes have the same chemical properties because they have the same number of electrons. Deuterium bonds with oxygen to form heavy water, D_2O. This liquid is used in nuclear reactors and in chemical experiments.

A third isotope of hydrogen is called tritium, T. It is made in nuclear reactors and is radioactive.

The main use of hydrogen is for the manufacture of a wide range of chemicals. It is also used as a propellant for space rockets and as a fuel for welding.

USES OF HYDROGEN

The Haber process reacts hydrogen with nitrogen to make ammonia, NH_3, which is used to manufacture fertilizers, explosives, dyes and plastics. Hydrogen is also used to change vegetable oils into margarine.

Hydrogen is a good fuel for rockets and welding. It burns in air to produce large amounts of energy and pure water, which does not cause pollution. Hydrogen is being tested as an alternative fuel for cars.

The problem with hydrogen as a fuel is that it is difficult to store. Hydrogen gas occupies too much space to be useful, and liquid hydrogen must be kept in insulated containers at temperatures below –253°C.

Evaporators change liquid hydrogen to gas

Engine

Fuel tank

This experimental car burns hydrogen fuel in an ordinary piston engine. Its exhaust gases cause no pollution, because they contain only water vapour.

SEE ALSO

110–1 The elements, 112–3 Atoms, 122–3 Chemical compounds

THE HALOGENS

The halogens are the elements that form group 17 of the periodic table. They are reactive non-metals and include fluorine, chlorine, bromine and iodine.

Dry ski slopes are made with PTFE-coated tiles.

Fluorine — Carbon

PTFE molecules contain only carbon and fluorine.

▲ Polytetrafluoroethene, or PTFE, has a surface that is more slippery than ice. Its many uses include coatings for cooking utensils and ski slopes.

The halogens are five non-metallic elements that form group 17 of the periodic table. Molecules of these elements have two atoms joined by a single covalent bond. Fluorine, F_2, is a pale yellow gas; chlorine, Cl_2, is a greenish yellow gas; bromine, Br_2, is a dark red liquid; iodine, I_2, is a purple-black solid. Astatine is a radioactive metallic solid that can only be made in nuclear reactors.

The halogens are all poisonous and will attack skin. Their reactivity decreases in the order F > Cl > Br > I. Fluorine reacts explosively with many substances; iodine reacts only slowly, if at all. The halogens combine with metals to form ionic salts called halides. Examples include sodium chloride, NaCl, which is common salt or table salt. Most halides are soluble in water and many are found in sea water. The halides of hydrogen are acids, and include hydrochloric acid, HCl. Halogens also form compounds with some non-metals, such as carbon and sulphur.

Some refrigerators contain chlorofluorocarbons, or CFCs. CFCs can cause holes in Earth's ozone layer. They must be removed before the refrigerators are scrapped.

EXTRACTION AND USES

Fluorine is made by electrolysing a liquid mixture of hydrogen fluoride and potassium fluoride at 100°C. Fluorine is used to make fluorocarbons, such as the plastic PTFE, and to purify nuclear fuel. Small amounts of sodium fluoride, NaF, in drinking water help prevent tooth decay.

Chlorine is made by the electrolysis of sodium chloride solution. Chlorine is used to sterilize drinking and bathing water, and for the manufacture of bleach and plastics.

Bromine is made by reacting chlorine with magnesium bromide from sea water. Similarly, chlorine reacts with iodide salts from seaweed. These two halogens are used in the manufacture of photographic film, medicines and antiseptics.

9 **F** 19.0	Fluorine Atomic number 9
17 **Cl** 35.5	Chlorine Atomic number 17
35 **Br** 79.9	Bromine Atomic number 35
53 **I** 126.9	Iodine Atomic number 53
85 **At** 210.0	Astatine Atomic number 85

Halogen atoms are just one electron short of a complete outer electron shell. They react readily to accept one electron each from other atoms.

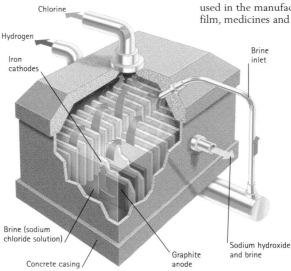

Chlorine
Hydrogen
Iron cathodes
Brine inlet
Brine (sodium chloride solution)
Concrete casing
Graphite anode
Sodium hydroxide and brine

Hooker cells are used to manufacture chlorine. Brine flows past electrodes inside a concrete casing. Bubbles of chlorine form at the graphite anodes and rise up through the brine. Chlorine leaves the cell through a pipe at the top. Hydrogen gas forms at the iron cathodes and leaves the cell through side pipes. The brine leaving the cell contains sodium hydroxide, a useful chemical substance.

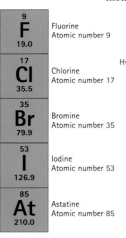

SEE ALSO

112–3 Atoms, 114–5 The periodic table

METALS

Most metals are extracted from ores that are mined from Earth's crust. The method of extraction depends on the chemical reactivity of each metal.

Every eight hours, workers wearing protective clothing open the hole at the bottom of a blast furnace. White-hot molten iron runs out at a temperature of 1,600°C.

REACTIVITY SERIES OF METALS	
Potassium	**K**
Sylvite, KCl	
Sodium	**Na**
Rock salt, NaCl	
Calcium	**Ca**
Limestone, $CaCO_3$	
Magnesium	**Mg**
Dolomite, $MgCO_3.CaCO_3$	
Aluminium	**Al**
Bauxite, Al_2O_3	
Carbon	**C**
Zinc	**Zn**
Zinc blende, ZnS	
Iron	**Fe**
Haematite, Fe_2O_3	
Tin	**Sn**
Cassiterite, SnO_2	
Lead	**Pb**
Galena, PbS	
Copper	**Cu**
Copper pyrites, $CuFeS_2$	
Mercury	**Hg**
Cinnabar, HgS	
Silver	**Ag**
Free metal	
Gold	**Au**
Free metal	
Platinum	**Pt**
Free metal	

The reactivity series of metals. The non-metal carbon occupies the space between aluminium and zinc. Carbon is more reactive than the metals below it, but less reactive than the metals above it.

About 80 of the elements are metals. They are to the left and in the centre of the periodic table. All metals are shiny solids at room temperature except mercury, which is a liquid. Metals are malleable and ductile which means that they can be hammered or stretched into different shapes. They are also good conductors of heat and electricity because their outermost electrons can move from one atom to the next.

METAL REACTIVITY

Gold and platinum are examples of metals that rarely react with other elements to form compounds. Potassium and sodium are extremely reactive metals. They even react violently with relatively unreactive substances, such as water. The reactivities of most metals lie between these two extremes. Iron, for example, rusts slowly in moist air. Copper is almost unaffected under the same conditions.

The reactivity series lists the common metals in order of reactivity. The most reactive metals are normally at the top of the list. Non-metals carbon and hydrogen are often included in the list to give a comparison of their reactivities.

Aluminium oxide is mixed with cryolite so that it melts at 850°C. Electrolysis between graphite electrodes produces aluminium metal at the cathode and oxygen gas at the anodes. The electrolysing current keeps the mixture hot enough to stay liquid. Oxygen gradually oxidizes the anodes, so they have to be replaced from time to time.

ORES AND METAL EXTRACTION

Most metals occur as compounds in Earth's crust. Only the least reactive metals, such as gold and platinum, are found as pure elements. Rocks that contain metal compounds are called ores. Haematite, a type of iron ore, contains iron oxide, Fe_2O_3. The main lead ore is galena, or lead sulphide, PbS.

Many metals are extracted by heating their ore in a furnace with a substance that removes the elements attached to the metal. This process is called smelting. Smelting iron ore with carbon produces iron metal and carbon dioxide gas. Carbon removes the oxygen from iron oxide.

Carbon can be used to smelt metals that lie below it in the reactivity series. Zinc, iron, tin and lead are all produced by smelting their ores with carbon.

Carbon cannot smelt metals that are more reactive than itself. As a result, the metals from aluminium to potassium in the reactivity series are produced by electrolysis of their molten compounds.

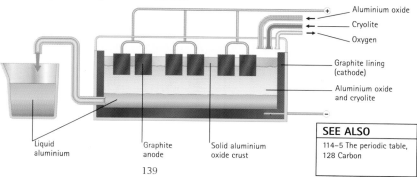

Aluminium oxide — +
Cryolite
Oxygen
Graphite lining (cathode)
Aluminium oxide and cryolite

Liquid aluminium

Graphite anode

Solid aluminium oxide crust

SEE ALSO
114–5 The periodic table, 128 Carbon

ACIDS

Acids are compounds of hydrogen that produce hydrogen ions with water. The hydrogen ions make the solution acidic. Acids turn litmus paper red.

Small pieces of zinc metal dissolve rapidly in hydrochloric acid. Bubbles of hydrogen gas are given off and the zinc dissolves to form a solution of zinc chloride.

Many everyday things contain acids. Lemons contain citric acid and vinegar contains ethanoic acid, which is also called acetic acid. Different acids give lemons, vinegar, sour apples and sherbet their sharp taste. Car batteries contain sulphuric acid and digestive juices in the stomach contain hydrochloric acid.

Acids are solutions of substances that produce hydrogen ions, H^+, when they dissolve in water. Many acid substances can be obtained as pure solids, liquids, or gases, but they act as acids only when they are dissolved in water.

ACID REACTIONS

Acids can be detected by adding a special dye called an indicator. Litmus paper, for example, contains an indicator dye that changes from purple to red in acidic solutions. Many acids produce hydrogen gas when mixed with reactive metals, such as zinc or magnesium. In another test, an acid is mixed with sodium bicarbonate, $NaHCO_3$. Strong acids make the mixture fizz as carbon dioxide gas is produced.

▲ Bottles of laboratory acids carry a warning label that shows they are corrosive and harmful (top). Organic acids in fruit taste sharp and are harmless, but the acid in wasp stings causes a painful reaction. Car batteries (bottom) contain harmful sulphuric acid.

▶ Concentrated sulphuric acid is a dehydrating agent. It removes hydrogen and oxygen from substances to form water. This picture shows sugar being turned to carbon by sulphuric acid. The acid converts sugar, $C_{12}H_{22}O_{11}$, into 11 molecules of water and 12 carbon atoms. The heat of the reaction turns some of the water into steam, which forms a froth of black carbon.

ORGANIC ACIDS

Plants and animals produce a variety of acidic carbon compounds called organic acids. Most are harmless and many give flavour to fruit and other food. Oils and fats are compounds of organic acids with glycerol. Soaps are salts of organic acids that are made from oils and fats. DNA, deoxyribonucleic acid, is an extremely complex acid that carries genetic code.

A few naturally-occurring organic acids are harmful. Stinging nettles and some ants defend themselves with methanoic acid, HCO_2H, which causes stings. An old name for this acid – formic acid – comes from the Latin word for ant, *formicus*.

The leaves of some plants, such as rhubarb, contain poisonous oxalic acid. Animals soon learn not to eat these plants.

INORGANIC ACIDS

Acids made from minerals and non-metals are called inorganic acids. The common inorganic acids include sulphuric acid, H_2SO_4, hydrochloric acid, HCl, nitric acid, HNO_3, and phosphoric acid, H_3PO_4.

Industry produces millions of tonnes of these acids every year. They are used to make plastics, fibres, fertilizers, dyes and other chemicals. Concentrated inorganic acids are often highly corrosive. They can damage skin and rapidly dissolve most metals. Hydrofluoric acid, HF, dissolves even glass. Other inorganic acids are not at all dangerous. Boric acid, H_3BO_3, is the main ingredient in soothing eye lotions.

SEE ALSO

120–1 Chemical reactions, 122–3 Chemical compounds

BASES AND ALKALIS

A base is a substance that can neutralize an acid by reacting with hydrogen ions. An alkali is a base that dissolves in water. Bases turn litmus paper blue.

Sodium bicarbonate solution with indicator

Vinegar

Indicator changes colour

▲ A litmus indicator turns blue in a solution of sodium bicarbonate (top). This is because the solution is alkaline. If drops of vinegar (above centre) are added to the solution (above), the acid in the vinegar turns the litmus red for a moment until it is neutralized by the alkali. When all the bicarbonate has been neutralized, the solution becomes purple. If more acid is added, the solution turns red.

▶ Crops do not grow well in acidic soil. Soils that are slightly alkaline are ideal for plant growth and for the spread of beneficial soil organisms. Farmers add powdered lime, calcium hydroxide, to acid soil. The lime neutralizes the acid and makes the soil slightly alkaline.

Most bases are minerals that react with acids to form water and a salt. Bases include the oxides, hydroxides and carbonates of metals. Examples are sodium hydroxide, NaOH, calcium carbonate, $CaCO_3$, and potassium oxide, K_2O.

Bases react with the hydrogen ions in an acidic solution to make water. Copper oxide, CuO, is a typical base. It neutralizes sulphuric acid, H_2SO_4, to produce copper sulphate, $CuSO_4$, and water. The chemical equation shows this change:

$$CuO + H_2SO_4 \rightarrow CuSO_4 + H_2O.$$

If black copper oxide powder is added to the colourless acid, the powder dissolves and the solution becomes coloured with blue copper sulphate. After a while, no more copper oxide dissolves. The acid is no longer present and the solution does not turn indicators to their acid colour. This is an example of a neutralization reaction. The salt produced in the reaction consists of a copper ion, Cu^{2+}, and a sulphate ion, SO_4^{2-}. The positive copper ion is from the base and the negative sulphate ion is from the acid. The other product of the reaction is water. The part of an acid that forms salts in neutralization reactions is called the acid radical. The acid radical of sulphuric acid is SO_4^{2-}.

Acids from rotting vegetation and acid rain harm aquatic organisms. This lime-dosing column adds calcium hydroxide to a flowing river to neutralize acids.

ALKALIS

Alkalis are bases that dissolve in water. A solution of alkali can be detected by adding an indicator. A solution of sodium hydroxide, for example, changes litmus from purple to blue. Alkaline solutions contain hydroxide ions, OH^-. When acidic and alkaline solutions mix, hydroxide ions from the alkali react with hydrogen ions from the acid to form water. A salt is the other product of the reaction.

SEE ALSO

120–1 Chemical reactions, 122–3 Chemical compounds

FACTS AND FIGURES

BRANCHES OF CHEMISTRY

Analytical chemists explore the types and proportions of substances in a sample.
Astrochemists identify substances found in stars and other bodies in space.
Biochemists study the compounds and chemical reactions in living organisms.
Electrochemists investigate the relationship between the flow of electricity and chemical reactions.
Environmental chemists study how changes in the natural environment affect living organisms.
Geochemists analyse the chemical composition of Earth.
Inorganic chemists study the chemistry of all the elements and their compounds, except for those compounds that contain mainly carbon and hydrogen.
Nuclear chemists investigate changes that happen in atomic nuclei.
Organic chemists study hydrocarbons – compounds of carbon and hydrogen – and other related compounds.
Photochemists investigate the relationship between light and chemical reactions.
Physical chemists use the principles of physics to explain observations about chemical substances and their reactions.
Radiochemists study the radioactive isotopes of the chemical elements.

THE ELEMENTS

Elements are substances that ordinary chemical reactions cannot break down into simpler substances.

An **element** is a substance that consists of atoms that all have the same number of protons. There are 90 naturally occurring elements and 21 artificial elements.
There are 16 non-metals, 5 metalloids and 90 metals. Under normal conditions of room temperature and pressure, 11 elements are gases and 98 are solids. Only two elements – bromine and mercury – are liquids under normal conditions.
An **atom** is the smallest part of an element that can exist separately. An atom consists of electrons orbiting around a nucleus. The **nucleus** contains two types of nucleons: protons and neutrons. The number of protons equals the number of electrons in a neutral atom.
The **atomic number** of an element is equal to the number of protons in its nucleus. The atomic number is sometimes called the proton number.
The **mass number** of an element is the sum of the numbers of protons and neutrons. An atom of iron, for example, consists of 26 protons, 26 electrons and 30 neutrons. The atomic number of iron is 26 and the mass number is 56.

LAWS AND PRINCIPLES OF CHEMISTRY

Avogadro's Law

At the same temperature and pressure, equal volumes of different gases contain the same number of molecules.
(*Amedeo Avogadro, 1811*)
By this law, a cubic metre of hydrogen contains the same number of molecules as a cubic metre of carbon dioxide.

Law of conservation of mass

During a chemical reaction, matter is neither created nor destroyed.
(*Antoine Lavoisier, 1774*)
According to this law, the total mass of the products of a chemical reaction is equal to the total mass of the substances that react together.

Law of constant composition

No matter how a substance is made, it will always contain the same elements in the same proportions.
(*Joseph Proust, 1779*)
By Proust's law, carbon dioxide exhaled in breath and carbon dioxide from a car exhaust both consist of molecules that contain one atom of carbon and two atoms of oxygen.

Heisenberg's uncertainty principle

It is impossible to specify both the precise position and the momentum of a particle at the same time.
(*Werner Heisenberg, 1927*)
The more accurately the position of a particle is measured, the less accurate the knowledge of its momentum becomes, and vice versa. The effect only becomes noticeable for subatomic particles, such as electrons and protons.

KEY DATES

BC
c.450 Leucippus of Miletus introduces the idea of atoms. Empedocles of Akraga introduces the four elements: earth, air, water and fire.
430 Democritus of Abdera develops the idea of atoms and suggests they explain the properties of matter.
340 Greek philosopher Aristotle proposes that substances are all combinations of the four elements.

AD
750 Arabian alchemist Geber describes how to prepare acids and their salts.
1473 Democritus' theory of atoms becomes known to alchemists in Europe through its Latin version.

1597 German chemist Andreas Libavius writes *Alchemia*, the first important chemistry text book.
1610 French chemist Jean Béguin publishes the first chemistry book not to be based on alchemy.
1661 Irish chemist and physicist Robert Boyle publishes *The Sceptical Chymist*, introducing the concept of chemical elements.
1766 British chemist Henry Cavendish discovers hydrogen, which he names 'inflammable air.'
1777 French chemist Antoine Lavoisier suggests air consists of two gases.
1781 British chemist and clergyman Joseph Priestley makes water by burning hydrogen in oxygen.
1803 British chemist and physicist John Dalton formulates atomic theory.
1807 British chemist Humphry Davy uses the recently invented electric battery to isolate the elements sodium and potassium.
1811 Italian chemist and physicist Amedeo Avogadro proposes that equal volumes of different gases contain the same number of molecules.
1828 German chemist Friedrich Wöhler makes urea, an organic compound, from inorganic ammonium cyanate.
1833 French chemist Anselme Payen discovers diastase, an enzyme.
1856 British chemist William Perkin makes the first synthetic dye.
1869 Russian chemist Dmitri Mendeleyev publishes the first form of the periodic table.
1893 German chemist Felix Hoffman synthesizes aspirin.
1911 British physicist Ernest Rutherford discovers the proton.
1913 Danish physicist Niels Bohr proposes a theory of atomic structure based on electron orbits.
1926 Austrian physicist Erwin Schrödinger develops a wave equation of atomic structure.
1932 British physicist John Cockcroft and Irish physicist Ernest Walton build the first particle accelerator to change one element into another.
1935 US chemist Wallace Carothers develops nylon, the polymer of the first totally synthetic fibre.
1938 German chemist Otto Hahn splits atoms of uranium.
1942 Italian-born US physicist Enrico Fermi creates the first controlled nuclear chain reaction.
1971 US company DuPont starts production of Kevlar, a polymer that is stronger than steel.
1985 Ball-shaped carbon molecules are discovered and named fullerenes.

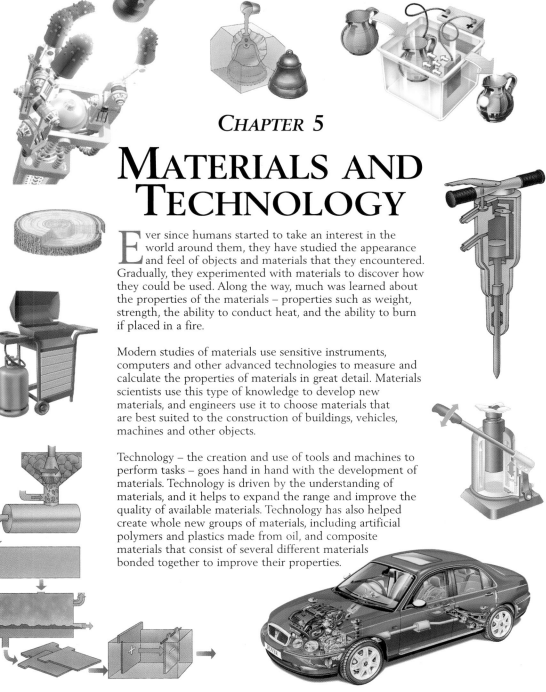

CHAPTER 5

MATERIALS AND TECHNOLOGY

Ever since humans started to take an interest in the world around them, they have studied the appearance and feel of objects and materials that they encountered. Gradually, they experimented with materials to discover how they could be used. Along the way, much was learned about the properties of the materials – properties such as weight, strength, the ability to conduct heat, and the ability to burn if placed in a fire.

Modern studies of materials use sensitive instruments, computers and other advanced technologies to measure and calculate the properties of materials in great detail. Materials scientists use this type of knowledge to develop new materials, and engineers use it to choose materials that are best suited to the construction of buildings, vehicles, machines and other objects.

Technology – the creation and use of tools and machines to perform tasks – goes hand in hand with the development of materials. Technology is driven by the understanding of materials, and it helps to expand the range and improve the quality of available materials. Technology has also helped create whole new groups of materials, including artificial polymers and plastics made from oil, and composite materials that consist of several different materials bonded together to improve their properties.

PROPERTIES OF SOLIDS

Materials such as rock, wood, rubber and diamond are all examples of solids. How they can be used depends on their individual properties.

The Mohs scale lists ten minerals in order of their hardnesses. Talc is the softest mineral on the Mohs scale. It has a value of one, and can be scratched by a fingernail.

Diamond is the hardest known substance. It has a value of ten on the Mohs scale. Diamond is pure carbon that has formed into crystals.

The first materials used by people were those that grew or lay around them. Rocks and stones were chipped to form spearheads and crude tools. Plant fibres were used to make rope, string and thread, and the thread was used to sew clothes from animal skins and furs.

Over many centuries, people have discovered new natural substances and found ways to make new materials. For example, it was discovered that clay could be hardened by heat to make ceramic pots, vessels and storage containers.

Materials that are used to make other materials that are called raw materials. The products of this processing are called manufactured or synthetic materials. For example, clay is a raw material that can be made into pottery, a manufactured material. Wood can be processed to make manufactured materials such as paper, card and textiles. Glass is a synthetic material that is made by heating sand with salts and other substances.

Materials such as dry leaves and wood can easily catch fire. In dry conditions, a discarded match can start a major inferno. Forests like this on Australia's east coast are especially vulnerable to fires.

PROPERTIES

The properties of a material include its hardness, strength and flexibility. Some materials are combustible, which means they can be catch fire. Others allow heat or electricity to pass through them and are called conductors. The properties of a material determine how it can be used.

▲ Solders melt at lower temperatures than most common metals. They are used to join metal surfaces together. Hot solder melts and then flows around and into the gap between the surfaces. When it cools, it solidifies, forming a strong bond.

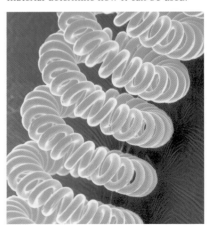

◀ Many metals can be drawn and spun to make thin wire. The device pictured winds niobium metal wire into coils for superconducting magnets.

▲ This lightbulb filament (magnified 68 times) is called a coiled coil. This type of filament is made by bending a tightly coiled wire into a looser coil.

Mohs value of 10. Talc, which is soft and crumbly, has a value of 1. Other materials can be given Mohs values by comparing them with the minerals in the Mohs scale. A typical fingernail has a hardness of 2½, while glass has a hardness of 6.

Hardness does not necessarily equal strength. A piece of rubber might feel soft and flexible yet it is extremely strong. Rubber is an example of an elastic material. Elastic materials return to their original shapes after being stretched. Materials that can be formed into different shapes have the property of plasticity. Materials that stretch easily without cracking or breaking are ductile.

Clay can be hardened by heat. Bricks and ceramic items are made by firing soft clay objects in a kiln.

A material's hardness is its ability to resist being scratched or dented. Hardness depends mainly on how tightly the atoms in a material are bound together.

German mineralogist Friedrich Mohs (1773–1839) drew up a scale of hardness using ten natural materials called minerals. He numbered them from one to ten in order of increasing hardness. Diamond is the hardest known material and has a

When designers set about planning how to make an object, one of their first decisions is which materials to use. In addition to choosing materials that have a good combination of hardness, flexibility and formability, the designer might have to choose materials that can withstand extreme temperatures or corrosive chemicals. Often, cost is the factor that decides which material is used out of a range of possible options.

Wood is a common material that is light and strong. It can be fashioned into a variety of shapes.

MATERIALS USED IN A CAR

A variety of materials is used to build a modern car such as this Rover 75. The engine is made from a metal alloy that can withstand heat and force. The windscreen is made from layers of glass and plastic that will not shatter over the passengers if the screen breaks. The seat covers are made from leather, which is comfortable and easy to cut, colour and stitch. Soft foam bumpers reduce the severity of injuries in collisions with pedestrians.

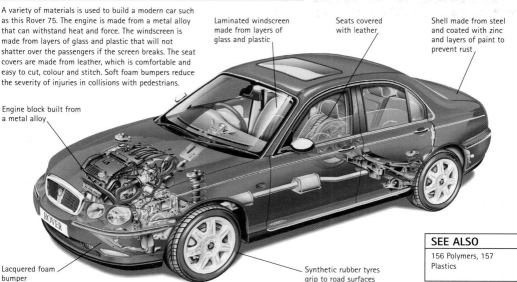

Laminated windscreen made from layers of glass and plastic

Seats covered with leather

Shell made from steel and coated with zinc and layers of paint to prevent rust

Engine block built from a metal alloy

Lacquered foam bumper

Synthetic rubber tyres grip to road surfaces

ROVER

SEE ALSO

156 Polymers, 157 Plastics

IRON

Iron is one of the most common metals in the Earth's crust. It has been worked with for thousands of years and is now mainly used to make steel.

Haematite is an ore of iron. It often forms kidney-shaped lumps. These give the ore its nickname of kidney ore.

▼ The steamship *Great Eastern* had a 211-metre-long hull made from riveted iron plates. The ship was designed by British engineer Isambard Kingdom Brunel (1806–1859) and was launched in 1858.

Iron makes up about 5 per cent of Earth's crust and some 35 per cent of Earth as a whole. Most of this iron is in Earth's core.

Iron is the cheapest, most commonly used metal. In its natural state, iron is normally combined with oxygen as iron-oxide ores. Haematite and magnetite are the two main iron ores.

Since the 14th century, huge ovens called blast furnaces have been used to turn iron ore and coke into pig iron. Pig iron is iron that contains a small amount of carbon left over from the blast furnace coke. Pig iron was used to make tools, weapons and many other objects.

Since the 1850s, an increasing amount of pig iron has been converted into steel. Steel contains less carbon than pig iron and is more flexible. Steel is made by blowing air or oxygen into pig iron.

Blacksmiths work with a type of iron called wrought iron. Wrought iron can be beaten and bent into shape when it has been softened in a red-hot furnace.

PROPERTIES OF IRON

Pure iron is a shiny, silver-white metal. It melts at 1,535°C, and it is ductile and malleable. Forms of iron that contain a small amount of carbon, such as steel, are harder than pure iron. This hardness makes steel more useful than pure iron for many uses.

Iron forms compounds with elements such as chlorine, oxygen and sulphur.

When unprotected iron is exposed to moist air it corrodes. A reddish-brown, flaky oxide forms. This oxide is rust.

BLAST FURNACE

In a blast furnace, measured amounts of iron ore, coke and limestone are loaded into the main chamber. Iron ore contains iron oxide, coke is a form of carbon made from coal and limestone is calcium carbonate.

Hot air is blasted into the bottom of the furnace. This causes the coke to burn and form carbon monoxide. The burning coke heats the contents of the bottom of the furnace to more than 1600°C. At this temperature, oxygen in the iron oxide reacts with carbon monoxide, freeing iron from its ore.

The liquid iron flows to the bottom of the furnace and is run off every three or four hours. The limestone reacts with impurities in the iron ore and forms a product called slag. A layer of liquid slag forms on top of the liquid iron and is removed from time to time.

Iron ore, limestone and coke

Waste gases

Air heater

Blast furnace

Dust catcher

Hot, dry air fed into furnace

Liquid iron

Iron

Slag

SEE ALSO

162–3 Chemical reactions, 170 Carbon, 178 Oxidation and reduction, 213 Coal

COPPER

Copper is a soft reddish-brown solid metal that has been widely used since 3000BC for its flexibility and, more recently, for its ability to conduct electricity.

Malachite is an ore of copper. Its dramatic bands of dark green make it popular in jewellery.

Electrical cables often consist of fine strands of copper wire woven together and encased in a plastic sleeve.

I n nature, copper occurs both as the free metal and as compounds in copper-bearing ores and minerals.

Pure copper melts at 1,083°C, which makes it easy to cast in moulds. Copper is also very malleable, which makes it easy to hammer into shape. As a result, copper has been used to make a wide range of objects, including coins, cooking utensils and ornaments.

Although pure copper is too soft for many uses, it forms hard, strong alloys, such as brass and bronze, when it is mixed with other metals. Brass is an alloy of copper and tin. Bronze contains zinc.

PROPERTIES AND USES OF COPPER

Copper is widely used in the electrical industry because it is an excellent conductor of electricity and can be extruded and drawn out to make wires as fine as 0.025 mm in diameter.

Copper wire is used for household electrical circuits and for the wiring found in electrical goods. Electromagnets, generators and motors also frequently contain coils of copper wire.

This copper statue, believed to be the world's oldest metal sculpture, is an image of Egyptian pharaoh Pepi I. This Old Kingdom pharaoh reigned from 2289 to 2244BC.

The flexibility of copper and brass make them ideal materials for making pipes for plumbing systems.

Copper reacts with other elements less quickly than iron. However, copper will corrode slowly in moist air. Over time, reddish-brown copper statues and roofs become coated in a green patina. This coating is copper carbonate. It forms when copper reacts with moisture and carbon dioxide in the air.

Copper ore is crushed into small pieces

Water is added to make a slurry

A ball mill grinds the copper ore into a fine powder

Air and chemicals are added. These help to concentrate the slurry.

Waste gases

Furnace

Slag

Liquid copper is cast into slabs

Electrolytic refining

Copper is formed into pipes, wires and ingots

MAKING COPPER

Copper ore is mined and then crushed into granules. The ore is mixed with water and then ground in a ball mill to form a slurry of fine particles. The slurry is heated in a furnace to form crude metallic copper. Electrolysis then purifies and refines the copper to over 99.9 percent purity. The final processing stage consists of melting and casting the copper into cakes, bars, ingots and billets. Billets are blocks of copper that are used to make copper tubing and piping. The photograph below shows copper piping being bent through a right angle.

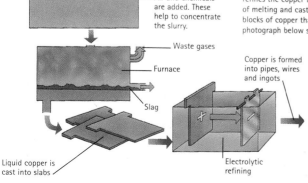

SEE ALSO

148–9 Alloys,
242 Electrochemistry

ALLOYS

Alloys are mixtures of metals with other metals and some non-metals. The properties of alloys are often superior to the properties of their components.

Most alloys are mixtures of two or more metals. Some are mixtures of metals with small amounts of other elements, such as carbon. They resemble metals in many ways: they are lustrous, or shiny, and conduct heat and electricity.

Alloys are normally designed to keep the good properties of a metal while eliminating its bad properties. This has been true since the first alloy, bronze, was made more than 6,000 years ago.

Bronze is an alloy of copper and tin. Like copper, it can be hammered into shape. It was found to be much harder than either pure copper or pure tin, however, which made it more useful for armour, tools and weapons. Bronze is still used to make heavy-duty machinery.

Today, many different alloys are manufactured for specific purposes. Some are designed to withstand extreme temperatures or strong chemicals; others are made to be light but strong. The most widely used alloy by far is steel.

TYPES OF STEELS

There are numerous types of steels. All are alloys of iron and carbon. Some steels also contain other elements.

More than 90 per cent of all the steel that is made is a form of carbon steel. Carbon steel contains amounts of carbon with small traces of manganese, silicon and

This Celtic helmet, shield and sword are made of bronze – an alloy of copper and tin. All are around 2,000 years old.

Brass instruments, such as this trumpet, are made of an alloy of copper and zinc.

▼ This photo shows a cross-section of a cylinder made from an alloy of titanium and aluminium. This alloy is widely used in the aerospace industry.

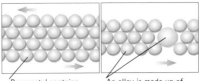

Pure metal contains one type of atom

An alloy is made up of different types of atoms

Pure metals are soft because their atoms can easily glide past each other. Different-sized atoms in an alloy resist this motion and make the alloy harder.

copper. Carbon steels are used to make a variety of objects, including springs, car bodies and girders for construction.

Alloy steels and tool steels contain larger amounts of manganese, silicon and copper than carbon steels. They also contain elements such as the metals molybdenum, tungsten and vanadium. Alloy steels are used where a hard-wearing material is needed, such as in truck transmissions and machine tools. They are more expensive than carbon steels because of the higher cost of their ingredients.

High-strength low-alloy (HSLA) steels are a new class of steel. They are stronger than ordinary carbon steels but more economic to produce than alloy steels. This is because they contain less of the more expensive elements such as vanadium. HSLA steels are used for the same applications as carbon steels.

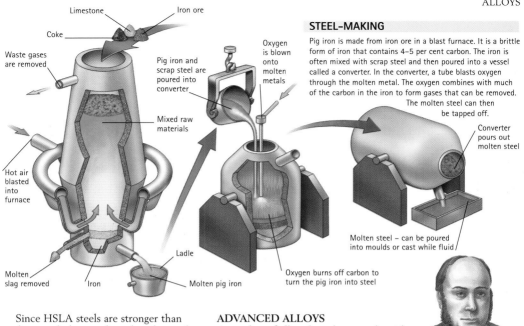

Limestone
Iron ore
Coke
Waste gases are removed
Pig iron and scrap steel are poured into converter
Mixed raw materials
Oxygen is blown onto molten metals
Hot air blasted into furnace
Molten slag removed
Iron
Ladle
Molten pig iron
Oxygen burns off carbon to turn the pig iron into steel
Molten steel – can be poured into moulds or cast while fluid
Converter pours out molten steel

STEEL-MAKING

Pig iron is made from iron ore in a blast furnace. It is a brittle form of iron that contains 4–5 per cent carbon. The iron is often mixed with scrap steel and then poured into a vessel called a converter. In the converter, a tube blasts oxygen through the molten metal. The oxygen combines with much of the carbon in the iron to form gases that can be removed. The molten steel can then be tapped off.

Since HSLA steels are stronger than carbon steels, less steel needs to be used. An HSLA girder, for example, would be thinner and lighter than a normal girder of the same strength.

Stainless steels contain chromium and nickel, which make them shiny and resistant to rusting and stains. Stainless steels are used to make a variety of objects from kitchen sinks and cutlery to surgical instruments such as scalpels.

In industry, stainless steel is shaped into pipes and containers to house corrosive chemicals safely. It is also used to make ball bearings.

ADVANCED ALLOYS

A number of alloys have been made with very special properties. Among them are superconductors and shape-memory alloys, or SMAs.

Superconducting alloys have almost no electrical resistance when cooled to extremely low temperatures. An example is an alloy of tin and niobium metals.

Shape-memory alloys 'remember' their shape and return to it. When cold, the alloy can be twisted out of shape. When heated, it returns to its original shape. SMAs are useful for spectacle frames and orthodontic wires and braces.

British engineer Henry Bessemer (1813–1898) invented a vessel called a converter for making large quantities of steel.

▼ Many different types of stainless steels exist. The type used for cutlery is sometimes called 18-8. This is because it contains 18 per cent chromium and 8 per cent nickel.

Stainless-steel fork

1 per cent carbon and other metals
8 per cent nickel   18 per cent chromium
73 per cent iron

◄ Completed in 1889, the Forth Bridge in Scotland was the first cantilever railway bridge. Around 55,000 tonnes of steel were used to make the Forth Bridge.

SEE ALSO

118–9 Solutions,
128 Carbon, 146 Iron,
147 Copper

WOOD AND PAPER

Wood is a tough material that forms the trunks and branches of trees. It is used as a building material, as fuel and as a raw material for making paper.

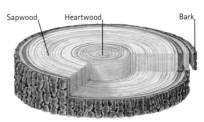

Sapwood   Heartwood   Bark

The layer of wood beneath the bark of a tree carries moisture and is called sapwood. The hard, solid wood in the centre of the trunk is called heartwood.

Wood is a versatile material. It is tough, flexible, easy to fashion and relatively strong. The logging and lumber industries are concerned with felling trees and preparing planks, beams and panels for use in construction and in making furniture and other goods.

Different woods have different properties. Balsa, for example, is extremely light, while ash is tough and can withstand sudden and repeated bending and stretching.

Prepared wood, or lumber, is used to make window, door and roof frames.

Wood is turned into lumber in a lumber mill. After the bark has been removed, mechanical saws cut the wood into thin planks or thicker beams, depending on what the lumber is to be used for.

PAPER AND RECYCLING

Paper is made mainly from the cellulose in wood. Paper is used to make books and stationery, waxed paper cartons for liquids and filter papers for car engines. Paper can be made using cellulose from sources other than wood. Bank notes and expensive writing paper contain cellulose fibres from cotton plants. These fibres make the paper tough and very smooth.

On average, each person in the United States uses about 300 kilograms of paper a year. About half that amount is recycled to make newspapers, toilet rolls and other low-quality papers. The rest is incinerated or dumped in landfill sites.

Bark removed

Wood chipped into small pieces

Water and chemicals added and wood chips cooked into pulp

Pulp beaten to break down the fibres

Pulp cleaned and bleached to make it white

Pulp drained on fine mesh belt

Pulp from recycled paper

Heated rollers dry paper and press fibres firmly together to form a sheet (of paper)

Finished paper is wound onto a reel

MAKING PAPER

Modern paper is largely made from coniferous trees, such as pine, spruce and fir. The fibres in wood consist of a strong material called cellulose. This makes the paper very strong, so it does not fall apart easily when pressed, folded or stretched. In a modern paper-making machine, wood chips are first boiled with caustic soda or another chemical in a tank. This releases the strong fibres of cellulose. Liquid is removed from the fibres to leave a pulp. The cellulose pulp is then spread on a conveyor belt and squeezed between hot rollers to remove moisture, resulting in the finished paper.

SEE ALSO

52–3 Trees, 144–5
Properties and solids,
151 Fibres

FIBRES

Fibres are used to make fabrics for clothing, bedding and many other products. Some fibres are natural products; others are made by chemical processes.

Silk

Cotton

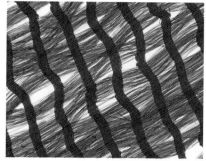

This close-up of crepe de Chine is magnified 70 times. The fabric is made by weaving fine silk fibres with thicker silk thread. The thread is made by spinning silk fibres together.

Natural fibres are obtained from animals and plants. These fibres can be spun to make yarn, which can then be woven on a loom to make cloth.

Wool, the most common animal fibre, is made from the fleece of sheep. The quality of wool depends on the breed of sheep.

Silk is made by a type of caterpillar called a silkworm, which spin a cocoon to prepare for their transformation into moths. In silk farming, the long fibres are collected before the cocoon forms.

Many fibres are obtained from plants. Cotton comes from a clump of fibres that forms around the head of cotton plants. Jute, sisal and hemp are strong plant fibres that are used to make rope and canvas. Linen is made from the stem fibres of flax.

SYNTHETIC FIBRES

The first synthetic fibres were made from cellulose early in the 20th century. Rayon is prepared from cellulose by dissolving wood pulp in an alkali. This mixture is then treated with a chemical that turns it into a sticky liquid called viscose. The viscose solution is sprayed through tiny holes into a bath of sulphuric acid. The acid makes the viscose solution harden and form fibres that resemble silk. These fibres can be spun to make yarn and then woven to make a silky cloth.

Fibres such as nylon, polyester and acrylic are produced from oil by chemical processes. They are similar to plastics.

Synthetic fibres are often stronger than natural fibres. They can be woven to form crease-resistant cloths and are also used to make ropes and carpets.

Wool

Flax and cotton come from plants. Wool is from the fleece of sheep. Silk is produced by silkworms.

Pressure rollers flatten cotton fibres

Teasing wires form loose bundles of fibres called slivers

Cotton fibres are formed into fine threads by a process called spinning. Rollers and teasing wires flatten and divide the cotton fibres. The fibres are then gathered into slivers, which are stretched between more sets of rollers before being spun to make yarn.

Cotton fibre

Slivers stretched and spun to form cotton yarn

Dividers

Rollers that move at different speeds stretch the slivers

Weft

◀ A weaving loom interlaces threads to make cloth. Two sets of warp threads are stretched across the frame of the loom. A shuttle pulls the weft thread between the two sets of warp threads. The loom then reverses the upper and lower warp threads before the shuttle passes back between them.

Slivers wound onto bobbins

Yarn bobbin

Shuttle

Warp

SEE ALSO

141 Bases and alkalis,
156 Polymers, 157 Plastics

OIL AND REFINING

Oil or petroleum is a naturally occurring liquid that can be refined to make fuels and lubricants as well as raw materials for the chemical industry.

This pump in Bakersfield, California, is an example of a nodding donkey. It draws oil to the surface as it 'nods' up and down.

Oil is the decomposed remains of tiny organisms that lived in the sea many millions of years ago. After they died, they were covered with layer upon layer of sediment. Over time, the weight of those layers turned the organic remains into crude oil. The appearance of crude oil varies from a pale yellow liquid to a black sticky tar. More than half the world's known oil reserves are in the Middle East.

Crude oil, or petroleum, is a complex mixture of chemical compounds that consist mainly of hydrogen and carbon. This mixture is separated and treated in oil refineries to produce materials that are used to make a huge range of materials called petrochemicals or oil derivatives.

The main derivatives of oil are fuels such as diesel, jet fuel and petrol. Other derivatives include dyes, lubricants, medicinal drugs, nylon and polyester fabrics, plastics and polymers, solvents, synthetic rubber and waxes.

Oil rigs are designed to provide a stable platform for the incredibly powerful drill bit to bore deep down into the Earth to tap into a reserve of oil. Life on oil rigs is dirty and sometimes dangerous, but usually well paid.

FINDING AND PRODUCING OIL

The search for oil is called oil exploration. The extraction of oil from natural reserves is called oil production.

As oil forms underground, it tends to seep up through porous rock towards Earth's surface. If it hits a 'roof' of non-porous rock, it becomes trapped and starts to form a reservoir of oil.

Oil and natural gas collect in domes or ridges of porous rock that lie under solid rock. When the rock above the reserve is punctured by a drill, natural pressure forces the gas and oil to the surface.

Many oil reserves lie out at sea. One type of oil rig floats on large pontoons and is held in place by computer-controlled engines. Huge drills are sunk into the ground in search of deposits of oil.

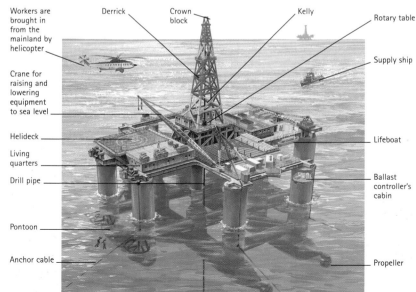

Workers are brought in from the mainland by helicopter

Crane for raising and lowering equipment to sea level

Helideck

Living quarters

Drill pipe

Pontoon

Anchor cable

Derrick

Crown block

Kelly

Rotary table

Supply ship

Lifeboat

Ballast controller's cabin

Propeller

▲ The Trans-Alaska Pipeline carries crude oil 1,284 kilometres through Alaska. The pipeline is 1.2 metres in diameter and can transport 318 million litres of crude oil a day.

▶ Oil is frequently transported by ships called tankers. The larger tankers, such as this Arco Alaska oil transporter, are known as supertankers. They can be hundreds of metres long.

Geologists know the patterns of rock where oil is likely to collect and have developed surveying techniques that help to locate potential oil deposits.

In one method, explosives are set off to send vibrations underground. The echoes of the explosions are then detected and analysed to form a picture of the rock structure that lies underground. If the rock structure is likely to hold oil, a number of test wells are drilled. If the test wells find oil, production oil wells are built.

Oil production is measured in units called barrels. One barrel is equal to 159 litres. The current world production of crude oil is about 25,000 million barrels a year. Experts believe that there are between 1.5 and 2 billion barrels of oil still remain in underground reserves and could be extracted in the future.

OIL REFINING

Crude oil is transported to oil refineries by pipelines or by large tanker ships. There, a process called fractional distillation is used to separate it into mixtures of products that have similar boiling points. Those mixtures are then treated further to produce a range of fuels and a multitude of raw materials for the chemical industry.

▲ An enormous number of products are derived from oil, including polymers and solvents for paints, lubricating oils, waxes and fuels.

◀ Crude oil is processed in large refineries such as this one in Wales. Fractionating columns up to 75 metres tall are used to separate the components of the oil.

SEE ALSO

132–3 Organic chemistry, 156 Polymers, 157 Plastics

NATURAL GAS

Natural gas is an important energy resource. It is used to provide heat and energy, and is an important raw material for the chemical industry.

Hot-air balloons burn propane or butane from cylinders to warm air inside the balloon and lift the balloon into the sky.

The composition of natural gas varies depending on where it is extracted. North-Sea gas contains 92 per cent methane, 3.5 per cent ethane, 2.5 per cent nitrogen and 1 per cent propane.

Bottled gas is a portable source of fuel for appliances such as gas-fired barbecues.

Natural gas, like crude oil, formed over millions of years from the remains of marine organisms. Natural gas also collects in the types of rock structures where crude oil is found. For this reason, gas reserves are often found with oil reserves.

The main component of natural gas is methane, the simplest hydrocarbon (compound of carbon and hydrogen). Mixed with the methane are smaller amounts of other hydrocarbon gases such as ethane, butane and propane.

Natural gas provides almost one fifth of the world's energy supplies. Also, the components of natural gas are important raw materials for chemical processes.

Most of the world's natural gas comes from wells in Canada, Siberia and the United States.

PRODUCING AND PROCESSING GAS

Natural gas is extracted through wells that are similar to oil wells. Many gas reserves lie offshore, and gas is piped from offshore gas-production platforms to an onshore collection point and then on to a refinery where it is purified.

In the first stage of purification, water and any other liquids are allowed to settle out from the gas under the action of gravity. The dry gas then passes through a

Surplus natural gas is sometimes burned off by flares, such as this one on a gas-production platform in the Bruce field in the North Sea.

cooler, where butane and propane liquefy and are collected. These gases, called liquefied petroleum gases (LPGs), can be sold as raw materials for the manufacture of chemicals or bottled and used as fuel for heaters and cookers.

The remaining natural gas can be piped through a supply network or cooled and pressurized to form liquefied natural gas (LNG). Liquefied natural gas takes up much less space than natural gas and is a convenient way to ship the gas in tankers.

This view along the deck of a liquefied natural gas (LNG) carrier shows the tops of its large, insulated steel tanks. The tanks contain liquefied gas at −162˚C.

SEE ALSO

128 Carbon,
236 Power stations

COAL

Coal is an impure form of carbon that formed from the remains of prehistoric plants. Coal burns very easily, releasing large amounts of heat.

Anthracite is the hardest form of coal. It releases more heat than any other type of coal when it burns. Anthracite contains about 95 per cent carbon. Another form of carbon is the graphite that is used in pencils.

Coal, like oil and gas, is a fossil fuel. While oil and gas formed from the remains of living organisms, coal formed from the remains of decaying plants from prehistoric forests. These remains were compressed and changed by the layers of rock above them.

There are three main types of coals, each containing a different amount of carbon: anthracite, bituminous coal, and lignite. Anthracite is the most valuable form of coal as it contains approximately 95 per cent carbon. Bituminous coal contains around 70 per cent carbon, while lignite, or brown coal, contains less than 50 per cent carbon. Most coal is found in underground bands or seams.

The coal-mining industry extracts more than 4–5 million tonnes of coal every year. More than half of this total is mined in China and the United States. The amount of coal that could be mined from known reserves is some 1,200 billion tonnes.

USES OF COAL

Coal was the main fuel for the Industrial Revolution at the end of the 18th century. It powered steam engines and was used for making iron and steel. Today, most coal is burned in power stations to generate electricity. Coal-fired power stations use filters and other devices to trap soot and other pollution formed when coal burns.

When coal is heated in a chamber called a retort, it releases gas, oil and tar. The gas can be burned as fuel. The oil and tar contain chemicals that can be used for making products such as dyes, perfumes and artificial fibres. The solid that is left behind in the retort is coke. It is used as a smokeless fuel and for making iron.

MINING COAL

Different types of mines and mining techniques are used depending on the land conditions and how deep the coal seam is. Deep deposits are reached through vertical shafts and horizontal tunnels are dug into the coal seam. Coal deposits near the surface are mined in drift, opencast and slope mines, which cost less to construct and operate than shaft mines.

Excavator working an opencast mine

Ventilation fans

Shaft mines extract coal from deep seams

Slope mines extract coal from shallow seams

Ventilation shaft

Drift mine built where coal seam reaches surface

Coal seam

Cage travels up and down main shaft

Tunnel

Main shaft

SEE ALSO

128 Carbon, 152–3 Oil and refining, 154 Natural gas

POLYMERS

Polymers are very large molecules made up of thousands of smaller molecules all joined together. Polymers can be natural or synthetic.

Polymers occur frequently in nature. DNA, wood and protein are all polymers. Natural fibres, such as wool and silk, are also polymers.

Synthetic polymers have been made since the late 19th century. They are the basis of all synthetic plastics and fibres.

Polymers consist of long, chain-like molecules that are made by joining together smaller molecules called monomers. Polyethylene, for example, is made by joining together many thousands of molecules of the monomer ethylene.

Polymers can have one of three types of structure: linear, branched or cross-linked. Linear polymers consist of long simple chains of monomers. Nylon and polyvinyl chloride (PVC) are linear polymers. Some linear polymers have kinked chains. The chains straighten out when stretched but spring back when the force is removed. This makes some polymers elastic.

Branched polymers have shorter chains attached along their main chains, rather like the teeth of a comb.

Cross-linked polymers have links between their chains that form a net and make the polymer harder and less flexible.

This is just a small part of a polymer chain.

Raw materials for nylon are heated to form a stream of liquid polymer, which forms a filament of nylon when it cools in a water bath. The filament is dried, chopped, melted, then sprayed through a spinneret, which resembles a shower head. Cold air turns the spray into fine fibres that are then stretched and crimped.

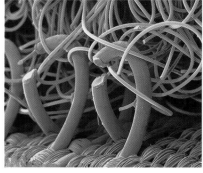

Velcro, seen here magnified 24 times, consists of two textures of nylon. The hooks at the bottom catch in the looped whiskers at the top.

NYLON

Nylon was the world's first synthetic fibre. It was developed by US chemist Wallace H. Carothers (1896–1937) in the 1930s.

Nylon was originally designed as a cheap alternative to silk. Nylon polymer makes a fibre that is stronger than cotton and wool. It can be mixed with natural fibres or used on its own to make yarn for weaving textiles.

Different types of nylon can be made by varying the raw materials used for making the polymer. Some nylons are hard enough to make gear wheels and bearings for machines. These types are also resistant to heat and chemical attack.

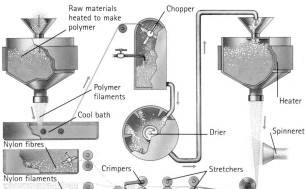

Raw materials heated to make polymer

Chopper

Polymer filaments

Cool bath

Nylon fibres

Nylon filaments

Crimpers

Heater

Drier

Spinneret

Stretchers

▲ This fly-fisherman in the River Dee, Scotland, uses a nylon line on his rod. The strength of nylon makes it ideal for this use.

SEE ALSO

128 Carbon, 132–3
Organic chemistry, 157
Plastics

PLASTICS

Plastics are materials that can be easily stretched or moulded into shape. Most plastics are made from chemicals derived from oil.

Many familiar objects, from casings for electrical goods to bags, drinks containers and sports helmets, are made from different types of plastics.

Articles such as brushes and combs are made by moulding plastics. The bristles of toothbrushes are nylon fibres.

Plastics are a form of polymer. The first plastics, such as celluloid, were made from naturally occurring polymers. The first completely synthetic plastic was Bakelite, invented in 1907 by US chemist Leo Baekeland. Since then, hundreds of different plastics have been developed. Nearly all are made using chemicals derived from oil.

The wide range of uses for plastics stems from their properties. They can be rigid or flexible and can be coloured, shaped and formed in a number of ways. Plastics are good electrical insulators and many are resistant to chemical attack.

Plastics do not all behave in the same way when they are heated. Some plastics, called thermoplastics, soften when they are heated. Polythene is a thermoplastic.

Other plastics, called thermosets, become harder when they are heated. Once they have set during manufacture, they cannot be reshaped. Electric plugs are made of thermosets, which is why they do not melt if the wires inside them overheat.

The walls of this exhibition squash court are made using perspex, which is transparent but much more resistant to high-speed impacts than glass.

ENVIRONMENTAL ISSUES

The chemical resistance of plastics is a good property for most of their uses. That same resistance becomes a problem in refuse landfills. The bacteria that cause wood, paper and natural fibres to biodegrade, or rot, cannot break down most plastics. As a result, plastic materials remain unchanged in landfills.

Scientists are developing plastics that are biodegradable, some of them based on plant materials. However, recycling as much plastic as possible is considered the most practical way to avoid the build up of plastics in landfill sites.

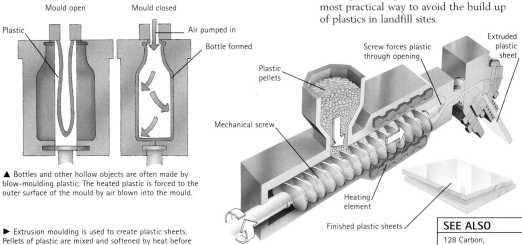

▲ Bottles and other hollow objects are often made by blow-moulding plastic. The heated plastic is forced to the outer surface of the mould by air blown into the mould.

▶ Extrusion moulding is used to create plastic sheets. Pellets of plastic are mixed and softened by heat before being forced through an opening by a mechanical screw.

Mould open   Mould closed

Plastic   Air pumped in

Bottle formed

Plastic pellets

Screw forces plastic through opening

Extruded plastic sheet

Mechanical screw

Heating element

Finished plastic sheets

SEE ALSO
128 Carbon,
156 Polymers

PETROL AND DIESEL ENGINES

Petrol and diesel engines burn fuel to create mechanical energy. This is used to drive machinery in everyday use, including road vehicles and boats.

In 1861, German engineer Nikolaus Otto (1832–91) built the first practical four-stroke petrol engine. In 1876, he invented the four-stroke internal combustion engine. He patented a four-stroke combustion engine in 1876.

Petrol and diesel engines are types of internal combustion engines. They take this name because they combust, or burn, fuel inside their cylinders. Some of the energy released when the fuel burns and is converted directly into mechanical energy. The rest is lost as heat.

Internal combustion engines are used to power most road vehicles and boats. They also power some aircraft and railway locomotives. Petrol or diesel engines are frequently used to turn emergency electricity generators.

The designs of four-stroke diesel and petrol engines are very similar. In both cases, a mixture of fuel and air burns inside cylinders that are fitted with pistons. As the burning mixture expands inside a cylinder, it forces the piston to slide towards the mouth of the cylinder and push against a crank on a crankshaft, just like a cyclist pushing against a pedal. The exhaust valve then opens and the waste combustion gases start to escape.

FOUR-STROKE PETROL ENGINE

German engineer Gottlieb Daimler (1834–1900) worked with Nikolaus Otto to develop the design of Otto's engine. Daimler patented a high-speed version of the four-stroke engine in 1887. Typical petrol engines have up to eight cylinders. A piston in each cylinder drives the crankshaft round when fuel and air ignite, driving the piston down. Springed valves allow the fuel–air mixture to enter the cylinder and the exhaust gases to leave.

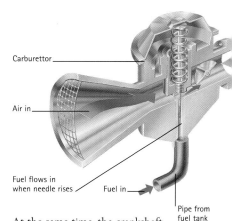

Carburettor

Air in

Fuel flows in when needle rises

Fuel in

Pipe from fuel tank

At the same time, the crankshaft pushes the piston back into the cylinder, driving out the remaining gases.

When the piston reaches the top of the cylinder, the exhaust valve closes and the inlet valve opens. As the crankshaft pulls the piston down, it draws fuel and air into the cylinder. This is called induction.

The inlet valve closes and the piston is pushed into the cylinder by the crank. This compression, or squeezing, causes the mixture of fuel and air to become hot.

In the case of a petrol engine, a spark from the spark plug then ignites the hot mixture and starts a new power stroke.

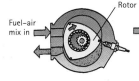

Fan blows air over engine to cool it

Air filter stops dust and dirt from being sucked into engine

Drive shaft

Disk brake

Spark plug

Twin rotors

German engineer Felix Wankel (1902–88) built a rotary engine in 1957. A triangular piston turns inside a chamber through the combustion cycle.

Fuel–air mix in

Rotor

Compressed fuel–air mix

Spark plug

Exhaust out

1 Induction: turning rotor sucks in mixture of petrol and air.

2 Compression: fuel–air mixture is compressed as rotor carries it round.

3 Ignition: compressed fuel–air mixture is ignited by the spark plug.

4 Exhaust: the rotor continues to turn and pushes out waste gases.

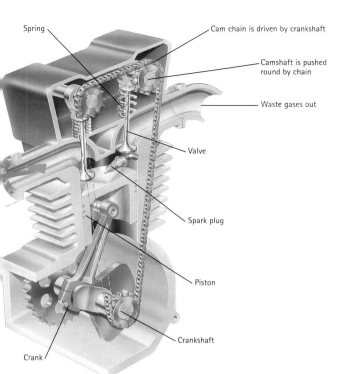

Spring

Cam chain is driven by crankshaft

Camshaft is pushed round by chain

Waste gases out

Valve

Spark plug

Piston

Crankshaft

Crank

1 **Induction:** the piston falls, drawing air and fuel into the cylinder.

2 **Compression:** the rising piston compresses the fuel–air mixture.

3 **Power:** a spark ignites the fuel and air, which forces the piston down.

4 **Exhaust:** the piston rises and forces the exhaust gases out.

▲ The four-stroke cycle is used in most car engines. The four piston strokes, or movements, are called induction, compression, power and exhaust. Only the power stroke drives the crankshaft round. In a four-cylinder engine, each cylinder will be at a different stage in the sequence at any one time.

DIESEL ENGINES
The diesel engine was invented in 1896 by German mechanical engineer Rudolf Diesel (1858–1913).

A diesel engine compresses the fuel–air mixture to roughly twice the pressure found in a petrol engine. This makes the fuel–air mixture hot enough to ignite without the need for a spark plug. Diesel fuel is usually less expensive than petrol. Also, diesel engines burn fuel more efficiently and so use less fuel.

IMPROVEMENTS
Petrol and diesel engines can be improved by increasing their fuel efficiency and reducing the pollution that they emit.

In many new engine designs, the carburettor has been replaced by electronic fuel injectors that pump fuel into the cylinder during the induction stroke. A microprocessor controls the amount of fuel that is injected as well as the timing so that the fuel burns efficiently with less pollution.

◄ The diesel engine of a diesel–electric locomotive drives an electricity generator, which powers the locomotive's electric traction motors. The turbocharger pumps air into the engine, which produces more power.

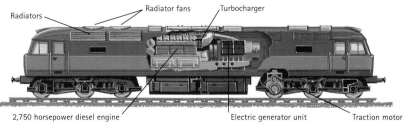

Radiators

Radiator fans

Turbocharger

2,750 horsepower diesel engine

Electric generator unit

Traction motor

SEE ALSO
152–3 Oil and refining, 172 Combustion, 200–1 Work and energy

JET ENGINES AND GAS TURBINES

Jet engines drive aircraft forwards by forcing gases backwards at high speed. Gas turbines convert the force of expanding gases into mechanical energy.

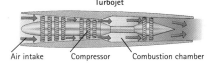

Turbojet

Air intake   Compressor   Combustion chamber

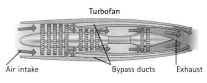

Turbofan

Air intake   Bypass ducts   Exhaust

Turbofan engines are quieter and more efficient than simple turbojet engines. Turbofans drive air around the combustion engine as well as through it.

British engineer Frank Whittle (1907-1996) designed the first jet engine. The first aircraft to be successfully powered by one of his engines flew in 1941.

Concorde reaches speeds greater than the speed of sound under the power of jet propulsion.

The rotor blades that keep Chinook helicopters in the air are driven by a form of gas turbine.

In a gas-turbine system, a compressor forces air into a combustion chamber. There, it mixes with fuel. The mixture is ignited by a spark. Hot gases are produced when the fuel burns. They expand and drive a series of fan blades called a turbine. The rotation of the turbine can be used to drive a generator as well as driving the compressor at the turbine's air intake.

Jet engines can power aircraft at greater speeds than propeller engines by burning fuel and compressed air in a combustion chamber and expelling the combustion gases backwards.

Gas turbines are similar to jet engines, but the hot gases that they produce are used to make a turbine rotate.

In either case, air enters the engine at the front and is compressed by a series of turbines. It enters the combustion chamber where it is mixed with fuel. The air-and-fuel mixture is ignited and burns fiercely to produce hot, expanding gases.

The hot gases pass through another turbine as they leave the combustion chamber. This turbine drives the compressor at the front of the engine.

In an aircraft's jet engine, these gases escape from the rear at high speed and drive the aircraft forwards.

Simple jet engines are called turbojets. They are capable of driving aircraft at great speeds and are used to power military aircraft. Turbojets, however have the disadvantages of being noisy and not very fuel efficient.

Jet airliners are usually powered by jet engines called turbofans. The compressor of a turbofan engine sucks in more air than is needed for the fuel that it burns. The extra air passes around the combustion chamber and is forced out of the rear of the engine. Turbofans are less noisy and more efficient than turbojets.

In a gas turbine, the force of the expanding hot air escaping from the combustion chamber is used to make a turbine rotate. The rotation of the turbine shaft can be used for example to drive an electricity generator.

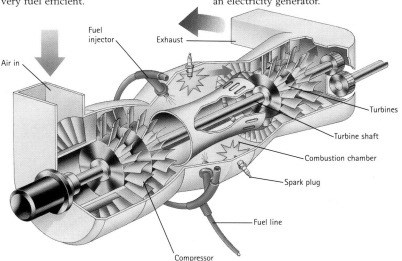

Air in

Fuel injector

Exhaust

Turbines

Turbine shaft

Combustion chamber

Spark plug

Fuel line

Compressor

STEAM ENGINES

Steam engines use pressurized steam from a boiler to drive pistons. They convert heat energy from burning fuel into mechanical energy.

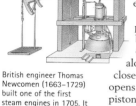

British engineer Thomas Newcomen (1663–1729) built one of the first steam engines in 1705. It was used to pump water from mines.

British inventor James Watt (1736–1819) developed the first practical steam engine. In 1764, he invented a condenser that turned waste steam back into water to be used again.

When water boils in an open pot, it produces about 2,000 times its own volume of steam. If water boils in a sealed tank, the steam cannot expand. Instead, its pressure increases.

Steam engines work by letting pressurized steam from a heated boiler expand in their cylinders. This expansion forces a piston to move along the cylinder. When the piston is close to one end of the cylinder, a valve opens and pressurized steam forces the piston in the opposite direction. While this happens, the steam on the other side of the piston escapes. A connecting rod and crank turn the back-and-forth motion of the piston into the rotation of a shaft.

STEAM ENGINE HISTORY

Some of the first steam engines were built by Thomas Newcomen in the early 18th century. They were used to pump water from mines. In the 1760s, James Watt made improvements on the early designs. By the end of the 18th century, steam engines were widely used to power machinery in factories.

In 1804, British engineer Richard Trevithick (1771–1833) built the first steam locomotive. Steam locomotives were widely used until the 1960s.

STEAM TURBINES

In 1884, Irish engineer Charles Parsons (1854–1931) invented a turbine that used expanding steam to make a shaft rotate. By 1897, Parsons had built a steam-turbine-powered ship called *Turbinia*. It was the fastest boat of its time.

Steam turbines have an advantage over piston engines in that they produce rotational motion directly, rather than needing a connecting rod and crank to turn a shaft. Steam turbines run more smoothly than piston engines, they are more efficient and they occupy less room than piston engines of equivalent power.

Today, the main uses of steam turbines are in power stations and for the propulsion of nuclear-powered ships.

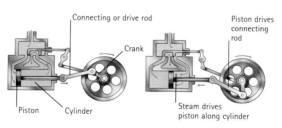

Connecting or drive rod

Crank

Piston

Cylinder

Steam drives piston along cylinder

Piston drives connecting rod

Crank converts motion to turn wheel

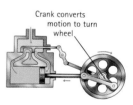

▲ Steam from a boiler drives a piston in a cylinder. The piston is connected to a crank that converts its back-and-forth motion into a circular motion.

◄ Locomotives such as these were capable of hauling express passenger trains at speeds of around 160 kilometres per hour using steam power.

SEE ALSO

155 Coal, 173 Expansion and contraction, 213 Pressure, 236–7 Power stations

AUTOMATION

Automation is the use of machines to perform tasks without human assistance. Automated machines work according to sets of programmed instructions.

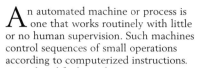

Invented in 1793, the hand-turned cotton gin simplified the removal of seeds from cotton. Until then, seeds had to be picked out by hand.

Henry Ford (1863–1947) pioneered automation in the production of the Model T car. About 15 million Model Ts were built between 1908 and 1927.

An automated machine or process is one that works routinely with little or no human supervision. Such machines control sequences of small operations according to computerized instructions.

Modern life depends on automation. Electronic cash registers recognize goods by a pattern of stripes and numbers called a bar code. This indicates the prices of items and send information to a central computer that controls stocks and orders goods. Traffic lights, central-heating controls and aircraft autopilots are all examples of automated systems.

STEPS TOWARDS AUTOMATION

The term automation was coined in 1946, by which time telephone exchanges and other equipment were already automatic. Moves towards automation had started almost two centuries before then.

Mechanization is the use of machines to perform tasks that would otherwise have to be done by humans. The first examples of mechanization took place in the textile factories of the late 18th century, where machines were first used to spin yarn and weave cloth.

Early machines needed humans to control each step of their operations. Later machines were programmed to perform sequences of tasks.

The first programmable machine was a loom built in 1801 by French inventor Joseph-Marie Jacquard (1752–1834). The Jacquard loom used a system of punched cards to control the pattern of threads woven by the loom.

▲ By 1913, the assembly time for a Model T had fallen from twelve hours to 90 minutes. This time saving was mainly due to the use of assembly lines.

▶ Automated guided vehicles, or AGVs, operate in many factories. They ferry goods and materials along carefully marked routes. Many AGVs are guided by signals from electrical loops buried under factory floors.

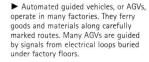

◀ These robots are welding a car body on a Honda production line in Ohio, USA. Robots are ideal for performing simple, repetitive tasks accurately. This welding line can be programmed to weld different models of car.

▲ The movements of this paint-spraying robot are programmed to make sure that paint coverage is even. The use of robots to spray paint protects workers from harmful paint fumes.

The spread of mechanization was encouraged by the availability of new power sources. Waterwheels and steam engines were used to power machines in factories via systems of belts and drive shafts. Later, electric motors were used to power machinery.

Throughout the 20th century, the growing car industry was a major driving force for automation. In 1913, Ford's Model T car was the first to be built on an assembly line. The line moved car bodies through all the stages of construction, and operators at each stage would repeat a number of tasks on each car as it passed.

The introduction of assembly lines reduced the cost of cars by increasing the efficiency of production. Other industries soon adopted the same approach.

ROBOTS AND AUTOMATION

The first robot started work in a car factory in 1961. The use of robots has grown continuously since then.

Robots are well suited to doing repetitive tasks with a great deal of accuracy. They can be programmed to perform a variety of tasks, often by 'learning' the actions of human operators. Robots can also send information about their own rate and quality of production to a central computer.

The use of computer-aided design, or CAD, saves many hours of producing detailed technical drawings. Even more work is saved by combining CAD with computer-aided manufacture, or CAM. CAD/CAM was used to design and build this French submarine.

SEE ALSO
158–9 Petrol and diesel engines

FACTS AND FIGURES

MAJOR MINERALS

A mineral is a naturally occurring inorganic material. Many minerals are metal salts.

Alabaster Hydrated calcium sulphate – translucent, fine-grained form of gypsum used for making ornaments.
Apatite Calcium fluoride–phosphate – former ingredient of phosphate fertilizers and source of phosphorus, now replaced by rock phosphate.
Asbestos Various magnesium silicates – natural fibre and good thermal insulator. Exposure to asbestos dust can cause lung diseases, so its use has been prohibited.
Azurite Hydrated basic copper carbonate – used as a pigment for its blue colour.
Bentonite A type of clay that swells in water – used as a filler in making paper.
Calcite Calcium carbonate – the second most common mineral after quartz.
Clay Fine-grained aluminosilicates – used to make bricks, pottery and fine ceramics.
Dolomite A mixture of calcium and magnesium carbonates.
Fluorite The mineral form of calcium fluoride – used in making glass and ceramics, and as a source of fluorine.
Graphite A soft form of carbon – used in nuclear reactors and in lead pencils.
Gypsum A soft form of calcium sulphate – used for making cement and plaster.
Kaolin A form of clay, also called china clay – used to make ceramics and in the manufacture of paper, rubber and paints.
Limestone Rock, consisting mainly of calcium carbonate – used as a building stone and in iron smelting.
Marble A highly crystalline form of calcium carbonate – used in architecture and sculpture.
Mica Various aluminium silicates – used in electrical capacitors and as an electricity insulator. Micas split into thin plates, which makes them useful as pearl-effect pigments for paints.
Oriental alabaster A variety of calcite – harder than true alabaster.
Rock phosphate Calcium phosphate – an ingredient in fertilizers and the main source of phosphorus.
Quartz Silicon dioxide. The most common mineral, quartz occurs in opaque and transparent forms that are sometimes coloured by impurities – used for making glass and some ceramics.
Saltpetre Potassium nitrate – used to make gunpowder and fertilizers.
Silica A hard mineral form of silicon dioxide with a high melting-point.
Slate Mixed minerals that naturally split into thin sheets – used for roofing.
Talc A soft white or greenish material – used in paints, ceramics and products of the cosmetics and health-care industries.

MOHS HARDNESS SCALE

German mineralogist Friedrich Mohs devised a scale of hardness based on ten minerals. The greater the number, the harder the mineral. A typical fingernail has Mohs hardness 2–3, since it scratches gypsum but is scratched by calcite. Materials up to Mohs hardness 4 can be scratched using a coin.

1 Talc
2 Gypsum
3 Calcite
4 Fluorite
5 Apatite
6 Orthoclase
7 Quartz
8 Topaz
9 Corundum
10 Diamond

Diamond is the hardest of all minerals. It will scratch all other materials. Silicon carbide, or carborundum, is a synthetic material with Mohs hardness 9.5. It is manufactured for use as an abrasive.

MAJOR ORES

An ore is a mineral source of a metal.

Bauxite An impure form of aluminium oxide from which aluminium is extracted by electrolysis.
Cassiterate An impure tin oxide, this mineral is the chief ore of tin.
Chalcopyrite Gold-coloured mixed copper–iron sulphide, also known as copper pyrites. The main ore of copper.
Chromite A mixed chrome–iron oxide, the main source of chromium.
Galena Lead sulphide. The principal source of lead.
Hematite The mineral form of iron oxide and one of the main ores of iron.
Halite Sodium chloride, also known as rock salt. A source of chlorine, sodium hydroxide, and sodium metal.
Ilmenite A mixed iron–titanium oxide. The main ore of titanium.
Malachite Copper hydroxide–carbonate. A bright green mineral used as an ore of copper and as a semi-precious stone for making ornaments.
Pentlandite Iron–nickel sulphide. The main ore of nickel.
Pitchblende Uranium oxide. The main source of uranium for making fuel for the nuclear industry.
Rutile Titanium oxide. A minor source of titanium metal.
Sphalerite Zinc sulphide. The main ore of zinc, also known as zinc blende.
Zincite Zinc oxide. A minor source of zinc, sometimes called spartalite.

TOP TEN METALS PRODUCTION

Iron	973,000,000 tonnes
Manganese	23,600,000 tonnes
Aluminium	17,700,000 tonnes
Chromium	12,500,000 tonnes
Copper	9,000,000 tonnes
Lead	3,300,000 tonnes
Nickel	895,000 tonnes
Tin	219,000 tonnes

REACTIVITY SERIES OF METALS

The reactivity series of metals lists metals in order of decreasing ease of reactivity, so potassium is more reactive than platinum. The more reactive metals are less easy to extract from their compounds.

Potassium
Sodium
Calcium
Magnesium
Aluminium
Zinc
Iron
Tin
Lead
Copper
Silver
Gold
Platinum

TEN COMMON ALLOYS

Coinage bronze – copper (95%), tin (4%), zinc (1%) – used for making coins and tokens for vending machines.
Coinage silver – silver (90%) copper (10%) – used for making coins.
Dental gold – gold (58%), silver (14–28%), copper (14–28%) – used for dental repair work.
Duralumin – aluminium (95%), copper (4%), manganese (0.5%), magnesium (0.5%) – used for making structural components for aircraft.
Manganin – copper (82.5%), manganese (16%), nickel (1.5%) – used in the circuits of electrical meters.
Nichrome – nickel (80%), chromium (20%) – used for making electrical resistors and heating elements.
Pewter – tin (65–80%), lead (20–35%) – used for making utensils and decorative drinking vessels and ornaments.
Solder – lead (20–70%), tin (30–80%) – used to join metal objects, particularly electrical wiring and circuit components.
Stainless steel – iron (60–80%), chromium (10–20%), nickel (8–20%) – used in kitchen utensils, sinks and machinery fittings.
Tool steel – iron (90–95%), molybdenum (6–7%), chromium (2%–4%) – used in tools such as chisels and saws.

CHAPTER 6
LIGHT AND ENERGY

The Sun is Earth's main energy source. Electromagnetic radiation carries its light and heat to Earth through 150 million kilometres of space. Light helps plants to grow on land and in the sea, providing food for Earth's animal life. Heat energy keeps the Earth at a suitable temperature for life to exist. It evaporates water to make clouds, makes winds blow and makes waves move across oceans. Fossil fuels contain energy from sunlight that fell on Earth millions of years ago. These fuels provide most of the energy for industrial societies. Without the Sun's light and heat, life on Earth would not exist.

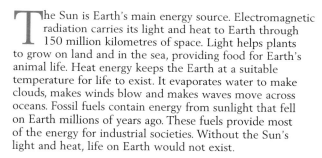

Many machines depend on the flow of heat energy to make them work. Scientists and engineers have formulated laws of thermodynamics that describe the flow of heat in chemical and mechanical systems. These laws help them to design heating and cooling systems as well as engines and motors. Scientists have developed kinetic theory to explain the influence of heat on the behaviour of the particles that make up materials.

Light is a form of energy. Electric lamps and lasers are two sources of artificial light. Precisely engineered lenses and curved mirrors in microscopes and telescopes bend light rays to form magnified images. Instruments that focus and detect electromagnetic radiation outside the visible spectrum give information about the temperature, composition and speed of motion of distant objects in space.

HEAT AND LIGHT FROM THE SUN

The Sun is the star at the centre of our solar system that includes Earth. It is 150 million kilometres from Earth and is its main source of heat and light energy.

During the day, sunlight provides heat and light, which are two types of energy. Life exists on Earth because the Sun provides just the right amount of energy. Planets that are closer to the Sun, such as Venus, receive more energy and are too hot for life. Planets that are further away, such as Mars, are too cold.

Energy is needed to do work and make things happen. Practically everything that happens on Earth depends on energy that originally came from the Sun. Plants depend on energy from sunlight for their growth, and pass on part of this energy to animals that eat them. Carnivorous animals gain their energy from the meat of other animals that eat plants. Also, coal, oil and natural gas are all formed from the remains of plants and animals that grew thanks to energy from sunlight.

Heat from sunlight evaporates water from Earth's oceans, warms the air and makes the winds blow. These winds carry heat and water vapour around the world and control its weather patterns.

Filtered by Earth's atmosphere, the Sun's rays keep the temperature on a sunny day at around 25°C. If the temperature were 15°C lower, the children in this photograph would be shivering; if it were 15°C warmer, they would be uncomfortably hot.

The leaves of plants absorb energy from sunlight. Plants use this energy to make complex chemicals from simple ones. Plants are the source of food for most life on Earth.

NUCLEAR FUSION

The Sun consists of about 70 per cent hydrogen and 30 per cent helium. The Sun's energy comes from nuclear fusion reactions its core, where atoms of hydrogen join together to make helium atoms. The mass of a helium atom is less than the mass of the hydrogen atoms that form it. This lost mass turns into energy: one kilogram of lost matter gives about 100,000 million megajoules – a large power station's total output in 25 years.

ENERGY FROM THE SUN

The diameter of the Sun is 100 times that of Earth, and its volume is one million times greater. The immense force of gravity produces enormous pressure in the core, and raises the temperature to about 15 million °C. At this temperature, hydrogen exists as a soup of electrons and protons, called plasma, that is denser than lead. Four hydrogen nuclei combine to create a helium nucleus and release a large amount of energy.

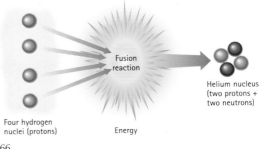

Four hydrogen nuclei (protons)

Fusion reaction

Energy

Helium nucleus (two protons + two neutrons)

Energy for living comes from food. Food comes from plants that use energy from the Sun, or from animals that eat plants, so human life is powered by sunshine.

SUNLIGHT AND FOOD

Plants use energy from sunlight to create complex chemicals from water and carbon dioxide. These complex chemicals have energy locked up in the bonds that hold their atoms together. Animals that eat plants are called herbivores. Meat-eating animals, called carnivores, eat the herbivores. In this way, energy from sunlight is trapped by plants, and then is passed on to different animals. Animals digest their food and release energy by breaking down the chemicals in it. The animals use this energy to move around and carry out all the internal processes needed to maintain their lives.

KEEPING COMFORTABLE

Earth's atmosphere acts as a filter that absorbs energy from the Sun's rays. Temperatures near the Equator are as high as 45°C because it is closest to the Sun. Heat and light pass straight down through the atmosphere. Around the North and South poles, temperatures can sink to around –40°C. For most of the year, the Sun is low on the horizon. The Sun's rays are spread over a wider area, and much heat and light is absorbed as they travel at a shallow angle through the atmosphere. Humans thrive at 25°C, the average temperature midway between the poles and the equator.

ENERGY IN THE FUTURE

Most of the energy used for heating and powering transport comes from fossil fuels that formed millions of years ago. Reserves of these fuels are running out and some may be exhausted within 50–100 years. Scientists are working to produce power using nuclear fusion, the process that makes energy in the Sun. They hope to build a fusion reactor that uses two isotopes of hydrogen called deuterium and tritium. Experimental Tokamak fusion reactors use powerful electromagnets to compress a plasma of deuterium and tritium. When the temperature reaches 100 million °C, scientists expect the isotopes to fuse together to produce helium and enormous amounts of energy.

Most deserts are near the equator, where the Sun is high in the sky for most of the day. The Sun's heat evaporates water and splits rocks into parched and lifeless grains of sand.

The Arctic Sun stays close to the horizon for most of the year. All the surface water is frozen and plants cannot survive. Most polar animals are carnivores that depend on creatures that live in the sea.

◀ Russian tokamak fusion reactors such as this are used to research nuclear fusion. Magnetic fields heat and compress hot gases, turning them into plasma and keeping them away from the walls of the container. The reactor has reached the necessary temperature for nuclear fusion, but not at the same time as the other conditions needed to produce energy. Scientists hope to have achieved controlled fusion by 2025.

SEE ALSO

14–15 Earth's atmosphere,
190–1 Light energy,
200–1 Work and energy

THE ELECTROMAGNETIC SPECTRUM

The electromagnetic spectrum is the full range of electromagnetic radiation and includes radio waves, heat and light rays, X-rays and gamma rays.

Mobile telephones send and receive very-high-frequency (VHF) radio signals.

Electromagnetic radiation is a form of energy that travels at the velocity of light, which is 300,000 km per second. As it travels, its energy switches back and forth between electric and magnetic fields. As one field increases in strength, the other decreases. The rate at which this exchange happens is called the frequency of the radiation. Different sorts of electromagnetic radiation have different frequencies. Radio waves have lower frequencies than light rays, for example, and blue light has a higher frequency than red. The frequency of electromagnetic radiation, measured in hertz (Hz), is the number of times in one second that the electrical field reaches its maximum value.

Scientists say that electromagnetic radiation travels in waves. This is because the strengths of the electric and magnetic fields vary up and down continually as they travel through space. The wavelength is the distance the wave travels in the time it takes the electric field to fall from its maximum value to its minimum value and then rise back to its maximum value. Because of this, the wavelength is the speed of light divided by the frequency. The signal from a radio station whose broadcast frequency is 1,200 kilohertz, or 1,200,000 Hz, has a wavelength of around 250 metres, for example.

RADIO AND MICROWAVE
Radio stations broadcast using frequencies in a range from 150,000 Hz to around 20 million Hz. Each station uses a particular frequency, so receivers tune to a given station by only accepting waves at the correct frequency for that station. Land-based television

Wavelength is the distance between two neighbouring peaks of an electromagnetic wave.

Travelling ray of light

Magnetic wave
Electric wave

▲ An electromagnetic wave is a combination of electric and magnetic fields that vibrate in planes at right angles to one another.

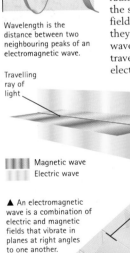

Gamma rays used to detect cracks in metal

X-rays used to look inside bodies

Ultraviolet rays used in a sunbed

Visible light from red to violet

Infrared rays in a heat-sensitive camera

High-frequency microwaves used to heat food in a microwave oven

Low-frequency microwaves used in radar

Ultra-high-frequency (UHF) radio waves for TV transmissions

Radio waves used in radio broadcasts

The electromagnetic spectrum stretches from radio waves at the lowest frequencies to gamma rays at the highest frequencies. Visible light occupies a narrow range of frequencies in the middle of the electromagnetic spectrum. The wavelength variations in this artwork are not to scale.

transmitters send signals between about 70 MHz and 800 MHz. (One megahertz is one million hertz.)

Satellite televisions work at even higher frequencies. These electromagnetic waves are captured by dish-shaped antennae that point towards the satellite.

Radars bounce radio waves off planes, ships and clouds to show their positions, which can be many kilometres away. They use wavelengths of a few centimetres. Doppler radar measures the speed of moving objects from the minute change in the frequency of the reflected waves.

Microwave ovens use wavelengths of a few millimetres, which correspond to frequencies of thousands of millions of hertz. The radiation heats food by causing water molecules to vibrate.

INFRARED LIGHT AND BEYOND

Infrared radiation has frequencies just lower than those of visible light. Its wavelength ranges from 1 millimetre to 750 nanometres. (A nanometre, or 1 nm, is one thousand-millionth of a metre.) Hot objects give off infrared radiation, and infrared is felt as heat. Visible light is the tiny part of the electromagnetic spectrum that human eyes can sense. The spectrum of colours stretches from red light at 770 nm to violet light at 400 nm.

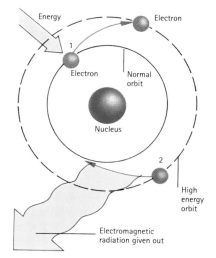

◄ The electrons of an atom move around its nucleus in orbits. The energy of each electron depends on its orbit. Energy can make an electron jump from its usual orbit (1) to an orbit with higher energy (2). When this happens, the atom takes in radiation of an energy that matches the energy gap between the two orbits. The atom gives out radiation if an electron falls from a high-energy orbit to a lower one.

Many aircraft have radar systems that bounce radio waves off clouds in the flight path ahead of the aircraft. A computer analyses the echoes to reveal the size and distance of clouds. The computer can also identify potential hazards, such as icy hailstorms.

The energy of electromagnetic radiation increases as the wavelength becomes shorter. Invisible ultraviolet rays cause sunburn and have shorter wavelengths (100–400 nm) than visible light.

X-rays have even shorter wavelengths, usually less than the diameter of an atom (0.1 nm). They penetrate flesh and bones.

Gamma rays have enormous energy and can be used to produce images of cracks deep inside pieces of metal.

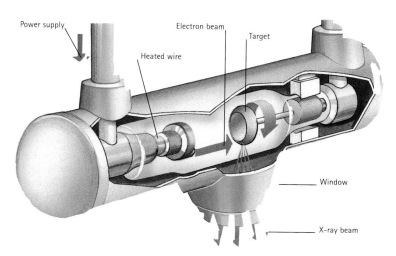

A hospital X-ray machine uses an X-ray tube to produce X-rays. A heated wire emits electrons, which are accelerated by electric fields towards a metal target. The collision ejects electrons from the metal atoms. Further electrons fall into the spaces left by the ejected electrons. As they do this, the electrons lose energy in the form of X-rays.

SEE ALSO
112–3 Atoms,
176–7 Light

HEAT TRANSFER

Heat energy moves around by conduction, convection and radiation. Transfer is from a region of higher temperature to a region of lower temperature.

Day-time: the land is warmer than the sea.

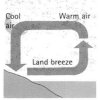

Night-time: the sea is warmer than the land.

Sea breezes are caused by convection currents. Warm air rises and is replaced by heavier, cooler air.

Heat travels from one atom to the next

Heated atoms vibrate and strike neighbouring atoms

Metal bar

Heat

Heat flows from the hot part of a solid object to the cold part. Rapidly vibrating atoms in the hot part speed up the vibrations of their cooler and slower neighbours.

One of the properties of heat energy is that it flows from hotter objects to cooler objects. If a hot drink at 80°C is placed in a room where the temperature is 25°C, the drink will gradually cool to 25°C. At the same time, the temperature of the air in the room will increase slightly.

CONDUCTION OF HEAT

Conduction carries heat through an object until its temperature evens out. Dip a metal spoon into a hot drink and the handle soon starts to feel warm. The hot liquid heats the part of the spoon that is in the drink. This increases the kinetic energy of its atoms and makes them vibrate with greater force. Heat travels up the spoon to the handle as vibrating atoms strike their neighbours and make them vibrate more vigorously.

CONVECTION OF HEAT

Convection carries heat through liquids and gases. Imagine heating a kettle of water on a stove. The heat source warms the water at the bottom of the kettle by conduction. As the water

gets hotter, it expands and becomes less dense. This causes the warm water rise, making room for cooler, denser water. The cooler water becomes warmer and less dense, so the convection process continues.

RADIATION OF HEAT

Radiation is a process that carries heat in straight lines through empty space. It is how the heat of a radiator can be felt without touching the radiator. This form of energy flow is also called radiant energy transfer. It is carried by infrared radiation.

All objects give off radiant energy. The hotter an object is, the more radiant energy it gives off. An object that is cooler than its surroundings gets warmer as it takes in more infrared than it gives out. At higher temperatures, the frequency range

The kettle absorbs heat directly from hot gases and radiant heat from the fire. The metal skin of the kettle conducts heat to the water inside it. Convection then spreads the heat through the water.

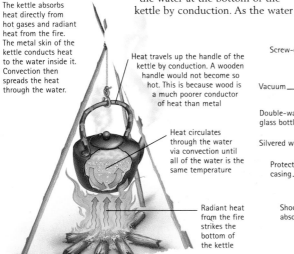

Heat travels up the handle of the kettle by conduction. A wooden handle would not become so hot. This is because wood is a much poorer conductor of heat than metal

Heat circulates through the water via convection until all of the water is the same temperature

Radiant heat from the fire strikes the bottom of the kettle

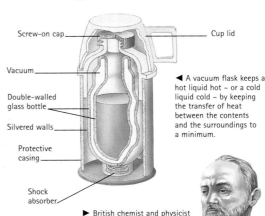

Screw-on cap

Cup lid

Vacuum

Double-walled glass bottle

Silvered walls

Protective casing

Shock absorber

◀ A vacuum flask keeps a hot liquid hot – or a cold liquid cold – by keeping the transfer of heat between the contents and the surroundings to a minimum.

▶ British chemist and physicist James Dewar (1842–1923) invented the vacuum flask, which is sometimes called a Dewar flask or Thermos flask.

of the radiation increases. The filament of a light bulb glows white-hot at around 2,000°C because some of its radiation is at frequencies higher than infrared.

HEAT TRANSFER EFFECTS

When the weather is hot and sunny, light-coloured clothing helps people to keep cool by reflecting sunlight and absorbing smaller amounts of radiant energy than dark colours. Dark objects also give off more radiant energy than light objects at the same temperature. This is why car radiators and the cooling panels on the backs of refrigerators are painted black: it helps them lose heat as rapidly as possible.

A cake taken from a refrigerator does not feel as cold as a bottle of milk at the same temperature. The reason is that heat flows from the hand into the bottle more rapidly than into the cake. The bottle is a better conductor of heat than the cake and so the hand cools more quickly. The cake is full of tiny air bubbles. Air is a poor conductor of heat and so lowers the rate of heat transfer from the hand. Air acts as an insulator.

THERMAL INSULATION

It is sometimes necessary to prevent heat from travelling from one place to another. Birds fluff up their feathers in winter, and people wear extra layers of clothing to trap an insulating layer of air next to their bodies. Cooks hold hot dishes with a cloth that insulates their hands from the heat. Metals are good conductors of heat, since they contain freely moving electrons that can carry heat energy. Plastic, glass and pottery are poor conductors of heat. This makes them good materials for insulating containers for hot drinks and food.

A computer's main circuit board becomes hot as electrical signals pass through its processor. While many processors are fitted with cooling fans, some have heat sinks with metal fins that disperse heat from the processor. The fins provide a large surface area for contact with the surrounding air, which convects heat away.

Solar panels — Thick turf on roof

Thick wooden walls

Small windows

Wood is a renewable fuel for heating and cooking

This energy-efficient house has an insulated roof and walls, as well as small windows that reduce heat losses. It uses renewable solar energy and wood fuel.

MODERN MATERIALS

Cups in drinks machines and containers from fast-food shops are often made from expanded polystyrene. This material is a mass of tiny plastic bubbles. The gas in the bubbles acts as a heat insulator. Plastic foam is often used to insulate cavity walls. This is because the foam prevents air convection in the space between the walls. People rescued from the sea are often wrapped in flexible foam blankets covered with a thin film of shiny aluminium. The foamed structure reduces heat loss, and the shiny surface on the inside reflects radiant energy back to the person's body. The shiny outer surface loses little heat by radiation.

▲ Friction raises the skin temperature of NASA's space shuttle to 1,500°C during re-entry. A layer of insulating ceramic tiles on the outside protects the cabin from overheating.

▲ This thermograph of a man's head shows small variations in temperature as differences in colour. Blue areas are the coolest. This type of image can be used to show where heat leaks from buildings.

Finned heat sink

Circuit board

SEE ALSO

112–3 Atoms,
157 Plastics

COMBUSTION

Combustion is another name for burning. It is a chemical reaction between a fuel and oxygen that releases energy as heat and sometimes light.

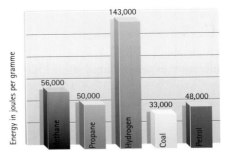

Different fuels release different amounts of energy during combustion. Comparisons are made by listing the amount of energy released by burning one gramme of fuel.

Mixture of gas and air

Air hole
Gas
Air

The blue inner cone of a Bunsen burner flame is a mixture of gas and air. Closing the air hole gives a bright yellow flame.

Candle wax melts easily to form a pool of liquid wax around the wick. Capillary action draws the wax up through the wick. The heat of the flame vaporizes the wax, which then burns as it mixes with air. Glowing particles of carbon make the flame bright yellow.

When a fuel burns, its molecules split into atoms that combine with oxygen to form other molecules. This process is called combustion. Most fuels are hydrocarbons, such as methane, CH_4. Hydrocarbons can burn in air, which is around 21 per cent oxygen. When they burn completely, the products are carbon dioxide gas, CO_2, and water vapour, H_2O.

If the supply of oxygen is limited, the products of combustion include toxic carbon monoxide gas, CO, and soot, which is a form of carbon.

Fuels burn extremely fiercely in pure oxygen. A mixture of ethyne gas, C_2H_2, and oxygen burns at 3,300°C, which is hot enough to melt and weld steel.

THE BURNING PROCESS

A mixture of fuel and air must be heated to a certain temperature before it ignites. The minimum temperature for burning depends on the type of fuel. When a spark lights a gas burner, the heat of the spark breaks fuel and oxygen molecules into atoms. These atoms then re-combine to form the products of combustion. As they do, the heat released by the reaction enables other fuel and oxygen molecules to break up and react together. This releases more heat, so the combustion process keeps itself going.

GAS, LIQUID AND SOLID FUELS

Various types of burners are used to burn fuel gases such as methane and propane. Most burners work in the same way as a laboratory Bunsen burner. The fuel gas enters the burner through a small hole called a jet. The stream of gas draws in air and forms a flammable mixture. The mixture combusts at one or more holes at the top of the burner.

Liquid fuels, such as paraffin, must vaporize before they can burn. Diesel engines, jet engines and oil-fired furnaces spray oil from tiny holes. The spray then vaporizes in the heat of combustion.

Solid fuels, such as coal and wood, burn by giving off flammable gases when they are heated. Chipped or powdered solid fuels burn faster than lumps.

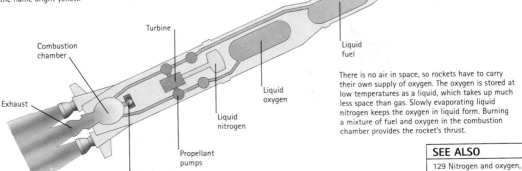

Combustion chamber
Turbine
Exhaust
Propellant pumps
Igniter
Liquid nitrogen
Liquid oxygen
Liquid fuel

There is no air in space, so rockets have to carry their own supply of oxygen. The oxygen is stored at low temperatures as a liquid, which takes up much less space than gas. Slowly evaporating liquid nitrogen keeps the oxygen in liquid form. Burning a mixture of fuel and oxygen in the combustion chamber provides the rocket's thrust.

SEE ALSO
129 Nitrogen and oxygen, 292–3 Rockets and the space shuttle

172

EXPANSION AND CONTRACTION

Substances expand when they are heated – their volume increases as temperature rises. This effect is reversible, since substances contract when they are cooled.

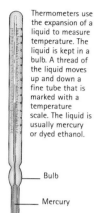

Thermometers use the expansion of a liquid to measure temperature. The liquid is kept in a bulb. A thread of the liquid moves up and down a fine tube that is marked with a temperature scale. The liquid is usually mercury or dyed ethanol.

Bulb

Mercury

At a constant pressure, doubling the absolute temperature of a gas doubles its volume.

The particles in a substance have more kinetic energy when the substance is hot than when it is cold. This means that the atoms, ions or molecules that make up the substance move faster as temperature rises. This motion tends to push the particles further apart, and increase the volume of the substance. At the same time, the density of a substance decreases, since the mass occupies a greater volume.

SOLIDS

A metre-long iron bar expands by around one-hundredth of a millimetre for each degree Celsius rise in temperature. One kilometre of railway line expands by almost 50 centimetres when a hot day follows a frosty night. This is why railway lines must have occasional sliding joints to prevent the rails from buckling. Rapid heating can cause glass dishes to crack as they expand unevenly.

LIQUIDS

For a given change in temperature, most liquids expand or contract around one thousand times more than solids. Water is unusual – it contracts as the temperature rises from 0°C to 4°C. This is due to a change in the structure of water.

This blacksmith is cooling a newly fitted iron tyre with water. When hot, the tyre fits easily around the wooden wheel. As the iron cools, it shrinks to form a tight fit.

GASES

Gases compress more easily than liquids and solids. This is because there is more space between their particles. For this reason, the volume of a gas is measured at a fixed pressure, usually the average pressure of Earth's atmosphere at sea level. When this is done, most gases expand by the same amount for each degree rise in temperature. In fact, the volume of a gas at constant pressure increases in proportion to its temperature above absolute zero, or –273°C.

As day and night alternate, Earth's atmosphere heats and then cools. Hot air rises because its density is less than cold air. Moving air makes the winds blow and controls Earth's weather patterns.

▶ For a 1°C change in temperature, solids expand or contract by different amounts. For this reason, engineers must make a careful choice of materials when they design a building or a machine.

◀ Flashing lights in cars use bimetallic strips made of two metals that expand at different rates. When the light is on, the strip gets hot and bends. This breaks the circuit that heats it. The strip then cools, closes the circuit and lights the bulb again.

Heating coil

Bimetallic strip

Contact open

Hot coil starts to cool

Lamp

Battery

Expansion ($^1/_{1,000,000}$ per cent per degree)

	26			
12				11
		8		
			4	
Tin	Aluminium	Glass	Pyrex	Concrete

SEE ALSO
116–7 States of matter,
166–7 Heat and light
from the Sun

THERMODYNAMICS

Thermodynamics is the study of the laws that control the direction in which heat will flow, and how energy changes from one form to another.

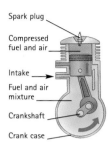

Spark plug
Compressed fuel and air
Intake
Fuel and air mixture
Crankshaft
Crank case

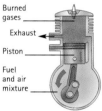
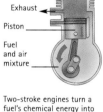

Burned gases
Exhaust
Piston
Fuel and air mixture

Two-stroke engines turn a fuel's chemical energy into mechanical energy and heat. The first stroke compresses and ignites the mixture. In the second stroke, burning gases push the piston down before more fuel and air blow them through the exhaust.

The study of thermodynamics began in the 19th century. Scientists used the results of experiments to draw up laws that describe how heat and energy behave in nature. These laws helped engineers to improve the design of machines such as steam engines, which change the chemical energy trapped in fuels into heat energy and then mechanical energy. As time went by, scientists realized that the same laws of thermodynamics apply to all processes, from the workings of diesel engines to the biological processes in living organisms.

THE FIRST LAW

The first law of thermodynamics states that energy can neither be created nor destroyed. One result of this law is that the amount of energy that flows into a device is matched by the amount of energy that flows out of it. Take the case of an electric lamp. Energy flows into the lamp in the form of electricity. The lamp produces heat and light as an electrical current flows through it, and the sum of the heat and light energy that the lamp gives out is equal to the amount of electrical energy that the lamp consumes. In other words, the amount of energy does not alter as the lamp glows – the energy simply changes from one form to another.

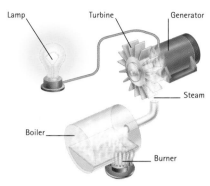

Lamp   Turbine   Generator

Steam

Boiler

Burner

This device uses heat from a burner to produce steam that drives a simple turbine and generator. Only a small proportion of the heat is converted to electrical energy.

THE SECOND LAW

The second law of thermodynamics states that all natural processes increase entropy. Entropy is a measure of the disorder of the Universe. One consequence of the second law is that heat flows from a hot place to a cooler place. In that way, the heat that was concentrated in a hot object becomes spread out and less ordered, so the process increases entropy. Heat does not flow naturally from a cold place to a hot place.

Entropy also plays a part in chemical reactions. Many reactions cause entropy to increase by turning chemical energy into heat, which spreads into the surroundings. Some reactions release gases, which are less ordered than liquids or solids.

▼ Only a small part of the heat from burning fuel in an engine turns into mechanical energy. The rest is lost in hot exhaust gases, through the radiator and to the air that flows around the engine. As a car moves, its tyres flex and become hot. Also, friction warms the air that passes over the car. When the brakes stop the car, they become hot and convert all the car's kinetic energy into heat.

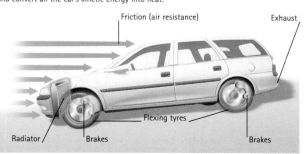

Friction (air resistance)   Exhaust

Radiator   Brakes   Flexing tyres   Brakes

THE THIRD LAW

The third law of thermodynamics states that there is a minimum temperature, called absolute zero. At this temperature, matter has its minimum possible heat energy and cannot become colder.

It is impossible to reach absolute zero, since any object at absolute zero would immediately absorb heat from any object around it. Nevertheless, calculations show that absolute zero is –273.15 degrees on the Celsius scale. Many thermodynamic calculations use temperatures on the thermodynamic scale, which sets absolute zero as 0K (zero kelvin). The equations that describe the properties of gases are an example of the use of thermodynamic temperatures. The volume of a gas at constant pressure varies in proportion to its temperature above absolute zero. Also, if the gas is kept in a fixed volume, its pressure increases in proportion to its thermodynamic temperature.

WORK AND ENERGY

Fuels such as petrol and diesel oil are called high-grade energy sources. They have this name because a small volume of fuel contains a large amount of useful chemical energy. If a motorist drives a car around a race track and returns to exactly the same place, all the chemical energy released by burning the fuel will have turned into heat. The engine wastes over 70 per cent of the fuel's energy as heat from the radiator and the exhaust. As the car travels down the road, friction converts

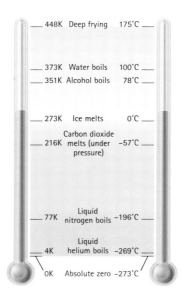

448K	Deep frying	175°C
373K	Water boils	100°C
351K	Alcohol boils	78°C
273K	Ice melts	0°C
216K	Carbon dioxide melts (under pressure)	–57°C
77K	Liquid nitrogen boils	–196°C
4K	Liquid helium boils	–269°C
0K	Absolute zero	–273°C

▲ In 1742, Swedish astronomer Anders Celsius (1701–1744) devised a temperature scale in which ice melts at zero degrees and water boils at one hundred degrees.

◄ The thermodynamic scale of temperature starts at 0K (zero kelvin), which is –273.15°C on the Celsius scale. Celsius temperatures are converted to kelvin by adding 273.15, so 0°C is equivalent to 273.15K.

kinetic energy into heat that warms the air and the tyres. Brakes change kinetic energy into heat energy. At the end of a journey, all the energy in the fuel has spread out into the surroundings and imperceptibly warmed up the world. This heat energy is called low-grade energy because it is spread out and cannot do useful work.

EFFICIENCY

The efficiency of a machine is the proportion of the energy input that it turns into useful work. The energy input to a car engine is the chemical energy released by its fuel, for example, and its useful work output is the kinetic energy that drives the car's wheels.

Thermodynamic calculations show that the maximum efficiency of an internal combustion engine cannot be greater than around 40 per cent. Electric engines are much more efficient: some convert more than 90 per cent of their electrical energy input into work. However, no more than 45 per cent of the heat produced in fossil-fuel or nuclear power stations is turned into electrical energy.

People become hot when they exercise. This is because the processes that cause muscles to move also produce heat.

Growing plants take part of the energy in sunlight and store it as chemical energy in their tissues.

The hull of a speedboat and the body of a dolphin are both streamlined. Their smooth lines help them to pass through water with little energy being wasted.

SEE ALSO

84–85 Muscles and movement, 158–9 Petrol and diesel engines, 200–1 Work and energy

LIGHT

Light is a form of electromagnetic radiation that can be detected by cells in the eyes of animals. Light can be scattered, reflected, refracted and diffracted.

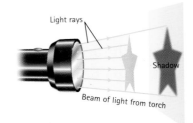

Light rays

Shadow

Beam of light from torch

When an object is placed in a beam of light, the object casts a shadow that matches its own shape. This is because light travels in straight lines through space.

Dutch physicist and mathematician Christiaan Huygens (1629–1695) developed techniques for making glass lenses for telescopes. He was the first person to describe light as a wave motion.

L ight is a form of electromagnetic radiation. The wavelengths of visible light are shorter than those of radio waves and infrared radiation, but longer than those of ultraviolet radiation and X-rays. Each colour of light has its own special wavelength. Visible light affects chemicals in the nerve endings in the back of the human eye. These nerve endings then send signals to the brain, which interprets the signals as light. Nothing travels faster than light. Its speed in a vacuum is about 300,000 kilometres per second, but it travels slightly slower through materials such as air, glass or water.

BEAMS AND RAYS

Light travels from its source as a series of electromagnetic waves. Scientists find it convenient to draw arrow-shaped lines, called rays, that show the direction of travel of these light waves.

A beam of light consists of groups of light waves that all travel in roughly the same direction. The beam of light from a torch spreads out slightly as it travels. The light from the torch appears fainter as distance increases, since its light waves are spread more thinly over a wider area.

The beam of light from a laser has almost parallel sides. This means that the light waves hardly spread out as they travel. This is why the beam from a laser can travel enormous distances before it becomes too weak to be seen.

SCATTERING

We only see light when it travels into our eyes. We cannot see a beam of light that travels past us through clear air. Light beams become visible when air is foggy. This is because the tiny water droplets in fog reflect some light out of the path of the beam. The light scattered to the side of the beam enters our eyes and makes the beam visible. The microscopic particles in smoke can also scatter light.

Solid substances that scatter light are said to be translucent. Greaseproof paper lets light pass through it, for example, but scattering disorganizes the light rays and prevents a clear image from being seen. Transparent materials, such as glass, allow light to pass through them without scattering, so a clear image can be seen.

Car headlights appear blurred in fog. This is because water droplets in the fog scatter light rays, so some light appears to come from around the headlights.

Visible light forms a spectrum of colours that ranges from red at longer wavelengths to violet at shorter wavelengths. The wavelength of infrared radiation is longer than visible light, and the wavelength of ultraviolet is shorter. The human eye cannot detect infrared or ultraviolet radiation.

← IR → Visible spectrum ← UV →

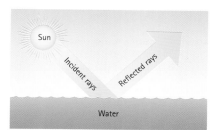

When the Sun shines on a choppy sea, it forms a broken, sparkling image. This is because the surface of the sea is in constant motion, and different parts of the surface reflect light at different angles.

When we look in a mirror, the image that we see is a true match of our appearance but with left and right reversed. The proportions and colour of the reflection are accurate.

REFLECTION

Few visible objects produce light; most are visible because they reflect some of the light that falls on them. Some objects appear bright or shiny because they reflect most of the light that falls on them. Objects that are bright but matte, scatter the light that they reflect. Dull objects absorb most of the light that falls on them and change it into heat. Mirrors produce clear images because they do not disturb the arrangement of the rays as they reflect.

REFRACTION

Light waves slow down when they move from air into a denser substance such as glass or water. The speed of light in water is three-quarters the speed in air. If a ray of light enters water or glass at an angle, it changes direction slightly. This effect, called refraction, causes lenses and other transparent objects to alter the appearance of things seen through them.

DIFFRACTION

When light waves pass through a hole or slit that has a similar width to their wavelength, they spread in all directions as they emerge. This effect is called diffraction. Waves that spread from separate openings can interfere with one another. Depending on the direction, some of the waves cancel each other out and some reinforce each other to produce patterns of light.

INFRARED AND ULTRAVIOLET

Over 200 years ago, scientists performed a simple experiment to show that light is a form of energy. They used a glass prism to split sunlight into a spectrum of colours ranging from red through green to violet. They then shone each colour in turn onto the bulb of a thermometer. The thermometer showed an increase in temperature as it absorbed energy from the light. The thermometer showed the presence of invisible infrared heat rays outside the spectrum in the dark space next to red light. Later, photographic film showed the existence of invisible ultraviolet radiation in the dark space next to violet light.

Dutch physicist Hendrik Lorentz (1853–1928) used mathematics rather than experiments to investigate light. He developed an electromagnetic theory of light to explain reflection and refraction effects.

A compact disc stores information as tiny pits in a sheet of aluminium. In a compact disc player, these pits reflect laser light of a single frequency onto a detector that converts the reflected pulses of light into an electrical signal. In normal light, reflected light rays interfere with one another. This makes some frequencies appear more strongly than others. This is why the surface of a compact disc sparkles with colour in white light.

Lighting effects help create atmosphere. Ultraviolet (UV) lamps produce a light that cannot be seen directly, but that makes a dye in clothing glow blue. The dye emits light at a lower frequency than the UV light that illuminates it. This is an example of an effect called fluorescence.

SEE ALSO

178–9 Reflection and absorption, 180–1 Refraction, 232–3 Electromagnetism, 256–7 Information technology

REFLECTION AND ABSORPTION

Light is reflected when it simply bounces off a surface. Absorption is when light changes into heat or some other form of energy when it strikes a surface.

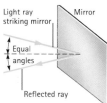

Light ray striking mirror | Mirror

Equal angles

Reflected ray

Light reflects off a mirror at the same angle at which the incoming ray strikes it.

The image that a mirror reflects is the opposite way round to the object. This right-handed woman appears to be left-handed. This reversal is called lateral inversion.

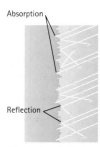

Absorption

Reflection

A light-coloured object reflects more light than it absorbs. Dark-coloured objects absorb more light.

▶ Soldiers use dull, dark clothing and paint lines on their faces to make themselves less visible in the battlefield.

A good mirror typically reflects around 98 per cent of the light that strikes its surface. The mirror absorbs the rest of the light energy and changes it into heat that slightly warms the surface. A dull black object absorbs almost all the light that falls on it. Most of the light energy changes to heat and very little reflects. A deep cave with soot-covered walls would be an almost perfect absorber: practically none of the light that entered the cave would come out again.

The Sun, stars and lamps are visible because they are hot and produce light. Most other objects are visible because they reflect some of the light that falls on them. White surfaces reflect light better than darker surfaces. Smooth, polished surfaces are the best reflectors. Rough surfaces tend to be poorer reflectors because they reflect light rays in all directions, and some light is absorbed in tiny cavities in the surface.

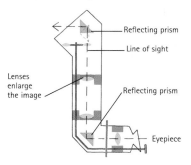

Reflecting prism

Line of sight

Lenses enlarge the image

Reflecting prism

Eyepiece

A simple periscope makes it possible to see over an obstruction. One type of periscope is used to see over protective lead walls in nuclear power stations.

MIRRORS
Ancient mirrors were made from sheets of polished metal, usually copper or brass. It was difficult to see a clear image in these mirrors because they were not perfectly flat and they absorbed much of the light that fell on them. Modern mirrors are made from flat sheets of glass with a thin reflective coating of silver or aluminium metal on the back. The glass protects the coating and ensures it is perfectly flat.

A ray of light that strikes a flat mirror behaves like a ball when it strikes a wall: the light bounces off at the same angle at which it made contact. The incoming ray is called the incident ray. The angle at which it strikes is called the incident angle. This is measured against an imaginary line sticking out from the surface at 90 degrees. The outgoing ray is called the reflected ray. The angle at which it leaves the mirror is measured in the same way and is called the angle of reflection.

REFLECTED IMAGES
Mirrors produce reflected images of objects that are placed before them. Plane mirrors, which are flat, form undistorted images of the objects they reflect. When a person looks into a mirror, they see an image of themselves. The image seems to be as far behind the mirror as the person is in front. The brain supposes that rays travel in straight lines and sees the image at the place the rays appear to come from.

FIBRE OPTICS

Fibre optics is a technology that uses strands of glass, called optical fibres, to channel light. Total internal reflection holds light within a glass fibre no matter how much the fibre is bent, and pulses of light can be used to carry telephone and data signals many kilometres through fibre-optic cables. Surgeons use fibre optics to look inside lungs and other body cavities. Tools called endoscopes have one bundle of optical fibres that leads light to the tip of the endoscope; a second bundle carries an image back.

Reflected light beam

Single glass fibre

Plastic sheath

Bundle of fibres

▲ This lamp uses fibre optics to produce a decorative effect. The optical fibres are bundled together at one end. Light passes through a coloured filter into the bundled end and emerges as pinpoints of brightness at the loose ends of the fibres.

TOTAL INTERNAL REFLECTION

Reflection can happen at the surface between two transparent substances, such as air and glass. This is how windows can reflect images of their surroundings. This type of reflection happens only for rays that travel at a shallow angle to the surface. Think about the surface of a swimming pool. When light from an underwater object hits the surface at right angles, or an angle of incidence of zero degrees, it passes straight through without changing direction. If a ray strikes the underside of the surface at an angle of 45 degrees, when it emerges into the air, the ray makes an angle of just 20 degrees to the surface. This is because an optical effect called refraction bends the ray closer to the surface. If a light ray hits the surface at an angle of incidence of 49 degrees, or 41 degrees to the surface, refraction bends it so much that the ray does not emerge

into the air. Rays that strike at angles less than 41 degrees to the surface do not pass into the air. Instead, they reflect downwards from the surface. This is called total internal reflection.

The minimum angle of incidence at which all light is reflected internally is called the critical angle and depends on the two materials that meet at the surface. For light passing from water into air, the critical angle is 48.8 degrees. From glass into air, the angle is 41.1 degrees.

An important feature of total internal reflection is that practically no light is absorbed by the reflecting surface. Fibre optics uses this property to carry light over great distances through optical glass fibres. As light passes through a fibre, total internal reflection can bounce it off the inside of the fibre's surface many millions of times without it becoming too weak to carry a signal or an image.

▲ The dark marks at the bottom of this photograph are reflections of markings under the surface of the water. From above, the surface sparkles as it reflects sunlight upwards.

The Moon shines because it reflects sunlight. The birds and tree appear as black silhouettes because they block the path of the rays of moonlight.

SEE ALSO

176–7 Light, 180–1 Refraction, 248–9 Telecommunications

REFRACTION

Refraction is the change of direction of light rays that can happen when the speed of light changes as it passes from one transparent substance into another.

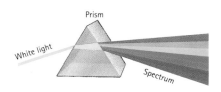

White light is a mixture of the colours, including red, orange, yellow, green, blue and violet. Dispersion of white light through a prism gives this spectrum of colours.

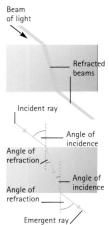

Beam of light

Refracted beams

Incident ray

Angle of incidence

Angle of refraction

Angle of incidence

Angle of refraction

Emergent ray

This ray diagram shows how refraction changes the path of light as it passes through a sheet of glass. When light passes from air into glass, the angle of incidence is greater than the angle of refraction. The direction changes again when the ray leaves the block, so the path of the emergent ray is parallel to the path of the incident ray.

Light rays travel in straight lines through any transparent substance, or medium. However, the speed of light depends on the medium. If light strikes the surface between two transparent media at right angles, it continues in the same direction. If it strikes at any other angle, the different speeds of light in the two media cause the light rays to change direction. This change of direction is called refraction.

REFRACTION AND WATER

When an underwater object is seen from outside the water, its appearance becomes distorted. This is because refraction changes the direction of the light rays that come from the object. When these rays enter the eyes of an observer, nerves in the eyes send signals to the observer's brain. The brain then constructs a picture based on where the rays appear to have come from. It does this without accounting for the effects of refraction, so the object's appearance is distorted. Looking at a straw in a glass of water, light rays from the part of the straw that is under water refract at the surfaces between the water and glass and between the glass and air. The rays appear to come from closer to the surface than they are, and the straw looks bent. If the straw were viewed from underwater, the part above water would be distorted.

DISPERSION AND PRISMS

Different colours of light have different wavelengths. The extent of refraction of light depends on its wavelength. Shorter wavelengths are refracted through greater angles than longer wavelengths. Because of this variation, violet light refracts more than green light, which refracts more than red light. The separation of wavelengths by refraction is called dispersion. Use of a prism, a triangular block of glass or plastic, maximizes the effect of dispersion. When white light enters a prism at the correct angle, dispersion splits it into a spectrum of different colours.

DISPERSION AND RAINBOWS

Rainbows appear when raindrops refract and reflect strong sunlight. Light passes into the raindrops, reflects off the back of them and passes out. Each wavelength of light is refracted through a slightly different angle on the way into the drop and then on the way out. Each raindrop reflects all the wavelengths of light, but an observer only sees colours from drops that are in the correct position to refract light into the observer's eyes. This is why the rainbow that each observer sees depends on that person's position.

DISPERSION IN LENSES

Refraction alters the appearance and position of objects. The images we see are not the same as the objects that cause them. Lenses use this effect to produce magnified images, and lenses are usually made from glass or transparent plastic.

More than 300 years ago, scientists discovered how to make telescopes and microscopes by fitting several glass lenses together inside tubes. They soon found

▲ The straw appears crooked because refraction bends light rays that come from the submerged part (see left). The shape of the glass increases the effect.

that dispersion limited the power of their lenses. The problem arose because lenses use refraction to produce magnified images. Each wavelength of light refracts differently, so the image that is seen is a combination of different coloured images, each of a slightly different size. Because of this, early microscopes and telescopes produced images that were surrounded by fuzzy coloured fringes. This effect is called chromatic aberration.

The greater the power of magnification, the more extreme the effects of chromatic aberration. The point is reached where magnified images become meaningless jumbles of overlapping coloured shapes. Modern optical instruments reduce the problem of chromatic aberration by using combinations of lenses made from different types of glass. One part of the lens combination produces a magnified image with chromatic aberration; the other part corrects the aberration. Some lenses are made from pieces of different glass that fit snugly together.

THE REFRACTIVE INDEX

Refraction happens because light travels at different speeds in different media. The greater the difference in the speed of light in two media, the greater the effect of refraction at the boundary between them. Scientists use a number called a refractive index to measure the refracting power of a

Medium	Refractive index
Acetone	1.36
Air	1.00
Benzene	1.50
Crown glass	1.52
Diamond	2.42
Ethanol	1.36
Flint glass	1.66
Polystyrene	1.59
Quartz	1.46
Sodium chloride	1.53
Water	1.33

The refractive index of a medium, or transparent material, is the speed of light in a vacuum divided by the speed of light in that medium. Refraction happens when light passes between two media with different refractive indices.

substance. The refractive index of a given substance is the speed of light in a vacuum divided by the speed of light in that substance. Refractive indices are normally quoted for yellow light, which is in the middle of the visible spectrum, but the exact value of refractive index depends on wavelength.

Light travels fastest through a vacuum, so the refractive index of a material is always greater than one. The refractive index of glass, for example, is 1.52, and the value for diamond is 2.42. The refractive index of air is only slightly greater than one.

Refractive indices are important in the design of optical equipment. This is because a lens made from a material with a large refractive index will magnify more than a lens of the same shape but of a lesser refractive index.

Under certain weather conditions, ice crystals in the upper atmosphere refract and disperse light from the Moon, causing a spectacular halo effect.

Refraction bends light rays from underwater objects as they pass through the water's surface. In this picture, refraction of light through ripples in the pool gives a distorted appearance to the parts of these people's bodies that are under the surface of the water.

▲ The speed of light in a diamond is less than half its speed in air. Strong refraction and dispersion of white light through the angled faces of a cut diamond cause it to sparkle with colour as it moves.

SEE ALSO

182–3 Lenses and curved mirrors, 190–1 Light energy, 192 The speed of light

LENSES AND CURVED MIRRORS

Lenses are curved pieces of transparent material. Like curved mirrors, they produce images that are larger or smaller than the object placed in front of them.

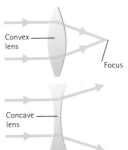

Convex lens

Focus

Concave lens

These ray diagrams show the path of light through lenses. A convex lens forms an image that is magnified by making light rays converge at a focus. A concave lens produces a diminished image by making light rays diverge from the lens's centreline.

▼ The centreline of a mirror or lens is called its principal axis. Any ray that strikes a concave mirror parallel to its principal axis reflects through a point called the principal focus of the mirror.

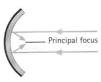

Principal focus

L enses and curved mirrors produce images that are larger or smaller than the object they come from. Ray diagrams help to visualize how this happens by showing the paths of a few light rays, particularly those that strike a mirror or lens parallel to its principal axis, or centreline.

CONCAVE MIRRORS

A concave mirror curves inwards and reflects light towards the inside of the curve. Rays that strike the mirror parallel to its principal axis all reflect through a single spot called the focal point. The distance of the focal point from the mirror is called the focal length. The shorter the focal length, the greater the magnification. The focal length depends on the curvature of the mirror. A tightly curved mirror has a short focal length and strong magnification.

CONVEX MIRRORS

Convex mirrors are curved like concave mirrors, but they reflect on the outside of the curve. When parallel rays of light strike a convex mirror, they are reflected outwards. The focal point of a convex mirror is at the point behind the mirror that the reflected rays appear to come from. The shorter the focal length, the smaller the reflected image.

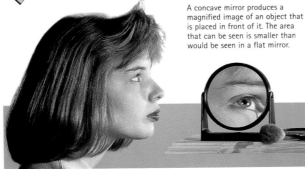

A concave mirror produces a magnified image of an object that is placed in front of it. The area that can be seen is smaller than would be seen in a flat mirror.

The reflective anti-glare coating on the lenses of these sunglasses make them act as convex mirrors. Parallel rays reflect outwards, forming a shrunken image.

MAGNIFICATION AND VIEW

Flat mirrors, called plane mirrors, form undistorted, unmagnified reflections. A concave mirror of the same size gives a magnified view of a smaller area – it is said to have a smaller angle of view. The magnification of a concave mirror is the number of times larger an image appears than the original object that is reflected. Concave mirrors are useful when a detailed view of a small area is required. This is why concave bathroom mirrors are helpful when shaving or applying make-up.

Convex mirrors have a wider angle of view than flat mirrors, but they produce diminished images. A convex mirror reflects more of the scene in front of it but with less detail. Many shops have large convex security mirrors fitted on the walls high above the customers. They give wide-angle downward views that include parts usually hidden from sight. They are often used in combination with video cameras, when they allow a wider angle of view to fit in frame. Car driving mirrors are slightly convex, so that the driver can see more of the scene behind the car.

LENSES

The optics of lenses work are similar to the optics of mirrors, but a lens affects the view of an object behind it while a mirror affects the view of an object in front of it.

Lenses work by refracting light at their surfaces. Refraction causes the rays to bend as they enter a lens from the air. The rays bend again as they pass out through the opposite side of the lens. Lenses are usually circular in shape and are made from glass or clear plastic. Convex lenses are thicker in the middle than at the outer edge. Concave lenses are thinner in the middle than at the outer edge.

CONCAVE LENSES

Concave lenses are also called diverging lenses because they bend parallel rays away from each other. The focal point of a concave lens is at the point behind the lens that light rays appear to come from. Objects appear smaller through a concave lens, just as they do in a convex mirror. Increasing the steepness of the curvature decreases the focal length of concave lenses and makes the image even smaller.

The most common use for concave lenses is in glasses for short-sighted people. Concave lenses are also used in compound lenses for cameras, telescopes and microscopes. They fit snugly against the

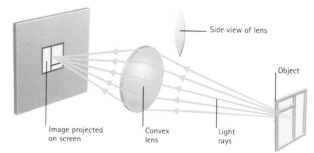

Side view of lens

Object

Image projected on screen

Convex lens

Light rays

▲ A convex lens can be used to focus light from an object onto a screen. The image is upside down and inverted from left to right.

outward curves of convex lenses that are made from other materials. Compound lenses reduce chromatic aberration, which is a blurring of images caused by different wavelengths of light being magnified to slightly different extents.

CONVEX LENSES

Convex lenses, or converging lenses, bend rays of light towards each other. They bend parallel rays towards a single point called the principal focus. The focal length is the distance from the principal focus to the lens. Convex lenses are used in magnifying glasses and glasses for long-sighted people. By concentrating sunlight at its focal point, a convex lens can be used to ignite paper and other flammable materials.

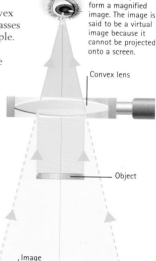

▼ The convex lens of a magnifying glass focuses light rays onto a point to form a magnified image. The image is said to be a virtual image because it cannot be projected onto a screen.

Convex lens

Object

Image

▲ A magnifying glass uses a convex lens to reveal details that would be difficult to see with the naked human eye.

The main mirror of the Hubble space telescope has a diameter of 2.4 metres. It is designed to collect as much light as possible from distant stars and galaxies.

183

SEE ALSO

178–9 Reflection and absorption

COLOUR

The sensation of sight is the result of light stimulating receptors in our eyes. The colour that is seen depends on the mixture of wavelengths present in light.

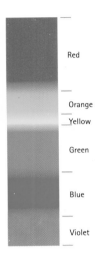

The visible spectrum is a continuous gradation of colours ranging from red through orange, yellow, green and blue to violet.

Red

Orange

Yellow

Green

Blue

Violet

▼ When viewed in white light, a white object reflects all wavelengths equally. A green object absorbs red, orange, yellow, blue and violet light; it reflects green light. A red object absorbs orange, yellow, green, blue and violet light; it reflects red light. A black object absorbs all wavelengths equally.

Visible light is electromagnetic radiation that has wavelengths in the 390 to 740 nanometre range. Receptors in the retina of the human eye detect the overall strength of light, and the balance of frequencies in light. These receptors produce nerve impulses that the brain interprets as brightness and colour. Light at 575 nanometres, for example, gives the sensation of greenish-yellow.

WHITE LIGHT

The light that comes from the Sun and from electric lamps is called white light because it does not appear to be coloured. When a beam of white light is shone through a prism, it emerges as bands of coloured light. These bands make up a spectrum of colours that are arranged in the order red, orange, yellow, green, blue and violet. The wavelengths of the colours decrease from red to violet. Mixing these colours of light together gives white light.

DESCRIBING COLOURS

Colours are described in terms of three qualities: hue, saturation and luminosity. Hue refers to the wavelengths present in coloured light. Saturation indicates the strength of a colour. Red, for example, is a more saturated version of pink. The third quality – luminosity – is the amount of energy in electromagnetic radiation. If the strength of the light falling on this page increases, for example, the luminosity of all the colours on the page will increase.

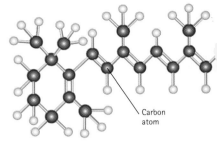

Carbon atom

A beta-carotene molecule has a chain of alternate single and double bonds between carbon atoms. This makes beta-carotene absorb blue light and reflect orange.

SOURCES OF COLOUR

A few light sources produce coloured light directly. The low-pressure sodium lamps that light some roads are an example. An electrical discharge causes the sodium vapour in the lamp to glow orange.

Sunlight and electric light appear white because they contain the full range of visible wavelengths in approximately equal proportions. Most sources of coloured light work by removing certain ranges of wavelengths from white light. A stop light glows red because it has a filter that absorbs all the wavelengths of light from a white lamp except red.

Coloured objects absorb certain ranges of frequencies from white light and reflect others. The reflected wavelengths cause the colour that is seen. A red tomato, for example, absorbs all the colours in white light except red. Some tomatoes are deep red and others are red-orange. These differences in hue result from variations in the mixtures of wavelengths reflected by each tomato. A piece of white paper reflects all wavelengths. A black object absorbs all the light that falls on it.

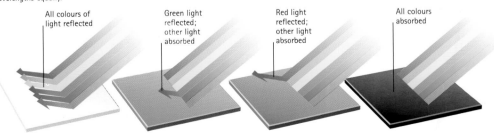

All colours of light reflected

Green light reflected; other light absorbed

Red light reflected; other light absorbed

All colours absorbed

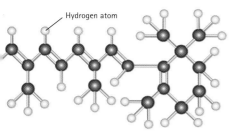

Hydrogen atom

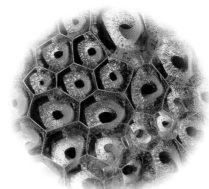

The skin of a soap bubble may be thinner than the wavelength of visible light. Some wavelengths of light cancel out when white light reflected from the back of the film interferes with white light reflected from the front of the film. The colour of the light depends on the thickness of the film and the angle from which it is viewed.

COLOUR AND ELECTRONS

The electrons in atoms and molecules move in orbits that each have a different energy. Light has an energy that depends on its wavelength. When an object absorbs light, electrons jump directly from lower energy levels to higher energy levels. These excited electrons then fall back down to lower energy levels in a series of smaller steps, releasing heat as they do so.

DYES AND PIGMENTS

Many living things contain coloured substances. The orange colour of carrots, for example, is caused by a natural chemical called beta-carotene. Molecules of such substances often contain chains of carbon atoms that are joined by alternate single and double bonds. The electrons in these bonds absorb visible light as they move between energy levels. Chemists use their knowledge of natural coloured substances to produce artificial dyes and pigments for dyeing textiles, and for the manufacture of coloured plastics, paints and inks.

OPTICAL EFFECTS

When sunlight falls on a patch of oil on a wet road, the colours of the rainbow are reflected by a thin layer of oil floating on the water. White light reflects from the upper and lower surfaces of the oil layer at the same time. When the reflected rays meet, some of their wavelengths join together and become stronger, while others cancel themselves out. This results in coloured light. The colour depends on the thickness of the oil layer and the angle at which it is viewed, so the colours swirl as the oil layer moves around. The same effect causes the colouration of soap bubbles and some insects' wings.

The lighting at a stage show helps create an atmosphere. A typical stage might be equipped with dozens of lights, each having a powerful electric lamp that shines white light through a coloured filter and a set of lenses.

A rainbow appears when strong sunlight shines from behind an observer onto rain falling in front of the observer. Raindrops act as prisms, and split white light into its component colours. The full spectrum of colours is usually seen, with red on the outside of the bow and violet on the inside. The exact position of a rainbow varies with the position of the observer.

SEE ALSO

168–9 The electromagnetic spectrum, 178–9 Reflection and absorption, 180–1 Refraction, 186–7 Colour mixing

COLOUR MIXING

Mixing colours together produces other colours. There are two different processes: one for mixing coloured light and one for mixing coloured substances.

Red, green and blue are the primary colours of light. Mixing those colours together gives white light.

Magenta, yellow and cyan are the primary colours of substances. Mixing them together gives black.

When a beam of light falls on a white screen, the colour that appears depends on the wavelengths of light in the beam. If a beam of light of another colour shines on the same spot, the colour that appears results from the combination of the colours of light in the two beams. This process is called additive mixing, since two colours of light form a third colour by adding their combinations of wavelengths together.

A different process happens when two coloured substances mix together. White light is a combination of all the visible wavelengths of light. When white light falls on a coloured powder, for example, the powder absorbs certain wavelengths of light. The colour that appears depends on the wavelengths that are reflected. If a second powder of a different colour is mixed with the first, the second powder will absorb some of the wavelengths that are not absorbed by the first. The colour

that results depends on what is left of white light after both powders have absorbed light. This process is called subtractive mixing, because the mixed colour is formed by each coloured substance in a mixture taking away a selection of wavelengths from white light.

PRIMARY COLOURS

There are two sets of primary colours that can be mixed to make any other colour. The primary colours for the additive mixing of light are red, green and blue. Mixing these three colours in equal amounts gives white light. Mixing pairs of these colours in equal amounts gives the secondary colours: red and green give yellow light, green and blue give cyan light, which is a similar shade to turquoise, and red and blue give magenta light. Any colour can be produced by the correct mixture of the three primary colours.

The primary colours for the subtractive mixing of paints and other coloured substances are magenta, yellow and cyan. The secondary colours for subtractive mixing are red, green and blue.

MAKING PAINT

Paints are coatings that are used to decorate and protect surfaces. A typical gloss paint is made by first mixing natural oils and resins called alkyds. These are the binder materials that gradually harden in air after the paint has been applied. Thinner is added to make the mixture easier to pump through a filter that removes any solid particles from the blended liquids. Pigment is mixed into the binder blend in a powerful mixer called a disperser. More thinner is added before the blend is pumped through a bead mill. Bead mills have rotating discs and glass beads that grind together and crush pigment particles between them. The final adjustments are done in a holding tank before the paint is packaged in cans or drums.

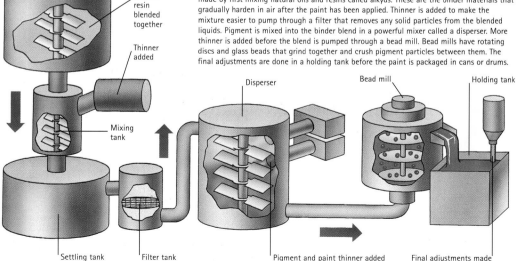

Oil and resin blended together

Thinner added

Mixing tank

Disperser

Bead mill

Holding tank

Settling tank

Filter tank

Pigment and paint thinner added

Final adjustments made

PAINTS AND PIGMENTS

Paints consist of white and coloured solid particles, called pigments, mixed into a sticky liquid called a binder. The binder hardens as the paint dries and acts as a protective layer that holds the pigments.

Most pigments are metal compounds. The most widely used pigment is titanium dioxide, TiO_2, which is a white powder. Titanium dioxide is used to make paints, toothpaste and many other products. Other pigments are strongly coloured because they absorb specific wavelengths of light. Yellow and red pigments are often compounds of iron, for example, while blue pigments are compounds of cobalt. An enormous range of colours can be made by combining a few pigments.

DYES

Dyes are strongly coloured substances that are used to give colour to products such as textiles and inks. Unlike pigments, dyes dissolve in water and other solvents. Dyes are complex organic compounds.

Textiles are coloured by dipping them into vats that contain dye solutions. The solutions soak into the fibres and form chemical bonds with them. Chemicals called mordants are often added to help the dyes to attach firmly to the fibres.

People have used natural dyes for thousands of years. In traditional leather dyeing, as seen here in Morocco, animal hides are soaked with dye solutions in pits in the ground. The hides can be soaked for many days.

COLOUR PRINTING

Colour pictures printed in books and newspapers seem to contain all the colours imaginable. However, most printing machines use just three colours – magenta, cyan, and yellow – and black. Magenta absorbs green light, cyan absorbs red and yellow absorbs blue. Black absorbs all wavelengths. Together, these three colours can produce any other colour.

A printed colour picture consists of millions of tiny dots of the three colours together with black. Varying the sizes of these dots creates different shades. Seen from a distance, the dots blend together to give the patches of different colours that make up the whole picture.

TELEVISION SCREENS

The picture on a television screen or computer monitor is made up of tiny points of red, green and blue light. The light comes from phosphors, which are chemicals that glow when they are hit by electrons. Three electron guns at the back of the tube fire beams of electrons at the phosphors. The beams scan across the screen from side to side and top to bottom. Each beam scans one colour of phosphors and varies in strength as it goes. The variations make the phosphors glow stronger or weaker, depending on how much of each primary colour is needed at each point. From a distance, the viewer's eye blurs the light from the individual phosphors, so the viewer sees the colour that results by mixing the light from the three colours of phosphor.

▲ Factories that use dyes often make their own dye mixtures from a standard range of colours. Quality-control testing ensures that the correct shade is achieved every time.

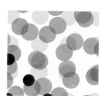

Colour printers use groups of magenta, cyan, yellow and black dots. From the normal reading distance, these dots appear as a blended shade of colour.

A Trinitron television screen produces colours by blending light from red, green and blue phosphor strips. Other types of screens use phosphor dots instead of oblong strips.

SEE ALSO

184–5 Colour, 250–1 Television and video

PHOTOGRAPHY AND FILM

Photography records permanent images of objects. Conventional cameras record images on light sensitive film. Digital cameras store image data in computers.

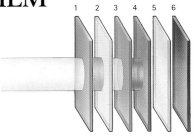

The six layers of colour film: layer 1 records blue light; 2 absorbs excess blue light; 3 records green light; 4 records red light; 5 is the plastic base; 6 is an opaque coating.

In 1826, French physicist Joseph Niépce (1765–1833) used bitumen-coated pewter plates to take the first-ever photographs.

British physicist William Fox Talbot (1800–1877) invented the negative-and-positive process that is still in use today.

In the 1880s, US inventor George Eastman (1854–1932) developed flexible-roll films to replace the glass plates that had been used earlier.

As early as 1515, Italian artist, engineer and scientist Leonardo da Vinci (1452–1519) described how an image could be made on the wall of a darkened room by letting light through a tiny hole in the opposite wall. This arrangement was called a camera obscura, meaning 'dark chamber' in Latin. The image could be recorded by sketching over it by hand.

THE FIRST PHOTOGRAPHS

During the 1820s, Joseph Niépce took the first photographs by placing flat, bitumen-coated pewter plates on the wall of a camera obscura. Although he would expose the plates to light for eight hours or more, the images were not clear.

In 1837, French painter Louis Daguerre (1789–1851) recorded images on metal sheets coated with silver iodide, a light-sensitive chemical. He would then treat these images, called daguerreotypes, with mercury vapour and common salt. Although mercury helped to prevent them from darkening in light, daguerreotypes had to be kept behind glass and viewed from a certain angle to see a clear image.

NEGATIVES AND POSITIVES

In 1841, William Talbot patented a process that steadily replaced Daguerre's method. Talbot's method used paper coated with silver iodide and a basic form of camera. The photographer would expose the paper to light for around 30 seconds. The paper would then be developed by washing it with chemical solutions. Silver iodide crystals that had been exposed to light would turn black, and unexposed crystals would be washed away. The image is called a negative because bright parts of the original scene appear dark. Several positive prints of each photograph could be made by shining light through the negative onto light-sensitive paper.

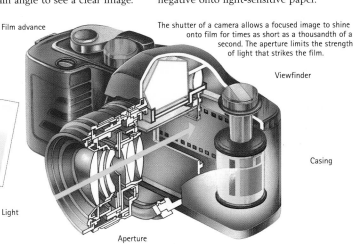

The shutter of a camera allows a focused image to shine onto film for times as short as a thousandth of a second. The aperture limits the strength of light that strikes the film.

Film advance

Viewfinder

Casing

▲ A colour negative shows the complementary colours of an image. When white light is shone through a negative onto photographic paper, a positive image develops.

Light

Lens

Aperture

Shutter

Film

MODERN PHOTOGRAPHY

In 1889, the American George Eastman introduced roll film for use in small hand-held cameras. Eastman's film consisted of a flexible strip of celluloid coated with a mixture of gelatine and silver bromide crystals. Winding the strip through the camera allowed several images to fit onto one roll. Enlarged positive images could be produced on specially treated stiff paper.

By the early 1900s, there were many professional photographers; by the 1920s, amateur photography was also becoming popular. Flash bulbs and films for colour transparencies were introduced in the 1930s. The first films for making colour prints were launched in 1941. By the 1980s, cameras could automatically focus the lens and expose the film to the correct amount of light. Photographic laboratories now use machines that develop film, and print colour photographs of any size.

MOVING PICTURES

Moving pictures for projection in cinemas are recorded on cine film. A cine camera takes 24 separate pictures every second on a long strip of film. The film is developed to produce transparent positive images. At the cinema, the film passes through a projector, stopping at each image for $\frac{1}{24}$th of a second. A powerful light shines through the film and lenses focus a large image onto the screen.

Laptop computer

Digital camera

DIGITAL CAMERAS

Digital cameras have a lens system and aperture just like ordinary cameras. Instead of film, they use a light-sensitive plate made of semiconducting materials to detect images. Electronic circuits scan the plate, and convert the information into a binary digital code. A computer memory chip records the data, which can then be passed to a larger computer that has software to adjust the colours and contrast of each image. Some can even move parts of an image around and add special effects, such as distortions and colouring. A printer can reproduce the image on any material – from paper to T-shirts.

Toshi Kaqnda created a TV station that broadcasts on the Internet. He uses a laptop computer to edit images from a digital camera before sending them to his file server in Japan for inclusion in the latest programme. This technique provides news that is just a few minutes old when broadcast.

▲ A digital camera looks similar to a conventional camera. The number of separate photos it can take depends on the size of its memory chip.

▲ The film for a typical feature movie is around 2.5 kilometres long. Each frame stays on the screen for only $\frac{1}{24}$th of a second. The human eye blurs these frames into a smoothly moving image.

▶ Futurescope at Poitiers in France, has a screen that surrounds the audience. Films can be shown as separate scenes side by side or as one gigantic image.

SEE ALSO

176–7 Light, 182–3 Lenses and curved mirrors, 244–5 Conductors, 254–5 Computers

LIGHT ENERGY

Light energy is a form of electromagnetic radiation.
Depending on circumstances, light can behave like
a wave or like a stream of particles.

When artists and scientists want to describe an object, many of them start with its appearance. What they see depends on how the object reflects light and how their eyes detect that light. Obviously, it is impossible to look at light in the same way. Instead, artists take note of the way that light illuminates objects, and scientists look at how light affects the substances that come into contact with it. When they do this, they find that light sometimes behaves in a similar way to waves on a pond. On other occasions, light affects matter as if it were a stream of particles.

▲ A solar cell consists of two wafers of silicon semiconductor that are sandwiched between two electrical contacts. The layers have slightly different compositions, and sunlight makes electrons jump from one layer into the other. These electrons flow out of the cell through one electrical contact. Other electrons flow into the cell through the other contact, and a current starts to flow.

Light energy spreads out from its source like the water ripples that spread out when a stone falls into a lake. This is an example of the wave-like behaviour of light.

THE PHOTOELECTRIC EFFECT

Photovoltaic cells, or solar cells, use the photoelectric effect to generate electricity from sunlight. The photoelectric effect occurs when visible light or ultraviolet radiation falls on the surface of certain materials. The light or radiation ejects electrons from the material, and can be measured by an electrical meter.

A curious detail of the photoelectric effect is that it only works with light that has a frequency above a certain value. Below that value, no electrons are emitted, no matter how strong the light that shines on the material. The minimum frequency for the photoelectric effect to happen depends on the type of material.

A single solar cell develops around 0.5 volts. The current available depends on the area of the cell. A solar cell 10 centimetres in diameter can supply a current of about 1.5 amps in bright sunlight. The solar cells in the panels pictured here are wired in series to provide greater voltage, and in parallel to give a stronger current.

PHOTONS – LIGHT AS PARTICLES

In 1905, German-born US physicist Albert Einstein (1879–1955) suggested an explanation for the photoelectric effect. His explanation was based on light behaving as if it were a stream of particles.

Electrons are held within any substance by their attraction to the nuclei within that substance. They need a shot of energy to break free from that attraction, just as a football needs a kick to get up a hill. In the photoelectric effect, that kick of energy comes from light. Einstein proposed that light is a stream of packets of energy, called photons. The energy of each photon depends on its frequency. If the frequency of light is too low, the 'kick' that a photon can give an electron is too weak, so the electrons cannot escape. Above a certain frequency, each photon has enough energy to eject electrons, as happens in the photoelectric effect.

LIGHT AS WAVES

Until Einstein suggested that light was made up of packages of energy, physicists had described light as being a type of wave motion. They came to this conclusion because of experiments that showed light behaving like sound waves or water waves.

In 1873, British physicist James Clerk Maxwell (1831–1879) drew up equations that described light as a combination of electric and magnetic fields that vibrate at right angles to each other, and also at right angles to the direction of travel of light.

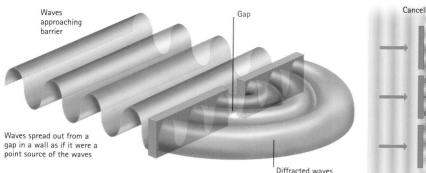

Waves approaching barrier

Gap

Waves spread out from a gap in a wall as if it were a point source of the waves

Diffracted waves

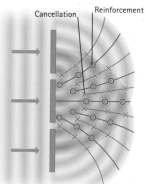

Cancellation   Reinforcement

DIFFRACTION

When water waves hit a wall with a small gap, waves spread out from that gap in a series of ever-widening circles, just as if the gap itself were a source of waves. This spreading from a gap is called diffraction. Sound waves also diffract. If a person walks past a house where music is playing and a window is open, the bass notes are heard first as the person approaches the window. The highest treble notes can only be heard close to the window because their high-frequency sound waves diffract less well. When light shines through a slit, it will not spread out unless the slit's width is similar to the light's wavelength.

INTERFERENCE

When water waves diffract through two side-by-side gaps, the gaps act as separate wave sources that are in step with each other. As the waves from these separate sources spread out, they interfere with each other. Where the crest of one wave meets the trough of another, the two waves cancel each other out, and the surface of the water remains calm. When two crests or two troughs meet, the result is an even larger crest or trough.

When light of a single wavelength shines through two tiny slits in a screen, it spreads out from the two slits and forms an interference pattern. Placing a screen in this pattern shows a series of light and dark lines where light waves reinforce each other in some places and cancel each other out in others. This is an example of light's wave-like behaviour.

PARTICLES AS WAVES

Shortly after Einstein proposed that light could behave as a particle, French physicist Louis de Broglie (1892–1987) suggested that the opposite could be true: that particles could behave like waves. He calculated that only the lightest particles would have wavelengths that could be tested in experiments. In 1924, he fired a beam of electrons at a crystal. The electrons diffracted through gaps between the atoms in the crystal, and left an interference pattern on a photographic plate. The electrons had behaved as waves.

Gaps in a barrier act as new sources of waves. The waves from different gaps interfere with each other as they meet. This diagram shows how the wave peaks reinforce each other.

▲ This solar-powered lighthouse buoy has four solar panels covered with photovoltaic cells. The cells charge a battery that powers the light at night.

◄ It is possible that future spaceships will have enormous sails. Photons from nearby stars will hit the sails, driving the ships forwards.

SEE ALSO
176–7 Light, 232–3 Electromagnetism

THE SPEED OF LIGHT

Light travels through a vacuum in straight lines at a constant speed of 299,792 kilometres per second. Nothing can travel faster than this speed.

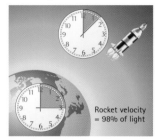

Rocket velocity = 10% of light

Rocket velocity = 98% of light

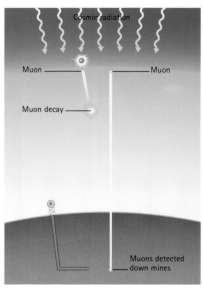

Cosmic radiation

Muon — — Muon

Muon decay ——

Muons detected down mines

Cosmic rays produce unstable particles, called muons, that travel near the speed of light. Without relativity, they would decay after only 600 metres. In fact, they reach deep underground because relativity slows their decay.

Strange things happen close to the speed of light. To an observer on Earth, a rocket travelling at nearly the speed of light would appear to have shrunk in the direction of flight and a clock on the craft would seem to move slowly. A passenger on the rocket would not be aware of these changes.

▼ The mass of an object increases rapidly as its speed approaches that of light. Einstein's equations predict that the mass of any object would become infinitely large at the speed of light. This makes it impossible to reach the speed of light, so faster-than-light craft will probably only ever exist in fiction.

Light travels 299,792 kilometres in one second through a vacuum. This speed has the symbol c. At this speed, light takes about eight minutes to travel 150 million kilometres from the Sun to Earth. It takes 4.3 years to reach Earth from the nearest star, Proxima Centauri.

EINSTEIN AND RELATIVITY

Einstein produced theories of relativity that predict strange effects near the speed of light. He did this by considering what he called 'inertial frames of reference.'

Imagine a fly buzzing forwards along a train at 2 metres per second. If the train were travelling at 100 metres per second, trackside observers would measure the fly's speed as 102 metres per second in their inertial frame of reference, while observers on the train would measure its speed as 2 metres per second in the inertial frame of reference of the train. At these speeds, the frames of reference tally.

Things would change if the train were capable of travelling at 1 metre per second slower than the speed of light. Observers in the frame of reference of the train would still see the fly moving forward at 2 metres per second. Observers by the track, however, would not see the fly moving at 1 metre per second faster than the speed of light. Many things would seem strange. The fly and the train would appear squashed to a fraction of their length, and the fly would be moving only slightly faster than the train – certainly slower than the speed of light.

Even at half the speed of light, a 100-metre-long train with a mass of 1,000 tonnes would seem to observers to be 87 metres long but 150 tonnes heavier. They would also say that the clocks on the train were running slow. The effects of relativity are bizarre, but they have all been demonstrated for small particles.

SEE ALSO

176-7 Light, 300-1 Space, time and relativity

LASERS

Lasers are devices that produce intense beams of light. The electromagnetic radiation in laser light has a single wavelength and the waves all vibrate in step.

In 1960, US physicist Theodore Maiman (b.1927) built the first working laser from a cylinder of artificial ruby.

Laser light

Ordinary light

Unlike ordinary light, photons of laser light all have the same frequency and are synchronized.

The letters that make up the word 'laser' stand for *l*ight *a*mplification by *s*timulated *e*mission of *r*adiation. An atom gives out a photon of light if an electron in the atom falls from a higher energy level, or excited state, to a lower one. In most cases, excited electrons give out light in this way of their own accord. This is called spontaneous emission. In a few cases, the properties of the excited state prevent electrons from giving out light unless they are triggered by another photon of light. This process is called stimulated emission. A stimulated photon has the same wavelength as the photon that triggered its emission and the two photons vibrate in step. Photons that have the same wavelength and vibrate in step are said to be coherent. It is the coherency of laser light that prevents a laser beam from spreading and makes it so intense.

TYPES OF LASERS

All lasers have two things in common. They contain a material that can be 'pumped' to an excited state but that does not emit light spontaneously. They also have a source of light or electrical energy to pump that material to an excited state.

A computer-controlled laser moves across a stack of fabric. The beam fires downwards, and cuts out the shapes that are later stitched together to make garments.

The first laser, built in 1960, was a ruby laser. This type of laser contains a rod of synthetic ruby with mirrored ends. Bursts of white light from a coiled flash tube around the rod excite atoms in the ruby. Once one of the excited atoms manages to emit a photon spontaneously, that photon stimulates other excited atoms to emit light as it reflects back and forth between mirrors mounted at the ends of the rod. One of the end mirrors is half-silvered so the laser beam can undergo multiple relections inside the tube and escape.

Other lasers use gas mixtures and dye solutions instead of ruby. In gas lasers, an electrical discharge provides energy to excite the gas atoms and start the laser action.

▶ The first laser produced light from a synthetic ruby. The ruby takes in ordinary light from a flash tube and emits it as laser light.

▲ Laser beams are extremely thin and straight. Here a laser beam is being used to check the accuracy of the direction of a long tunnel during its construction.

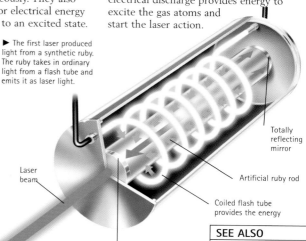

Laser beam

Totally reflecting mirror

Artificial ruby rod

Coiled flash tube provides the energy

Semi-silvered mirror

SEE ALSO

190–1 Light energy, 248–9 Telecommunications

FACTS AND FIGURES

THE SUN

Distance from Earth	149,503,000 km
Diameter	1,400,000 km
	(110 x Earth)
Mass	1.99×10^{30} kg
	(330,000 x Earth)
Surface gravity	38 x Earth
Composition	H 71%, He 27%,
	minute traces of
	other elements.

The **core** is within one-quarter of the Sun's total radius. Its temperature of nearly 16 million K results from the thermonuclear reactions that convert hydrogen into helium.
The **photosphere** is the Sun's visible surface that emits heat and light. Its temperature is 5,500 K.
The **chromosphere** lies outside the photosphere and is 10,000 km thick. Its temperature range is 4,000–50,000 K.
The **corona** is the outer part of the Sun's atmosphere. It is 70,000 km thick and its temperature is around 200,000 K.
The **solar wind**, consisting mostly of protons and electrons, flows into space from the upper part of the corona.

ELECTROMAGNETIC RADIATION

Broadcast radio		*low frequency*
Television		
Microwaves		
Radar		
Infrared radiation		
Visible light	Red	
	Orange	
	Yellow	
	Green	
	Blue	
	Violet	
Ultraviolet radiation		
X-rays		
Gamma rays		*high frequency*

WAVE–PARTICLE DUALITY

Electromagnetic radiation has properties of both waves and particles. Similarly, particles have wave properties. High-speed electrons have shorter wavelengths than light. Electron microscopes use this property to achieve magnifications greater than are possible with optical microscopes.

LIGHT AND MEDIA

Light can travel through a vacuum or through media such as air, glass and water. A clear image of an object can be seen through a **transparent** medium. Light passes through a **translucent** medium but a clear image of an object cannot be seen due to scattering. Light cannot pass through **opaque** substances.

OPTICS

Light waves travel in straight lines unless forced to deviate:
Diffraction is the spreading of waves as they pass through a narrow gap.
Dispersion causes white light to split into different colours during refraction.
Interference is the interaction of separate waves to give regions of high and low amplitude through mutual reinforcement and cancellation.
Reflection is the rebounding of light from a shiny surface or the boundary between two transparent media.
Refraction is the bending of the path of light as it moves across the boundary between two transparent media.

HEAT TRANSFER

Heat moves spontaneously from regions of higher temperature to regions of lower temperature. There are three mechanisms of heat transfer:
Conduction passes heat through a solid as vibrating particles bump into their neighbours. The solid itself does not move.
Convection moves heat through a fluid as changes in the fluid's density cause it to circulate in currents.
Radiation carries heat through a vacuum in the form of infrared light.

HEAT CAPACITY

The specific heat capacity of a substance is the energy required to increase the temperature of one kilogram of the substance by one kelvin.

Water	4,200 ($Jkg^{-1}K^{-1}$)
Aluminium	880
Iron	460
Lead	130
Glass	600

MELTING AND BOILING POINTS

The **melting point** (mp) of a substance is the temperature at which it changes from the solid to the liquid state. It is the same as the freezing point of the substance.
The **boiling point** (bp) of a substance is the upper fixed temperature at which it changes from liquid to gas. This phase change happens more slowly at lower temperatures by evaporation.

	mp (°C)	bp (°C)
Water	0	100
Aluminium	660	2,450
Lead	327	1,750
Oxygen	–219	–183
Helium	–270	–269
Ethanol	–114	78

KEY INVENTIONS

AD	
350	Gas lighting using marsh gas
1000	Camera obscura
1010	Optical lens
1550	Glass lens for a camera obscura
1590	Compound microscope
1608	Refracting telescope
1641	Liquid-in-glass thermometer
1663	Reflecting telescope
1714	Mercury thermometer and the Fahrenheit temperature scale
1758	A-chromatic lenses
1784	Bifocal spectacles
1792	Gas lighting using coal gas
1808	Practical arc lamp
1821	Modern hot-air central heating
1826	First photograph taken
1839	Calotype photography
1848	Kelvin temperature scale and the concept of absolute zero
1850	Photomicroscopy
1851	Mechanical refrigerator
1857	Silvered-glass mirror
1859	Battery-powered electric lamp
1878	Arc lamp for street lighting
1885	Incandescent gas mantle
1890	Telephoto camera lens
1892	Mercury-vapour lamp
1898	Osmium lamp filament
1899	Focusing camera lens
1900	Gas-fired room heater
1906	Colour film for movies
1906	Tungsten filament light bulb
1912	Ultraviolet microscope
1912	Modern colour film
1915	Gas-filled tungsten-filament lamp
1919	Flash photography
1920	Neon lighting
1932	Radio astronomy
1935	Fluorescent lighting
1935	Sodium-vapour lamp
1935	Colour transparency film
1936	Single-lens reflex camera
1938	Electron microscope
1945	Microwave oven
1951	Field ion microscope
1954	Solar-powered battery
1955	Fibre optics
1960	First practical laser
1966	Fibre-optic telephone cable
1974	Holographic electron microscope
1978	Scanning electron microscope
1981	Scanning tunnelling microscope
1984	X-ray laser
1985	Atomic force microscope
1986	Disposable camera
1990	Hubble space telescope placed in Earth orbit
1991	Nuclear fusion briefly achieved
1995	X-ray telescope in Earth orbit

Possible future developments will focus on using hydrogen as a pollution-free fuel and developing nuclear fusion as a 'clean' nuclear power source.

CHAPTER 7

FORCES AND MOVEMENT

orces are the pushes, pulls and twists that make things move faster or slower, change direction or change shape. Cyclists push on pedals to make their bikes move; birds pull worms out of the ground and cooks roll lumps of pastry into flat sheets. The force from a car's engine turns its wheels and makes the car accelerate; to slow down, friction in the brakes provides a force that decelerates the car.

The force of gravity makes objects fall towards the ground. Centripetal force makes the planets and satellites move in orbits and helps washing machines spin water out of wet clothes. Turning forces spin the shafts of motors and engines. They also twist doorknobs and tighten screws. Forces are often present even when nothing seems to be happening. A balanced seesaw does not move because there are two equal turning forces acting against one another.

When forces move, work is done as energy changes from one form to another. Muscles do work when a person lifts a box onto a shelf. The person converts chemical energy from food into potential energy as he or she lifts the box. The greater the rate of work, the greater the power. Forces, work and energy are what make things happen in the Universe.

FORCES

A force is a push or a pull that causes an object to accelerate, slow down or change shape. Forces can work in the same direction or against each other.

Load
Thrust

The load on a beam bridge is supported by vertical columns.

Load
Thrust

The load on an arch is carried by foundations on the river banks.

Cheetahs are the fastest ground animals, capable of bursts of speed of around 110 kilometres per hour when hunting. The muscles in cheetahs' legs provide the force that accelerates them forwards.

Distance

Time

Distance

Time

▲ The upper plot shows the distance covered against time for an object moving at constant speed. The speed is the gradient of the line. The lower plot is for an object that is decelerating. An upward curve would indicate an accelerating object.

The tension in stretched elastic, the pull of gravity on a raindrop and the thrust of a jet engine are all examples of forces. A force is an influence that can make an object start to move, slow down, change direction or change shape.

FORCES AND ACCELERATION

Acceleration is the most obvious effect of forces. Drop a ball from a height and the force of gravity will make it move faster and faster towards the ground. The ball accelerates because the force is acting downwards in the same direction as its motion. Throw a ball straight up into the air and the same force of gravity will make it move more and more slowly. This effect is called deceleration, which is negative acceleration. The ball decelerates because the force of gravity acts in the opposite direction to its motion.

Increasing the force on an object will increase its acceleration. The relationship between acceleration, force and mass is: force = mass x acceleration.

Forces can cancel one another out. A stone falling through water will reach a speed where the downward force of its weight is equal to the upward force caused by friction between the stone and water. The two forces cancel, so there is no acceleration. The speed remains constant.

PAIRS OF FORCES

Forces always occur in pairs. When a skier pushes backwards with ski sticks, for example, the backwards force on the sticks produces a force that pushes the skier forwards. The forces are equal in size but they act in opposite directions.

Sometimes, one of a pair of forces is less obvious than the other. The force that pulls a ball towards the ground is caused by gravity. Just as Earth attracts the ball, so the ball attracts Earth towards it. Because Earth's mass is so great, however, its motion towards the ball is negligible.

STATIC FORCES

There are situations where forces act without causing motion. If this book is lying on a table, the force of gravity is acting downwards on it. Although forces can make objects accelerate, the book is not moving because there is an equal and opposite force acting upwards on it. As the book presses down on the table, the table presses upwards on the book. The two forces are balanced and so the book does not move. The upwards force on the book is called the reaction force – it arises as a reaction to the weight of the book.

Backward thrust

Forward thrust

The miner's legs push back and downwards along the ground as they support his weight and produce a force that moves the wagon forwards.

▶ This man's legs push along the ground as he pulls the truck with his teeth. He sinks lower as the pulling force increases.

FORCES, WORK AND ENERGY

A force does work when it causes a mass to move. Work is the conversion of one form of energy into another. When a person walks, for example, muscles in their legs use chemical energy from substances in the blood to provide a force. The force does work as it makes the person move, and increases his or her kinetic energy.

THE UNIT OF FORCE

The unit of force is the newton, symbol N. Gravity exerts a force of 9.8 N on each kilogram of mass. A typical car engine can produce up to 4,500 N, while the four Rolls Royce RB211-524 jet engines of a Boeing 747 'Jumbo' jet develop a total thrust of more than 1,000,000 N when they are at full throttle for take-off.

MEASURING FORCE

Forces are measured by their effects on things. A spring gauge is a simple force-measuring device. It has a spring attached to a pointer and a hook. As a force on the hook stretches the spring, the pointer moves along a scale marked in newtons. The pointer stays still on the scale when the tension force in the stretched spring matches the force applied at the hook. The stronger the spring, the greater the range of forces this type of gauge can measure. Other devices use piezoelectric materials whose surfaces become charged when they are stretched or compressed.

◄ A potter's hands squeeze clay as it spins on a wheel. The force causes the clay to change shape.

▲ In this simple turbine of around AD50, the force of the escaping steam made the turbine rotate.

COMBINED FORCES

In most cases, several different forces act at the same time to produce what appears to be the result of a single force. Imagine a rower sitting still in a boat on a river. The boat floats because its weight is balanced by its buoyancy. The weight of the rower is matched by the reaction force of the seat. If nothing is moving, then none of these forces does any work.

When the rower grasps the oars and pulls them, the blades of the oars push against the water. This action creates a reaction force that pushes the boat forwards. At the same time, friction acts against the direction of travel. The acceleration or deceleration of the boat results from the difference between the push of the oars and the frictional force.

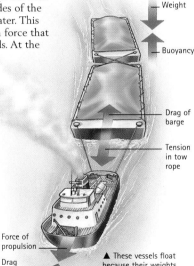

▲ These vessels float because their weights match their buoyancies. At constant speed, the tug's propulsive force matches the total drag on the tug and the two barges. Tension in the tow ropes pulls the barges forwards.

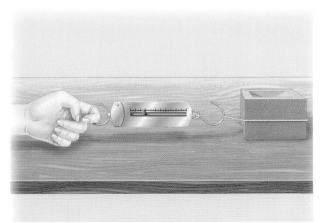

◄ The pointer of this spring gauge shows the tension in the hook as the brick is pulled along. When the brick moves at a constant speed, the spring gauge indicates the force of friction between the brick and the surface.

SEE ALSO

200–1 Work and energy,
202–3 Momentum,
218–9 Floating and
sinking

POTENTIAL AND KINETIC ENERGY

Potential energy is energy that is stored in an object as a result of its position or state. Kinetic energy is the energy that an object has due to its speed.

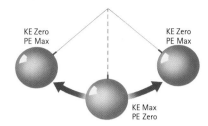

KE Zero
PE Max

KE Zero
PE Max

KE Max
PE Zero

A pendulum has maximum kinetic energy and minimum potential energy at the lowest point of its swing. Its kinetic energy falls to zero as its potential energy rises to a maximum at either extreme of the swing.

A ny moving object has kinetic energy. Moving cars, swinging hammers and spinning wheels are examples of objects that have kinetic energy.

Stationary objects also have a form of energy due to their height or state. The water in a mountain lake has potential energy, since it could flow to a lower level.

A hydroelectric power station uses the potential energy of high-level water to generate electricity. The potential energy first turns into kinetic energy as the water falls through pipes. Some of this kinetic energy transfers to turbo-generators that produce electrical energy as they turn.

During a dive, divers gain speed and lose height, so their potential energy becomes kinetic energy.

KINETIC ENERGY

The kinetic energy of an object depends on its speed and its mass. A parked car has no kinetic energy at all. It has kinetic energy when travelling along a road because it is moving. A train travelling at the same speed as a car has more kinetic energy because of its greater mass.

Doubling the mass of a moving object doubles its kinetic energy; doubling the speed increases the kinetic energy by a factor of four.

The standard unit of energy is the joule, symbol J. The kinetic energy (KE) in joules of an object is calculated using the equation $KE = \frac{1}{2}mv^2$, where m is the object's mass in kilograms and v is its speed in metres per second.

If a 40-kilogram child travels on roller blades at 5 metres per second, the kinetic energy is $\frac{1}{2} \times 40 \times 5^2 = 500$ joules. By comparison, a 4-gramme bullet travelling at 500 metres per second also has a kinetic energy of $\frac{1}{2} \times 0.004 \times 500^2 = 500$ joules. Although the child's mass in ten thousand times the mass of the bullet, the bullet need only travel one hundred times faster than the child to have the same kinetic energy. This is because kinetic energy depends on the square of speed.

Pulling back the plunger on a pinball machine compresses a spring. The force of the pull does work to store potential energy in the spring. When the plunger is released, the coiled spring expands and loses potential energy. It shoots a pinball and provides it with kinetic energy.

Spring   Plunger

▼ A bullet has a large amount of kinetic energy because of its high speed. On impact, this energy causes damage as it transfers to the target, here a chocolate bar.

POTENTIAL ENERGY

Potential energy is the energy that an object has due to its position or state. Objects that can fall downwards possess potential energy. Compressed, stretched or coiled springs also have potential energy.

One way to store energy as potential energy is to lift a brick up from the floor and place it on a table. The person or device that lifts the brick does work against the force of gravity to lift the brick upwards. Once on the table, the brick has gained potential energy equal to the amount of work done. If the brick were tied to a piece of string wound around the shaft of an electrical generator and then allowed to fall, its potential energy would turn into electrical energy during its fall.

When an object is lifted, the work done is equal to the downward force of gravity multiplied by the increase in height. Since the work done equals the increase in potential energy (PE), the equation for calculating the increase in potential energy in joules is $PE = mgh$, where m is the mass of the object in kilograms and h is the change of height. The constant g converts the mass of an object in kilograms into its weight in newtons. Its value is 9.8.

A one-kilogram bag of sugar on a two-metre-high shelf has a potential energy of $1 \times 9.8 \times 2 = 19.6$ joules more than it would have if it were on the floor. A five-kilogram bag of potatoes on the same shelf would lose $5 \times 9.8 \times 2 = 98$ joules of potential energy if it fell to the ground. It would have 98 joules of kinetic energy by the time it hit the floor.

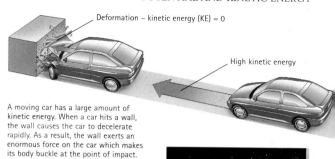
Deformation – kinetic energy (KE) = 0

High kinetic energy

A moving car has a large amount of kinetic energy. When a car hits a wall, the wall causes the car to decelerate rapidly. As a result, the wall exerts an enormous force on the car which makes its body buckle at the point of impact.

ENERGY CONVERSIONS

A 1,000-kilogram car moving at a speed of 108 kilometres per hour, or 30 metres per second, has 450,000 joules of kinetic energy. This energy has come from the chemical energy of the fuel, which is released during combustion in the engine. Without friction and air resistance, the car would be able to coast to the top of a hill almost 46 metres high. In reality, some of the energy is used in overcoming friction and air resistance. The brakes of a car work by changing kinetic energy into heat energy – 450,000 joules is enough to boil nearly five buckets of cold water.

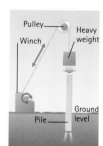
During a crash, a car stops quickly, but its occupants and any loose objects continue moving forwards.

Pulley

Heavy weight

Winch

Pile

Ground level

▲ A pile driver is a machine that rams steel girders into the ground. A simple pile driver has a winch that hauls a heavy mass to the top of a frame. When the mass falls, it gathers speed and kinetic energy. The mass delivers more than one million joules of energy to the pile with each blow.

◄ The three pile drivers in this picture are preparing foundations. The piles will carry the weight of the final construction.

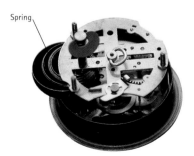
Spring
A mechanical clock is driven by the potential energy of a wound spring. An escapement mechanism releases this energy in precisely timed bursts that turn the hands.

SEE ALSO
200–201 Work and energy, 202–3 Momentum

WORK AND ENERGY

Work is done when a force moves through a distance.
Energy is the ability to do work. Power is the rate at
which work is done by converting energy.

A shot putter's throw does work on a heavy iron ball to give it kinetic energy. The energy to do that work comes from the chemical energy of substances in the blood that come from food.

Machines do work as they change energy from one form into another. For example, an electric fan does work as it changes electrical energy into the kinetic energy of moving air. The electric motor in the fan uses electrical energy to provide a turning force that spins the fan. The fan blades apply a force to the air that makes it move. At each stage, energy produces a force that moves and does work.

HUMAN MACHINES
When people climb stairs, they change food energy into potential energy and heat. It takes less energy to descend stairs because the force of gravity assists the descent. Some energy is still required to move forwards and keep the body upright.

Pulling a worm from the ground requires work. The exact amount of work depends on the friction between the worm and the soil, and the distance of the pull.

FIRST LAW OF THERMODYNAMICS
Thermodynamics is the study of energy and the ways that it converts from one form to another. The first law of thermodynamics states that the total energy content of the Universe is fixed. Energy cannot be created or destroyed: it can only be changed from one form to another. Machines are energy converters, so the first law of thermodynamics helps engineers to design effective machines.

WASTE HEAT
One of the problems with all machines is that they produce waste heat. Machines such as car engines have radiators that remove waste heat to prevent the engine from becoming too hot. Electric motors contain small fans that blow cold air across the hot coils of wire inside.

Machines take in one form of energy, and do work as they produce another form of energy. The waste heat does no useful work and is wasted energy.

DOING WORK AND USING FUEL
When a digger does 20,000 joules (20 kJ) of work on a load, it uses fuel that provides around four times as much heat energy when it burns in the digger's engine. To release 80 kJ of chemical energy as heat, the engine burns approximately 2 cm³ – almost half a teaspoonful – of diesel oil. Only 38 kJ of that heat is converted into mechanical energy. The rest is lost as heat in the exhaust and engine block. As the engine drives a hydraulic pump, the hydraulic fluid warms up and more energy is lost as heat. After a further small heat loss in the hydraulic rams, only 20kJ is left to do work.

Diesel fuel
80 kJ

Engine

Waste heat 52 kJ

Hydraulic pump

Waste heat 6 kJ

Hydraulic ram

Waste heat 2 kJ

Work lifting load 20 kJ

Machine

Energy

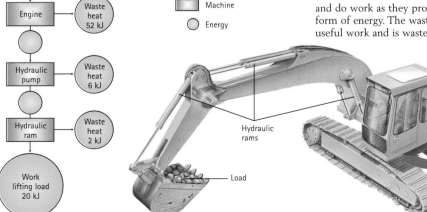

Exhaust waste heat

Engine

Hydraulic rams

Load

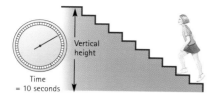

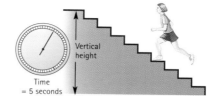

WORK AND ENERGY UNITS

Energy is measured in joules, symbol J. One thousand joules is one kilojoule (1 kJ), and one million joules is a megajoule (1 MJ). One kilojoule of heat energy can bring a teaspoonful of cold water to the boil or run a torch bulb for about 15 minutes.

A one-kilogram bag of sugar contains 21 megajoules of chemical energy. This amount of energy is the same as the increase in potential energy when three 70-kilogram people climb from sea level to the top of Mount Everest. In fact, each person eats the equivalent of more than one-third of a bag of sugar to do this work, because most of the chemical energy in food is converted into heat.

FORCE AND WORK

Work is done when a force acts on an object. One joule of work is done when a force of one newton moves a distance of one metre. If a force of 100 newtons is needed to push a pile of books across a two-metre-square table, the amount of work done is 100 x 2 = 200 joules. In this example, work is done to overcome the force of friction between the books and the surface of the table.

By walking upstairs, this girl uses half the power that she would use by running upstairs in half the time. The increase in potential energy is the same in both cases.

POWER

Sometimes people walk up flights of stairs; other times they run. In either case, the gain in a person's potential energy depends only on their mass and the vertical rise of the staircase. Climbing faster requires more effort: work is being done at a greater rate. Power is the rate of doing work or the rate of converting energy. It is measured in watts, symbol W. One watt is equivalent to doing one joule of work, or converting one joule of energy in the period of one second.

Power is energy divided by time. If a person who weighs 40 kilograms climbs a three-metre-tall staircase, the increase in potential energy is just under 1,200 joules. If the person makes this climb in five seconds, their power is 240 watts. By dashing up the steps in one second, the person's power is 1,200 watts – the same as a lawnmower engine. However, engines can develop power continuously. Human beings can only work at this level of power in short bursts.

Power is a useful measurement for comparing energy converters. For example, a 60-watt lightbulb takes in 60 joules of electrical energy every second. Its total output of heat and light is also 60 joules per second. A 120-watt television set converts energy at twice the rate of the bulb, and will cost twice as much to run for the same time.

Car
40,000 watts
(40 kilowatts)

Television
120 watts

Human
400 watts

A typical car engine is capable of producing around 40,000 watts of useful mechanical power. A television consumes around 120 watts of electrical power, while an average human being uses around 400 watts to run.

Work out

Heat out

Heat out

Food energy in

Heat out

◀ Bees obey the first law of thermodynamics. The total of the heat energy they produce and work they do is equal to the amount of energy they take in from food.

SEE ALSO

84–85 Muscles and movement, 102 Food and nutrition, 198–9 Potential and kinetic energy

MOMENTUM

Momentum is the mass of a body in kilograms multiplied by its velocity in metres per second. It features in Newton's three laws of motion.

British mathematician and physicist Isaac Newton (1642–1727) devised three laws to describe the motion of objects.

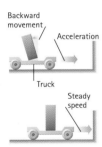

Backward movement

Acceleration

Truck

Steady speed

Forward movement

Rapid deceleration

▲ A load standing upright in a truck will be pulled forwards by its base if the truck accelerates. This can make a tall load tilt back or fall over. If the truck stops suddenly, the force of deceleration again acts through the base of the load, which can fall over.

► A dragster's engine develops enormous power to make it accelerate as quickly as possible. A vehicle's rate of change of momentum depends on its acceleration, which depends on the force the engine can apply to the road through the wheels.

More than 300 years ago, Isaac Newton used the ideas of momentum and inertia to formulate his three laws of motion. These laws describe and predict the effects of forces on objects. They give accurate predictions for most situations, but Einstein's relativity theory gives more accurate results for objects whose speeds approach the speed of light.

NEWTON'S FIRST LAW

The momentum of an object is equal to its mass multiplied by its velocity. If someone who weighs 50 kilograms runs at 10 metres per second, their momentum is 50 kg x 10 m/s = 500 kg m/s. In this case, the units of momentum are kilogram metres per second. When standing still, the person's momentum is zero.

Newton's first law of motion states that the momentum of an object stays constant until a force acts on it. For example, this page stays still until you turn it or the wind blows it. In both cases, a force makes it move and changes its momentum.

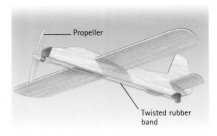

The second law predicts that maximum acceleration results when the plane has the smallest possible mass, and the propeller develops the greatest possible thrust.

SPEED AND VELOCITY

Momentum is defined in terms of velocity, not in terms of speed. It is important not to confuse the velocity of an object and its speed. Velocity joins together two pieces of information: the speeds at which an object is moving, and the direction in which it is moving. Velocity changes if either speed or direction changes.

If two cars travel side by side in a straight line at 50 kilometres per hour, they have identical velocities. If they travel at the same speed but in different directions, their velocities are not equal. If a third car travels in a circle at a constant speed, its velocity is constantly changing because it constantly changes direction away from a straight line.

NEWTON'S SECOND LAW

When a single force acts on an object, it makes it accelerate in the direction of the force. When throwing a ball, for example, force from the thrower's arm muscles accelerates the ball and increases its momentum. The greater the mass of the ball, the more difficult it is to accelerate.

Newton's second law of motion states that the rate of change of momentum of an object is proportional to the force that acts on the object. Since acceleration is the rate of change of velocity, mass times acceleration is the rate of change of momentum. Newton's second law is often written $F = ma$, where F is the force in newtons, m is mass in kilograms, and a is acceleration in metres per second squared.

NEWTON'S THIRD LAW

Newton's third law of motion states that whenever a force acts on one body, an equal and opposite force acts on some other body. The equal and opposite force is often called the reaction force.

When a spacecraft fires its rocket motor, the burning process in the combustion chamber causes hot gases to escape from the rocket's nozzle at high speed. Since the fuel and oxidant that feed the rocket have almost no momentum, the burning process must cause a 'backwards' force on the gas molecules to drive them out of the nozzle. The reaction force of the gases on the combustion chamber drives the spacecraft forwards. Since the mass of the craft is much greater than the mass of the rocket gases, the spacecraft accelerates much less than the gases for the same rate of change of momentum.

MOMENTUM AND INERTIA

Inertia is the tendency of an object to stay still or to move steadily in a straight line. Changing the momentum of an object requires that work is done against its inertia. It takes more effort to start cycling from a standstill than it does to keep moving at steady speed in a straight line. This is because cyclists must overcome their own inertia and that of the cycle to start moving. At constant speed, the cyclist need only overcome air resistance. Isaac Newton was the first person to realize that a force is needed to overcome inertia, and make objects accelerate or decelerate.

Skiers depend on Newton's third law to be able to move – a backwards push with the ski sticks produces an equal and opposite reaction force that pushes them forwards over snow.

CONSERVING MOMENTUM

When a gun fires a bullet, the force that acts on the bullet is equal and opposite to the recoil that acts on the gun. According to the second law, the rate of change of momentum must also be equal and opposite for the bullet and the gun. This means that the changes of momentum of the bullet and the gun must be equal and opposite, since both the firing force and the recoil force act for the same amount of time. In mechanics, 'opposite' is shown by a minus sign, so the sum of the equal and opposite momentum values for the bullet and gun is zero before and after firing. This is an example of the conservation of momentum.

When two balls collide and rebound from each other, their combined momentum before collision is equal to their momentum after collision.

THE THREE LAWS OF MOTION

Far away from Earth's gravity and its frictional forces, a spacecraft shows Newton's three laws of motion at work. The rocket motor fires gases backwards to produce a force that propels the craft forwards (3rd law). The spacecraft's acceleration is inversely proportional to the force from the motor and inversely proportional to its mass (2nd law). When the rocket switches off, the spacecraft continues at a constant velocity (1st law), flying in a straight line at a constant speed.

Acceleration proportional to thrust

Exhaust jet (backwards)

Acceleration proportional to mass

SEE ALSO

196–7 Forces, 198–9 Potential and kinetic energy, 292–3 Rockets and the space shuttle

RELATIVITY AND GRAVITY

Newton explained gravity as the force of attraction between masses. Einstein's theories state that masses distort the geometry of the space around them.

The theories of relativity developed by German-born US physicist Albert Einstein (1879–1955) explain observations of astronomy and physics that defy Newton's laws.

Earth's gravitational field attracts this falling car downwards. The car also attracts Earth upwards, but the effect is too small to be measured.

Acceleration due to Earth's gravity is 9.8 metres per second squared (9.8 m/s²). During one second, the speed of any free-falling object will increase by 9.8 metres per second. Acceleration due to gravity does not depend on mass: without air resistance, a half-brick would accelerate just as quickly as a whole brick.

All masses attract each other. This attraction is called the gravitational force or the force of gravity. The strength of this force between two objects depends on their masses: doubling either mass will double the force between the objects; doubling both masses quadruples the force. The force between two objects decreases in proportion to the square of the distance between them: doubling the distance between two objects reduces the force to a quarter of its original strength.

The force of gravity between two objects only becomes readily apparent when one or both of the objects has great mass. The force between two people at a distance of one metre from one another is only about one-millionth of a newton. Both people feel the gravitational pull of Earth, which has a mass of 6×10^{24} (six million-billion-billion) kilograms.

GRAVITY, MASS AND WEIGHT
At sea level, an object that has a mass of one kilogram is attracted to Earth by a force of 9.8 newtons. This attraction is the object's weight. The mass of the Moon is approximately one-sixth the mass of Earth, so a mass of one kilogram weighs one-sixth of its weight on Earth.

It is important to be aware of the difference between mass and weight. Mass is a measure of the amount of matter. It does not change from place to place. Weight is the force experienced by matter in a gravitational field. It varies with the strength of the gravitational field.

- Earth has a mass of six million-billion-billion kilograms. The gravitational pull of this mass is such that an object in free fall accelerates at 9.8 m/s² towards it.
- The Moon has a mass one-sixth the size of Earth's mass, so its gravitational pull is one-sixth as great.
- Jupiter is three hundred times as large as Earth. Its pull is three hundred times stronger, so free fall acceleration near Jupiter is an enormous 3 km/s².

The gravitational pull of a planet varies in proportion to its mass. Consequently, the rate of free fall acceleration varies in direct proportion to the mass of a planet.

ACCELERATION AND FREE FALL
The force of gravity on a mass of one kilogram is 9.8 newtons; the force of gravity on a mass of two kilograms is 19.6 newtons. By Newton's second law, either mass would accelerate at a rate of 9.8 metres per second squared if gravity were the only force acting. In fact, any mass would accelerate at this rate, since the force of gravity increases in proportion to mass. The constant *g* has the value of 9.8 metres per second squared. It is used to calculate the effects of gravity.

Free fall is when an object drops under the force of gravity alone. Free fall is rare on Earth, since air resistance opposes the pull of gravity on a falling object. This is why a feather will fall more slowly than a stone in Earth's atmosphere.

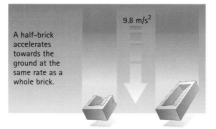

9.8 m/s²

A half-brick accelerates towards the ground at the same rate as a whole brick.

WEIGHTLESSNESS

An object is only truly weightless when it is in a zero gravitational field. There is a point between Earth and the Moon where Earth's gravitational field cancels out the Moon's field, and objects are weightless.

An orbiting spacecraft is accelerating towards the centre of its orbit under the force of gravity. This is why its occupants feel weightless – they are in free fall. This sensation can be felt closer to Earth in a plane that flies in a parabolic curve with downward acceleration g. A milder version of the same effect happens when a rollercoaster accelerates into a dip.

NEWTON AND GRAVITATION

In 1687, Isaac Newton published a law of gravitation that connected the force of gravity between two objects to their masses and the distance between them. It included a constant G, called the universal constant of gravitation. Newton's law is still used to predict the effect of gravity on objects, but it fails to explain how gravity works and why G has its value.

EINSTEIN AND RELATIVITY

In his Special Theory of Relativity of 1905, Albert Einstein stated that nothing – not even information – can travel faster than the speed of light. This created a problem with Newton's view of gravity, which required objects to exchange some sort of information at infinite speed in order to attract one another.

Ten years later, Einstein resolved this problem with his General Theory of Relativity. In this theory, Einstein

◄ A spacecraft in orbit around Earth constantly accelerates towards Earth under the pull of gravity. The craft's occupants feel weightless because they are in a type of free fall.

An astronaut, whose muscles are accustomed to the much stronger pull of the Earth's gravity, finds that it takes relatively little effort to jump high above the Moon's surface.

proposed that matter creates a distortion of the space that surrounds it. This is similar to the dip caused by placing a marble on a stretched sheet of rubber. In this distorted space, the shortest distance between two points is a curve. This is why a planet can bend the path of a passing object or even hold it in orbit – the object is simply following a straight line through space distorted by the planet.

Proof for the General Theory came from a total eclipse of the Sun in 1919. Astronomers observed stars that should have been obscured by the Sun. This proved that light from those stars had followed the curvature of space caused by the Sun's mass. Since light has no mass, Newton's law would not have predicted the path of light to be affected by gravity.

◄ Skydivers experience the full acceleration of gravity for only a few instants after they jump from an aircraft. As their descent becomes faster, the upward force of air resistance increases until it is equal but opposite to the downward pull of gravity. The speed of descent then remains constant at approximately 60 metres per second.

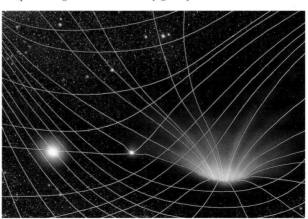

▲ Relativity describes gravity as a distortion of space, shown here as dips in a net. Although much heavier than Earth, the Sun (left) creates only a minor dip when compared with a smaller but more massive neutron star (centre), or an immensely massive black hole (right).

SEE ALSO

116–7 States of matter, 300–1 Space, time and relativity

RAMPS AND WEDGES

Machines are devices that make it easier to do work against a force. Ramps and wedges are examples of simple machines based on inclined planes.

Effort up ramp

Vertical effort

It takes less force to push a load up a sloping ramp than to lift the same load straight upwards to the same vertical height.

Machines do work as they move loads. The load moves when a force, called the effort, is applied to one part of a machine. It takes less effort to move a load with the help of a machine than without. The total amount of work done is the same, whether a machine is used or not.

RAMPS
Ramps are the simplest of all machines. They consist of a flat surface lifted up at one end to make an inclined plane. An example is a plank of wood with one end on the floor and the other end resting on a chair. It takes less effort to push a load up a ramp than to lift it straight upwards.

WEDGES
Wedges have two back-to-back sloping surfaces. The narrowest part of a wedge is where the two surfaces meet. Wedges are used for splitting objects made of wood or stone. A large hammer applies effort to the flat end of the wedge. The wedge directs this force outwards at right angles to its two sloping sides. The narrow end of the wedge drives its way into the object, and the force from each sloping face splits the object apart. An axe is a hammer and a wedge combined in a single tool.

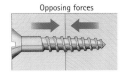

Opposing forces

When a screw holds two pieces of wood together, they are forced together as the screw tightens.

SCREWS
Screws are simple machines that hold things together. A screw consists of an inclined plane, called a thread, wrapped around a pointed cylinder. Effort is applied to a screw by using a screwdriver to turn its head. Rotating the head through a large angle moves the point of the screw forward by a short distance.

Bolts are similar to screws, but they are not pointed. The thread on the outside of a bolt fits into the inside thread of a nut. Some bolts multiply the turning force up to 40 times as they tighten with their nut.

MECHANICAL ADVANTAGE
When an effort is applied to one part of a machine, another part of the machine applies a force to a load. Most machines change a small effort into a large force that is applied to a load.

A number called the mechanical advantage of a machine is the ratio of the force on a load to the force of effort. The greater the mechanical advantage, the better the machine at multiplying effort. The total amount of heat and work that comes out of a machine is equal to the total work put into the machine.

Downward force

Sideways force

▲ A wedge converts the downward force of a hammer blow into opposing sideways forces that can be used to split logs and rocks.

The hairpin bends in this road allow it to climb the hill at a gentler gradient than if the road were straight. The course of the road is much longer than if it were straight, but it can be climbed more easily by cars and trucks.

SEE ALSO
196–7 Forces, 200–1 Work and energy

LEVERS AND PULLEYS

Levers and pulleys are simple machines. A lever is a combination of a bar and a fulcrum, or pivot. A pulley has a rope that passes over one or more wheels.

First–class lever

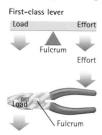

Pliers are pairs of first-class levers. The fulcrum is the pivot between the load in the jaws and the handles, where effort is applied.

Second–class lever

A wheelbarrow is an example of a second-class lever. The load is between effort and fulcrum.

Third–class lever

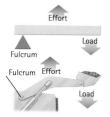

▲ The arm on the front of a mechnical digger is an example of a third-class lever. Effort from a hydraulic ram acts between the fulcrum and the load.

Effort applied to one part of a lever produces a force on a load at another part. Levers can act as force multipliers – the force that moves the load can be much greater than the effort applied to the lever. For example, a long thin metal bar called a crowbar can act as a lever and move enormous boulders that are too heavy to pick up unaided.

Two or more pulleys can also work together to multiply the force of the effort. Multiple pulleys not only multiply force, they also change the direction of the effort applied to one end of the rope.

LEVERS

A lever moves around a fixed point called the fulcrum. The distances of the load and the effort from the fulcrum affect how well the lever multiplies the force. A coin can be used as a lever to remove the lid of a can of paint. One edge of the coin fits under the lid and the rim of the can acts as a fulcrum. Pushing down on the free edge of the coin usually opens the lid. If the lid is too tight to open with a coin, the handle of a spoon will usually work. The effort acts further from the fulcrum, so the free end of the spoon moves a greater distance and applies a greater force at the end of the spoon under the lip of the lid.

SINGLE PULLEYS

A pulley is a wheel that turns on an axle. There is a groove around the rim of the pulley wheel that holds a rope. Pulling on one end of the rope moves a load attached to the other end. When one pulley wheel is used, the force at the load is the same as the effort pulling on the rope. A single pulley does not multiply the effort, so there is no mechanical advantage. The direction of the force moving the load is different to the direction of the effort.

MULTIPLE PULLEYS

When two pulleys work together, one is attached to a high fixed support, such as a beam, and the other is attached to the load. One end of the rope is attached to the fixed pulley. The rope passes under the load pulley and over the fixed pulley. Pulling on the free end of the rope supplies the effort to lift the load.

Using two pulleys halves the effort required to lift a load, so an effort of 100 newtons can lift a load of 200 newtons. Therefore the mechanical advantage of a two-pulley system is two – as the load moves one metre, the effort moves two metres. Since the work done is force times distance, the work put into the pulley – 100 x 2 = 200 joules – is the same as the work done on the load – 200 x 1 = 200 joules. In general, the mechanical advantage of a pulley system is equal to the number of pulleys the rope passes over.

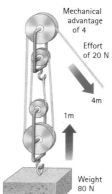

Mechanical advantage of 4

Effort of 20 N

4m

1m

Weight 80 N

Four pulleys working together multiply the force of the effort by four times. The distance moved by the effort is four times the distance moved by the load. Cranes and other machines use systems of pulleys and cables or chains to lift extremely heavy loads.

SEE ALSO
196-7 Forces, 198-9 Potential and kinetic energy

WHEELS AND AXLES

Wheels are simple machines. A wheel is a load-bearing cylinder that revolves on a rod-shaped axle at right angles to the centre of its circular face.

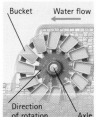
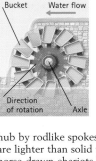

A wheel and axle act together as a rotating lever. The force of water falling on the rim of a water wheel is magnified to give a greater rotating force at the axle.

Bucket — Water flow — Direction of rotation — Axle

The wheel is possibly the single most important invention in the history of transport. Before its invention nearly 6,000 years ago, loads were dragged over ground on sledges. There are large forces of friction between sledge runners and the ground. Wheels rotate on their axles and the friction is much smaller, so less force is required to move a load.

Car wheels fitted with pneumatic tyres are the modern equivalent of tree-trunk rollers that were used thousands of years ago.

THE FIRST WHEELS

The first step towards the invention of the wheel was the use of log rollers to shift heavy loads. Rollers reduce friction by turning as the load on them moves. Simple rollers have to be collected from behind the load and placed in front of it as it moves along.

The earliest wheel was probably sliced from the end of a round log. A straight, round branch served as an axle. Each end of the axle fitted into a hole cut into the centre of two wooden discs. Carts using pairs of wheels on axles developed from the potter's wheel around 5,500 years ago in Mesopotamia, now Iraq. The soft clay pot rotated on the top wheel.

When turning a corner, the wheel on the inside of the curve travels a shorter distance than the outer wheel. A linkage called a differential lets the wheels turn at different speeds.

LATER DEVELOPMENTS

The first wagons for carrying heavy loads were fitted with solid wheels. Around 4,000 years ago, iron tools enabled carpenters in Mesopotamia and Egypt to shape pieces of wood to make spoked wheels. This type of wheel consists of a circular wooden rim attached to a central

hub by rodlike spokes. Spoked wheels are lighter than solid wheels, so fast horse-drawn chariots became possible. From Roman times, wheels were often fitted with an iron band, called a tyre. Tyres helped to reduce wear but did not improve the ride. The pneumatic tyres now used on cars and bicycles help to cushion bumps and provide a smoother ride. They were first patented in 1845.

WHEELS AS ROTARY LEVERS

A wheel and axle combination is a form of rotary lever. If a wheel is turned by a force at its edge, the force at the edge of the axle will be much greater.

Torque, or turning force, is force times the distance from the point where the force acts to the centre of the wheel. This is how a large steering wheel helps make steering a car easier – the force to turn the wheel is applied at a great distance from the centre of the wheel. Old ships had even larger wheels to turn their rudders against the force of moving water. Some valves are operated by turning a large wheel on a threaded shaft that controls the flow of a gas or liquid.

◄ Roller blades have several single wheels fitted in a row. The wheels are made of high-strength plastic. Moving along requires an action similar to that of ice skaters.

► The large diameter of a screwdriver's handle allows it to develop enough torque to shift even tight-fitting screws.

Large radius, small force

Small radius, large force

SEE ALSO

196–7 Forces, 216–7 Balance and turning forces

GEAR TRAINS

Gear trains consist of two or more gearwheels whose cogs mesh together. They transmit torque from one place to another, and can change the speed of rotation.

An 8-cog gearwheel meshes with a 16-cog gearwheel. Turning the smaller wheel causes the larger wheel to rotate with twice the turning force and half the speed.

A gearwheel has toothlike cogs that protrude from its surface. A gear train consists of two or more meshing gearwheels. A gear train can alter the direction, speed and torque of rotation from an input shaft to an output shaft.

ROTATIONAL SPEED

Two gearwheels meshed together rotate in opposite directions. The speed of rotation is usually measured in revolutions per minute, abbreviated to r.p.m. A gearwheel that makes one complete turn in a second, has a speed of rotation of 60 r.p.m.

The relative speed of two meshed gearwheels depends on the ratio of cogs on the wheels. They rotate at the same speed only if both of the gearwheels have the same numbers of cogs. If the numbers of cogs are different, the gearwheel with fewer cogs rotates faster than the wheel with the greater number of cogs. If the numbers of cogs is 8 and 16, then the gear ratio is 1:2. The speed of the gearwheels is 2:1. If the small wheel rotates at 50 r.p.m. then the larger will rotate at 25 r.p.m.

A flexible chain connects the gearwheels on a bike. The chain moves between different-sized gearwheels to change the ratios of the gears to suit the road gradient.

ROTATIONAL FORCE

A gearwheel is a combination of a wheel and levers. Each cog acts as if it were a lever attached to the centre of the wheel. In a gear train, the cogs on the gearwheel attached to the input shaft exert leverage on another gearwheel. When a small gearwheel turns a larger one, there is a mechanical advantage. The larger gearwheel turns with more torque than the smaller one, but more slowly.

GEARWHEELS AND CHAINS

Some machines have gearwheels that are joined by a chain. Bicycles are an example. The cogs of the gearwheels fit into slots on the chain. The gearwheels rotate in the same direction. As with gearwheels that mesh together, the change in speed and turning effort depend on the number of cogs on each gearwheel.

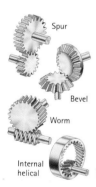

Spur

Bevel

Worm

Internal helical

▲ Meshing gearwheels can change the direction, speed or force of rotation. Different gear trains change the axis of rotation in different ways.

▶ A car's transmission uses gearwheels to match the speed and torque of an engine to the load. Climbing hills requires a high torque at low speed. High-speed cruising requires faster rotation and less torque.

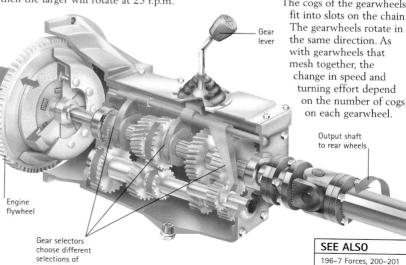
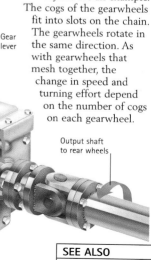

Gear lever

Output shaft to rear wheels

Engine flywheel

Gear selectors choose different selections of gear wheels

SEE ALSO

196–7 Forces, 200–201 Work and energy

FRICTION

Friction is a force that acts against the movement of surfaces that are in contact. Friction changes kinetic energy into heat energy as it resists motion.

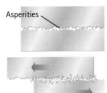

Asperities

Even a polished metal surface is covered in microscopic rough points called asperities. These points lock into one another and cause friction when the surfaces move.

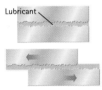

Lubricant

A film of lubricating oil holds the two metal surfaces apart. They slide past each other without making contact. Friction is reduced and movement creates far less heat.

When NASA's space shuttle enters Earth's atmosphere at more than 25,000 kilometres per hour, friction with air molecules brakes the craft and raises its skin temperature to around 1,500°C. Friction is at work everywhere. Some of its effects are unwanted; others are useful.

USEFUL EFFECTS OF FRICTION

Humans walk forwards by pushing backwards with feet. Without friction, floors, roads and pavements would be more slippery than an ice rink. People would fall over as they tried to walk or run. Objects with completely slippery surfaces would be impossible to pick up.

STATIC FRICTION

If a person tries to push a loaded crate along the floor, the force that resists the motion of the crate is friction. As the strength of the push increases, there comes a point when the crate starts to move. The force just before the crate moves is the limit of static friction. It depends on the combined weight of the crate and its contents. If the weight doubles, the limit of static friction also doubles.

The limit of static friction also depends on the materials of the surfaces in contact, in this case the crate and the floor.

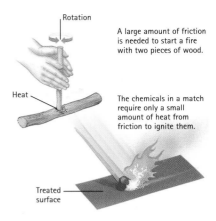

Rotation

A large amount of friction is needed to start a fire with two pieces of wood.

Heat

The chemicals in a match require only a small amount of heat from friction to ignite them.

Treated surface

MAKING FIRE

The simplest equipment for making fire depends on friction, and consists of two dry wooden sticks. Hands or a bowstring rapidly rotate one stick against another. Friction raises the temperature to around 300°C, when the wood starts to smoulder.

Matches also use friction. When the head of a match is rubbed against the strip on the side of a matchbox, friction makes the temperature rise. The heat causes chemicals in the match head and the strip to react together. As the temperature increases further, the match head burns in air and finally ignites the wood.

SLIDING AND ROLLING

Friction is a problem when moving loads. Early humans dragged loads on wooden sledges. The sledge runners helped to support the loads, but there was still a great deal of friction between the moving runners and the ground.

Humans later discovered that rollers made it much easier to move heavy objects such as blocks of stone. Rollers rotate and reduce friction because the load does not slide in contact with the ground as it travels. The drawback of rollers is that the load leaves them behind as it travels along. Around 5,500 years ago, this problem was overcome by the invention of the wheel and axle combination.

Heavy parcels move easily down a gentle slope on this roller conveyer. Each roller is fixed in position and rotates with little friction as a load travels over it.

BEARINGS

Bearings are devices that support moving parts and enable them to move with less friction. Bearings are used in most types of machines, including cars, bicycles, electric motors and roller skates. One type of bearing connects a moving shaft to a static support. Other bearings connect rotating objects to static shafts. Without bearings, friction between the two parts would slow the machine, waste energy as lost heat and rapidly wear the surfaces in contact.

A typical bearing consists of an inner and an outer metal ring. In ball bearings, steel balls run in grooves between the two rings. In roller bearings, steel cylinders roll between the inner and outer rings. High-quality bearings can run so smoothly that friction wastes less than one per cent of the energy consumption of a machine.

LUBRICANTS

Lubricants are fluids that hold sliding surfaces slightly apart to reduce friction. Mineral oils are the most common machine lubricants. Internal combustion engines contain pumps that supply oil to lubricate the pistons as they slide inside cylinders. Oil is also continuously supplied to the bearings inside these engines.

Some machines use high-pressure air to lubricate air bearings that support shafts that rotate at extremely high speeds; others use graphite as a solid lubricant.

▼ Friction between the bobsleigh's metal runners and the ice is low because the downward pressure melts the ice to form a thin lubricating layer of water.

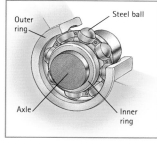

Each wheel of a roller skate is supported by a bearing. Steel balls roll inside grooves cut into the inner and outer rings.

SLOWING DOWN

Vehicles use bearings to reduce friction and help them move with the minimum amount of effort. Brakes cause deceleration by increasing the force of friction on wheels. Applying the brakes presses a hard pad of heat-resistant material against a steel drum or disk attached to each wheel. Friction between the material and the revolving part changes kinetic energy into heat and reduces the vehicle's speed. Heat is quickly lost to the surrounding air.

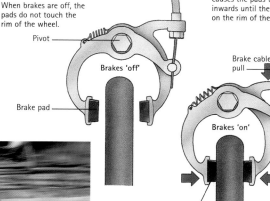

When brakes are off, the pads do not touch the rim of the wheel.

When brakes are on, the pads grip the wheel rim from either side. Friction between the pads and the rim converts the cycle's kinetic energy into heat as they reduce its speed.

▼ Bicycle brakes usually consist of a pair of hard rubber pads mounted at the ends of curved levers. These levers are attached to the bicycle frame by a pivot. Applying the brakes causes the pads to move inwards until they press on the rim of the wheel.

FLUIDS

Liquids and gases are classed as fluids because they are able to flow from one place to another. Finely powdered solids can also sometimes act as fluids.

There are three common states of matter – solid, liquid and gas. Liquids and gases are classed as fluids. Fluids are able to flow because the forces of attraction between their particles are weak. As a result, gases and liquids can easily change shape. Solids cannot flow under normal conditions because there are strong forces of attraction between their particles, and their shapes are fixed.

The temperature at the top of this waterfall is slightly lower than the temperature at the bottom. Frictional forces in flowing liquids cause them to slow down and become slightly warmer.

GASES

Gases flow from places of high pressure to places of lower pressure. For example, air flows from a bicycle pump into a bicycle tyre because the pressure inside the pump is greater than the pressure in the tyre. Weather patterns are caused by flowing air. Winds blow across Earth's surface because heat from the Sun causes differences in air pressure.

LIQUIDS

Liquids are denser than gases, so the force of gravity makes them flow downwards. Water, petrol, oil and treacle are all liquids, but they flow at different rates. There are stronger forces of attraction between the particles in treacle than there are in water. As a result, water flows more easily and quickly than treacle. Liquids that flow with difficulty are said to be viscous. Viscosity varies from one liquid to another and decreases with increasing temperature.

Fine powders can be made to behave like fluids by floating them on an upward air current. This arrangement is called a fluidized bed. A fluidized powder flows like a liquid.

Honey is a viscous fluid. Charged parts on the molecules in honey attract them to each other. They cling together as the liquid flows, slowing its motion.

FLUIDIZATION OF SOLIDS

Solids such as cement and flour are fine powders. Each powder particle is millions of times larger than the molecules in liquids and gases, but it is possible to make these powders flow and act like fluids. Blowing air up through a fine powder lifts its particles and lowers the friction between them. This is why cement and flour can be loaded into tankers and unloaded by fluidizing them with air.

Power stations burn powdered coal in fluidized bed furnaces. The pulverized coal flows into a layer of sand that is kept fluid by high-pressure air. The air helps the coal burn, and keeps the bed at the optimum temperature for efficient combustion.

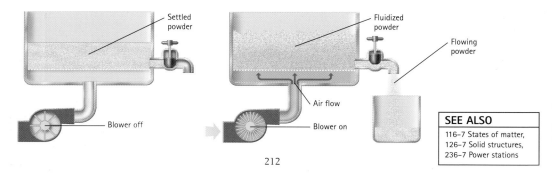

Settled powder

Fluidized powder

Flowing powder

Air flow

Blower off

Blower on

SEE ALSO
116–7 States of matter,
126–7 Solid structures,
236–7 Power stations

PRESSURE

Pressure is the result of a force pushing on a surface. Unlike solids and liquids, gases are easily compressed; their pressure increases as volume decreases.

A thousand metres below the ocean's surface, an armoured suit protects the diver from being crushed by pressures more than 100 times atmospheric.

D ivers feel the pressure of the water on their bodies. It pushes equally from all directions. The deeper the dive, the greater the pressure. Water pressure is the result of gravity pulling down on the water above the diver. On dry land, air pressure is the result of gravity pulling downwards on the gases in Earth's atmosphere.

MEASURING PRESSURE

The unit of pressure is the pascal, symbol Pa. One pascal is equivalent to a force of one newton acting over an area of one square metre. Atmospheric pressure at sea level is approximately 101,000 Pa.

Pressure increases as area decreases. A 70-kilogram person in stiletto heels exerts a higher pressure on the floor than a three-tonne elephant on its four large feet.

Pressure gauge   Thermometer

The pressure of a gas results from its molecules colliding with the walls of its container. As temperature increases, the collisions get more violent and the pressure increases.

PRESSURE AND COMPRESSIBILITY

Although gases fill the container in which they are placed, only a small proportion of that space is filled by molecules. The atoms or molecules of a gas are in constant motion; they exert pressure on their container as a result of their collisions with the container walls.

When a gas is compressed, its particles are contained in a smaller volume. They collide more frequently with the walls of the container, so the pressure increases. Liquids and solids have hardly any space between particles, so they do not compress as easily as gases do.

PUSHING WITH FLUIDS

Applying a force to the surface of a liquid increases the pressure at all points inside it. Hydraulic machines use this effect to move heavy loads. In a mechanical digger, the engine drives a pump that forces oil along pipes into cylinders. The pressure of the oil forces a piston to move along inside the cylinder. Oil can push on either side of the piston to make it move with immense force in either direction.

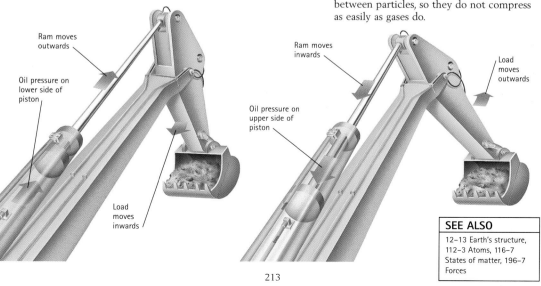

Ram moves outwards

Oil pressure on lower side of piston

Load moves inwards

Ram moves inwards

Oil pressure on upper side of piston

Load moves outwards

SEE ALSO

12–13 Earth's structure,
112–3 Atoms, 116–7
States of matter, 196–7
Forces

SOUND AS CHANGES OF PRESSURE

Sound consists of vibrations that travel through a medium such as air or water. These vibrations can be detected by the ears of animals.

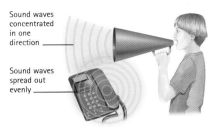

A simple cone makes the sound of a voice travel further by concentrating it in one direction. A telephone ringer spreads sound so that it can be heard from any direction.

Sound waves concentrated in one direction

Sound waves spread out evenly

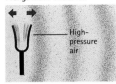

Prongs move outwards

High-pressure air

When the prongs move outwards, they compress nearby air.

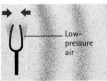

Prongs move inwards

Low-pressure air

When the prongs move inwards, they create a region of low pressure.

▲ The prongs of a tuning fork vibrate at a steady rate. They create regions of high and low pressure in the air that radiate away from the fork at the speed of sound.

When people speak, the vocal cords in their throats vibrate back and forth. As they vibrate, they produce sound waves that travel at around 340 metres per second. These waves are changes in air pressure of around one ten-thousandth the normal air pressure. The air as a whole does not move with a sound wave, it simply vibrates around an average position.

SOUND IN AIR

Sounds travel through air as a back and forth movement of air molecules. A vibrating surface makes sounds by alternately pushing and pulling at the layer of air surrounding it. This layer of air then pushes and pulls at the layer of air next to it, and so on. This is how vibrations travel through air.

SOUND IN OTHER MEDIA

Sound can travel through solids, liquids or gases. It cannot travel through a vacuum, because there are no particles to vibrate and carry sound waves. This is how the vacuum between the panes of a double-glazed window provides sound insulation.

A substance that carries sound is called a medium, and the speed of sound depends on the density of the medium. Sound travels five times faster in water than it does in air, for example, and more than three times faster in glass than in water. This is because the particles in a dense medium, such as glass, are closer together than in a less dense medium, such as air. The closeness of particles in a dense medium helps vibrations pass more quickly from one particle to the next.

Compressing a gas increases its density, so the speed of sound in a gas increases as the pressure of the gas increases.

Humans can hear sounds up to 20 kHz.

The ears of bats are sensitive up to 120 kHz.

▶ Dolphins and whales use high-frequency sound to communicate and to locate food. These sounds travel further in water than they would in air.

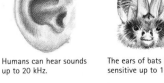

Crickets can hear sound frequencies up to 100 kHz.

Some frogs can detect sounds up to 50 kHz.

◀ Different animals can hear sounds within different ranges of frequencies. The range of human hearing is from around 25 Hz to around 15,000 Hz (15 kHz). The upper limit falls with increasing age.

SONAR

Sonar was developed in 1915 by French physicist Paul Langevin (1872–1946). Its invention was inspired by the disaster of the White Star liner *Titanic*, which was sunk by an iceberg in 1912. Sonar provides information about underwater objects through the echoes that return from them. Vertical echoes give information about the seabed; angled echoes can help locate submarines and shoals of fish.

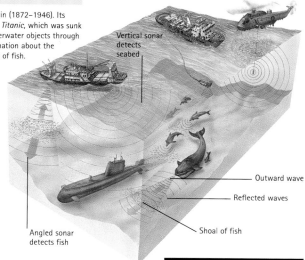

Vertical sonar detects seabed

Outward wave

Reflected waves

Shoal of fish

Angled sonar detects fish

USING SOUND

Some animals have ears that detect sounds as they travel through air or water. The ears convert vibrations into nerve impulses that travel to the brain. Other animals, such as snakes, sense ground vibrations.

Animals use sounds to communicate and to gather information about their surroundings. Sounds such as the noises of falling rocks or moving predators can provide warnings of imminent danger.

ECHOLOCATION

Some animals, such as bats, dolphins and whales, use sound to navigate and locate food. They send out short bursts of high-frequency sound and analyze the returning echoes. This is called echolocation; it helps these animals to detect prey and estimate its speed and direction of motion.

Ships use sonar – *so*und *n*avigation *a*nd *r*anging – to form images of underwater objects and landscapes. A device similar to a loudspeaker emits intense pulses of sound. A computer analyzes the timing and direction of returning echoes to calculate the position and size of objects.

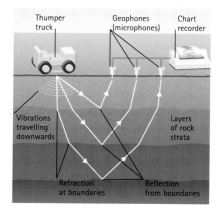

Thumper truck

Geophones (microphones)

Chart recorder

Vibrations travelling downwards

Layers of rock strata

Refraction at boundaries

Reflection from boundaries

Sound waves are refracted and reflected as they pass from one medium into another. Geologists analyze returning sound waves to build up pictures of rock layers.

INFRASOUND

Infrasound consists of sound waves that vibrate at frequencies below around 25 hertz, which is the lower limit of human hearing. Sounds in this range are said to be subsonic. Earthquakes send subsonic waves through the ground; explosions send them through air. Although infrasound cannot be heard, its pressure waves can sometimes be felt.

ULTRASOUND

Ultrasound consists of sound waves that vibrate at frequencies greater than around 14 kilohertz, which is the upper limit of human hearing for most adults.

Ultrasonic waves penetrate liquids and solids better than lower-frequency sounds. This is why ultrasound is used in some types of sonar equipment and in body scanners that give images of internal organs and growing babies. Ultrasound echoes can also detect flaws inside welded metal objects such as steel pipelines.

The energetic vibrations of ultrasound can be used to break kidney stones into pieces that are small enough to pass out of the kidneys in urine. Water-filled ultrasonic baths are used to dislodge encrusted dirt from laboratory equipment.

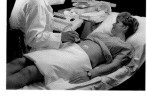

▲ An ultrasound scanner –much safer than X-rays, shows the outline of a baby growing inside its mother.

▼ Speed of sound depends on the medium it passes through. Speed increases with increasing temperature.

Medium	Speed
air at 0°C	331
air at 30°C	350
carbon dioxide	267
hydrogen	1,315
water	1,469
concrete	5,000
granite	3,950
brick	3,600
steel	5,121
aluminium	5,100
glass	6,000

Metres per second

SEE ALSO

22–3 Earthquakes, 92–3 Ears, hearing and balance

BALANCE AND TURNING FORCES

Turning forces can make an object rotate. An object is balanced when turning forces operate equally against each other so that rotation does not happen.

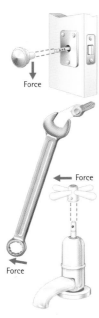

Force

Force

Force

Doorknobs, spanners and tap handles all increase the turning effect of a hand. In each case, a force that acts at a distance from a pivot point produces a twisting force.

Newton's third law of motion states that whenever a force acts on an object, an equal and opposite force matches it. A door opens when a person pulls its handle. The pulling force is matched by an equal force that acts through its hinges. This force points in the opposite direction to the force on the handle, but it acts along a different line. The result is a turning force, or torque, that makes the door turn on its hinges.

When a person sits on one side of a seesaw, their weight is matched by an upward force that acts through the pivot. This results in a torque that drives that side of the seesaw to the ground. If a second person of the same weight sits at the same distance on the other side of the pivot, they produce an equal and opposite torque, so the seesaw comes into balance.

CALCULATING TORQUE

A long-handled spanner will sometimes loosen a tight nut that will not respond to a smaller spanner. This is because a given force produces more torque when it acts further from a pivot, in this case the nut. It is pointless to try to loosen a nut by pushing or pulling along the handle of a spanner. This is because torque only arises from force that acts at right angles to the

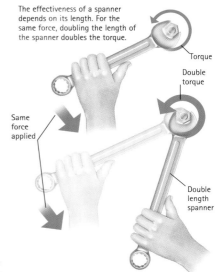

The effectiveness of a spanner depends on its length. For the same force, doubling the length of the spanner doubles the torque.

Torque

Double torque

Same force applied

Double length spanner

line between the pivot and the point where the force acts. At any other angle, at least part of the force will be resisted by a reaction force that acts along the spanner and has no turning effect.

Torque is calculated by multiplying the length of the line between the pivot and the point of action of the force by the part of the force that acts at right angles to that line. The mathematical expression is torque = force x distance. The usual unit of torque is the newton metre, symbol Nm, so the value of force in newtons must be multiplied by the distance in metres.

A torque of one newton metre is equivalent to a force of one newton pulling or pushing at right angles at the end of a one-metre-long lever. The motor of a food mixer produces a torque of around 1 Nm, the motor of an electric drill produces 2 Nm and the engine of a car generates up to 150 Nm.

The weight of a 40-kilogram cyclist is 40 x 9.8 = 392 newtons – the mass times the constant g. If that cyclist stands on a pedal attached to a 0.2-metre-long crank, the torque that acts on the chainwheel is 398 x 0.2 = 78.4 Nm – just over half the torque of a car engine.

▶ Pushing the pedals of a bicycle produces a torque in the chainwheel. The chain transfers torque to a gearwheel attached to the hub of the rear wheel. The torque at the rear wheel can be increased for hill climbing by changing to a lower gear ratio.

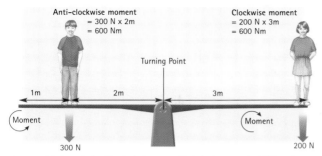

Anti-clockwise moment
= 300 N x 2m
= 600 Nm

Clockwise moment
= 200 N x 3m
= 600 Nm

Turning Point

1m   2m   3m

Moment

Moment

300 N

200 N

A seesaw is balanced when the clockwise moment equals the anti-clockwise moment. The boy's weight is 300 newtons (300 N) and he stands 2 metres (2 m) from the pivot. He causes the anti-clockwise moment of 600 newton-metres (Nm). The girl is lighter (200 N), but she stands further from the pivot (3 m). She causes a clockwise moment of 600 Nm, so the seesaw is balanced.

BALANCE AND MOMENTS

A seesaw is a lever that pivots on a central fulcrum. Two people sitting on opposite sides of the pivot of the seesaw will not move if the seesaw is exactly balanced. Gravity produces a downward force as it pulls on the mass of each person. The person on the right produces a torque that attempts to rotate the seesaw clockwise. This torque is called a clockwise moment. The person on the left produces a turning force that attempts to rotate the seesaw anti-clockwise: an anti-clockwise moment.

Moments are calculated in the same way as torque: by multiplying force and distance. If the people on the seesaw cause equal but opposite moments, there is no overall torque: the seesaw is in balance. The masses of the two people do not have to be identical to balance. The heavier person can reduce his or her moment by moving closer to the pivot.

EQUILIBRIUM

An object is in equilibrium if all the forces and moments that act on it are in balance. An object that is in equilibrium will not accelerate or start to rotate.

A food tin that stands on a flat surface is in equilibrium. The force of gravity acts on every particle in a tin. These separate forces act together as if they were a single force pulling downwards at one point. That force is the weight of the tin; the point where it seems to act is called the centre of gravity or centre of mass. The reaction force to the weight of the tin acts through its base.

If the tin is tilted through a small angle, its centre of mass will still be over its base. The reaction force acts at the rim of the tin, producing a torque that tends to set the tin upright. If it is tilted so that the centre of mass is no longer over the base, the torque reverses and pulls the tin over.

Standard weights are added to one side of the scales until they balance the objects on the other side. If the two arms of the scale are equal in length, the masses in the two pans are then equal.

This gymnast stays in balance while her centre of gravity is directly above the beam. If she leans to one side, she will topple.

Stable equilibrium

Centre of mass

An object in stable equilibrium can be tilted through a moderate angle without its centre of mass moving outside the footprint of its base. If tilted and then released, it will right itself.

Unstable equilibrium

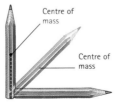

Centre of mass

Centre of mass

An object is in unstable equilibrium if a small angle of tilt takes its centre of mass outside the footprint of its base. If tilted, it will topple until it reaches a more stable equilibrium.

Neutral equilibrium

Centre of mass

An object is in neutral equilibrium if it can roll without its centre of mass straying outside the footprint of its base.

SEE ALSO

196–7 Forces, 202–3 Momentum, 204–5 Relativity and gravity, 209 Gear trains

FLOATING AND SINKING

A body floats in a fluid if the upward force from the fluid matches the downward force of gravity on its mass. This effect is called buoyancy.

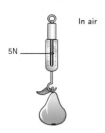

The mass of the pear is the same whether it is surrounded by air or by water. It appears to weigh less in water due to the upthrust it experiences by displacing water.

▼ A floating object, such as a boat, displaces water from the space that it occupies. The weight of the displaced water is the same as the weight of the boat. Buoyancy is the result of water pressure.

Wood and oil float on water, as do steel ships. Balloons full of hot air or helium float up into the sky. Submarines rise to the surface from the depths of the sea. All these floating effects are caused by buoyancy, which is the name given to the upward forces experienced by objects that are immersed in liquids or gases.

UPTHRUST
The weight of an object is the result of gravity acting on its mass. Weight always acts downwards. When an object is immersed in a fluid, it experiences an upward force called upthrust. This force acts in the opposite direction to weight.
Upthrust is equal to the weight of the volume of fluid the object displaces, or pushes out of the way. This effect is called Archimedes' principle.

SINKING
Suppose that a concrete block has a volume of 1 m³ and a mass of 3,000 kg. Its weight is almost 30,000 N. Immersed in water, the block displaces 1 m³ of water, which has a mass of 1,000 kg and a weight of almost 10,000 N. As a result, there is an upthrust of 10,000 N on the block that acts against its weight of 30,000 N. The apparent weight of the block when submerged is 20,000 N. This concrete block will sink in water because the resultant force acting on it is 20,000 N downwards. However, it will be easier to lift underwater because its weight is one-third less than on dry land.

THE CAUSE OF UPTHRUST
When an object is immersed in a fluid, the upthrust is the result of the pressure of the fluid on the object. Pressure in a fluid acts in all directions and increases with depth. As a result, the pressure acting downwards on the top of an object is less than the pressure acting upwards on the bottom of the object. The result of these two forces acting against each other is an overall upward force – the upthrust.
Imagine a one metre cube in water with its top face one metre below the surface. The forces on the sides cancel each other out. The force from pressure of the water above is 10,000 N. The force from below is 20,000 N. The net force is an upthrust of 20,000 N – 10,000 N = 10,000 N.

▼ Upthrust is the result of pressure increasing with depth. The pressure on the lower surface is greater than the pressure on the upper surface.

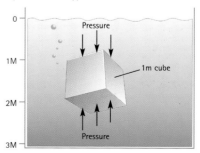

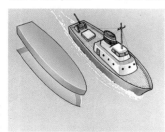

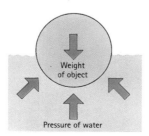

▶ The canopy of a hot-air balloon is made from synthetic materials that are light but tough. A wicker basket hangs underneath, carrying passengers and cylinders of liquefied butane gas for use in burners that heat the air in the canopy.

◀ A hot-air balloon holds around 1,500 cubic metres of hot air, which has a mass of about 1,500 kg. A balloon displaces 1,500 cubic metres of the surrounding cold air, which has a mass of about 2,000 kg. Buoyancy gives an upthrust of 500 kg.

Vent opens to allow hot air to escape for final descent

Opening in balloon

Gas burner

Flame

Butane gas in cylinders

FLOATING

Most types of wood float on water. Suppose that 1 m³ of wood has a mass of 500 kg. Its weight in air is 5,000 N. When totally immersed it displaces 1 m³ of water, which has a mass of 1,000 kg and a weight of 10,000 N. The upthrust on the wood is greater than its weight in air, so it floats to the surface. Floating objects are partly submerged so that the upward and downward forces are balanced. There is just enough of the object underwater to give an upthrust that is equal to the weight of the object in air. Placing weights on an object floating in water makes it sink lower into the water. It then displaces more water and the upthrust increases to match the greater downward force.

DENSITY AND FLOATING

The density of a substance is its mass per unit volume. The density of water is 1,000 kg/m³. The value for concrete is around 3,000 kg/m³, wood is around 500 kg/m³ and steel around 7,800 kg/m³. An object will float if its density is less than the density of water; it will sink if its density is greater. A steel ship floats because it is hollow. The average density of the steel hull and the air and fittings inside it is less than the density of water.

Hot-air balloons float because hot air is less dense than the cold air that surrounds them. Helium-filled balloons float because helium is less dense than air at the same pressure. As a helium balloon rises, it expands as the air pressure decreases.

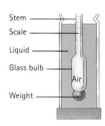

Stem

Scale

Liquid

Glass bulb

Air

Weight

A hydrometer is a floating device that measures the density of liquids. The more dense the liquid, the greater the upthrust and the higher the hydrometer floats.

TF
F
T
L O R
S
W

▲ The load line on the side of a ship shows how much load it can safely carry. How well a ship floats depends on the density of the water around it.

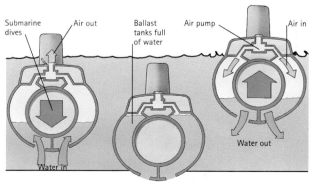

Submarine dives

Air out

Ballast tanks full of water

Air pump

Air in

▶ Submarines consist of an airtight compartment surrounded by ballast tanks. The submarine dives by filling these tanks with water. Submerged, its neutral buoyancy ensures that it neither floats nor sinks. It surfaces by using compressed air to force out the water.

Water in

Water out

SEE ALSO

196–7 Forces, 212 Fluids, 213 Pressure, 220–1 Principles of flight

PRINCIPLES OF FLIGHT

Machines fly by producing an upward force that overcomes their weight. They also generate thrust that drives them forwards and helps them to steer.

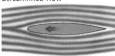

Streamlined flow

Streamlining cuts down drag, or air resistance, by helping to create a smooth airflow. Teardrop-shaped objects are streamlined.

Turbulent flow

Square objects are not streamlined. They have sharply angled edges that create the turbulent flow that results in drag.

Flight is the movement of insects, birds and aircraft through the air. There are forces acting on these objects that hold them up in the air. The principles of flight are the rules that govern the movement of objects as they fly through the air. They concentrate on the four forces – lift, thrust, drag and weight – that act on all flying objects.

LIFT

The top surface of an aircraft's wing is curved, while the bottom is flatter. This shape is called an aerofoil. Air that flows over the top of a moving wing has to travel a greater distance than the air flowing underneath. The air travels faster over the top than underneath. The pressure of air decreases when its speed increases. The pressure on the top surface of the wing is less than the pressure pushing up on the bottom surface. The overall force of lift pushes the wing upwards.

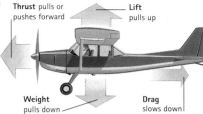

Thrust pulls or pushes forward — Lift pulls up

Weight pulls down — Drag slows down

Four forces act on an aircraft in flight. An aircraft will fly only when the lift, created by air flowing over the wings, is greater than its weight, and when the thrust of its engines is greater than the drag of the air.

DRAG

Aircraft are held back by air resistance. This force is called drag. The faster an aircraft flies, the greater the drag.

Drag is a particular problem when designing high-speed planes, since its effect increases greatly with small increases in speed. To reduce drag, all external surfaces must be streamlined.

STEERING AN AIRCRAFT

Aircraft have moveable parts on the wings and tail that are used to turn or tilt the aircraft in flight. Known as control surfaces, the ailerons, elevators and rudder produce vertical or horizontal forces by diverting the air that flows around them. Two or three control surfaces have to be moved at the same time to make even a simple turn. In larger aircraft, the control surfaces are moved by hydraulic rams. In fly-by-wire aircraft, such as the Airbus A320, the control surfaces are operated by computer systems.

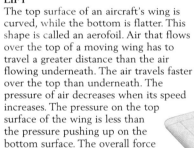

Aileron

Propellers
All aeroplanes had propellers until the 1940s. The shape of a propeller, or airscrew, is an aerofoil, similar to the wing. As the propeller is spun by the engine, it bites into the air to produce thrust.

Rolling
Moving one aileron up and the other down makes the aircraft bank, or roll to one side or the other.

Pitching
Lowering the elevators makes the nose dip and the tail rise. Raising the elevators makes the tail dip and the nose rise.

Yawing
Moving the rudder to the right turns the aircraft's nose to the right. Moving it to the left turns the nose to the left.

Rudder
Pushing the right rudder pedal moves the rudder to the right. Pushing the left rudder pedal moves the rudder to the left.

Control column
Pushing this column back and forth moves the elevators. Pushing it from side to side moves the ailerons.

THRUST AND PROPELLERS

All early aeroplanes were powered by petrol-driven piston engines that spun propellers. Each blade of a propeller has an aerofoil shape. As it moves through the air, it creates a forward force called thrust in the same way that a wing creates the upward force called lift. Many modern planes are fitted with variable pitch propellers that can alter the angle at which they bite into the air. The pilot changes engine speed and pitch in the same way that a cyclist changes gear and pedals faster or slower.

THRUST AND JETS

Many modern planes are powered by jet engines. These gas turbine engines burn kerosene fuel to produce a stream of exhaust gases that thrust the aeroplane forwards. The exhaust gases also spin a turbine at the rear of the engine that drives a compressor at the front. Turboprop engines use some of this power to drive a propeller. Turbofan engines have enormous compressors that shoot air around the engine to produce most of the engine's thrust.

HELICOPTERS

Helicopters use rotating blades to provide both lift and thrust. Two or more long rotor blades are attached to a shaft that sticks out of the top of the helicopter. Each blade is shaped like a long narrow wing that produces a downward force as it spins. This vertical force lifts the helicopter into the air. A smaller tail rotor provides a sideways force that prevents the helicopter from spinning. A helicopter can move in any direction by tilting the main rotors so that they are no longer parallel to the ground.

A glider's wings generate lift from the air that flows over them. A glider can rise in currents of warm air then dive gently to increase the airflow over the wings and provide lift.

Rising currents of warm air

Elevator

Lift

Thrust

Rudder

Ailerons
The ailerons are joined by wires so that when one goes up, the other goes down. Moving them makes one wing rise and the other one drop.

▲ A helicopter's rotor blades provide both lift and thrust. The rotor blade has the same aerofoil shape as a plane's fixed wing. As it spins, air flows over the blade, producing lift. Tilting the whole rotor provides thrust.

Nozzles point backwards for normal forward flight

Nozzles rotate for lift and forward flight

Nozzles point down for vertical takeoff or landing

Vertical Flight
The Harrier is one of the few fixed-wing planes able to rise vertically into the air and hover. It has four nozzles that direct thrust from the jet engines. When the nozzles point downwards, the plane moves vertically or hovers. When they point forwards, it flies conventionally.

SEE ALSO

61–3 Insects, 70–1 Birds, 160 Jet engines and gas turbines, 196–7 Forces, 213 Pressure

SUPERSONIC FLIGHT

Supersonic aircraft can fly at speeds greater than the speed of sound, which is around 1,200 km/h at sea level. Their designs differ from those of subsonic aircraft.

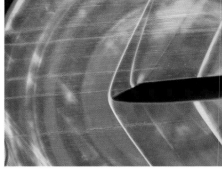

A model of NASA's space shuttle in a supersonic wind-tunnel test. A technique called schlieren photography reveals where shock waves form.

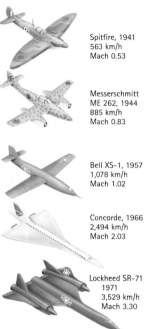

Spitfire, 1941
563 km/h
Mach 0.53

Messerschmitt
ME 262, 1944
885 km/h
Mach 0.83

Bell XS-1, 1957
1,078 km/h
Mach 1.02

Concorde, 1966
2,494 km/h
Mach 2.03

Lockheed SR-71
1971
3,529 km/h
Mach 3.30

Aircraft speed increased rapidly between the 1940s and the 1970s. Lift and drag increase with speed, so wings became smaller, thinner and more swept back to reduce drag.

In the early 1940s, propeller-engined fighter planes reached speeds of more than 1,000 km/h in steep dives. The pilots observed severe buffeting that threatened to tear the wings off their aircraft. Scientists realized that these effects were caused by the aircraft approaching the speed of sound.

THE SOUND BARRIER

A moving aircraft disturbs the air and sends out noise and pressure waves that travel away from it in all directions at the speed of sound. When a plane reaches the speed of sound, the pressure waves cannot outrun it. The sound waves build up in front of the aircraft as a layer of densely compressed air called a shock wave.

To travel faster than the speed of sound, the pilot must fly the plane through this barrier and overtake it. There is a jolt as an aircraft breaks through the sound barrier because drag increases and lift suddenly decreases. Once through, the plane leaves the shock wave behind and flight is smooth again.

A plane flying at supersonic speeds leaves a trail of shock waves in its wake. When these waves of highly compressed air reach the ground, they are heard as a deep bang called a sonic boom.

MACH NUMBER

The speed of sound depends on the temperature, pressure and humidity of air. Temperature can vary from 35°C near the ground to –55°C high in the stratosphere, so the speed required for a plane to break through the sound barrier depends on the local conditions around it. This local speed is given the value of Mach 1.0.

Any speed greater than Mach 1.0 is supersonic: from 1,240 km/h at sea level to approximately 1,060 km/h in the upper atmosphere at 13,000 metres altitude.

SUBSONIC AND HYPERSONIC

Subsonic speeds are below Mach 0.8, the approximate speed of a normal passenger jet. Transonic speeds range from Mach 0.8 to Mach 1.2. Hypersonic speeds are above Mach 5. Concorde and fighter planes are supersonic aircraft; NASA's Space Shuttle enters Earth's atmosphere at hypersonic speeds around Mach 20.

When an aircraft flies, sound waves spread out from it at the speed of sound. When the aircraft itself reaches the speed of sound – Mach 1.0 – these sound waves pile up in front of the aircraft. Above the speed of sound, these waves form a shock wave that reaches the ground. The shock wave is heard as a sonic boom that can be powerful enough to break windows.

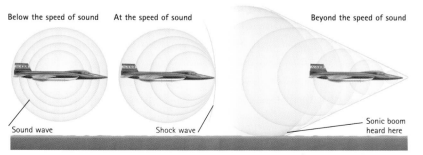

Below the speed of sound   At the speed of sound   Beyond the speed of sound

Sound wave   Shock wave   Sonic boom heard here

▶ In 1967, the US X-15 rocket plane flew at record-breaking speeds of around 7,200 km/h. Its wedge-shaped tail and thin stubby wings were designed for flight at altitudes of around 100 km.

Length 15.2 m

Wingspan 6.7m

Liquid-fuel rocket engine (more than 250,000 N thrust)

SUPERSONIC WINGS

The top speeds of the world's fastest aircraft increased by more than six times between 1940 and the end of the 20th century. The first plane to break the sound barrier was the Bell X-1 rocket plane in 1947. Scientists obtained information from early supersonic flights that helped them to design even faster aircraft.

In later designs, wings were swept back until they joined with the tail to form a single surface. This is called a delta wing; it helps make an aircraft more streamlined. The delta wing also fits snugly inside the supersonic shock wave, so the plane can pass through the sound barrier with the minimum of buffeting.

MOVEABLE WINGS

Some supersonic fighter aircraft have wings that can move forwards and backwards. Each wing is attached to a pivot in the plane's body. For takeoff and landing, the wings stick straight out from the fuselage to provide maximum lift. When flying at transonic and supersonic speeds, they move into a swept-back position to form a low-drag delta shape.

RECORD BREAKERS

The fastest winged aircraft is the rocket-propelled Bell X-15, built in 1959. It achieved a speed of Mach 6.72 in 1967. This record still stands because there is no longer any competition between countries to break the airspeed record. The fastest jet aircraft is the Lockheed SR-71, originally built as a spy plane. It achieved a speed of Mach 3.3 in 1971. Although no longer in military service, it is used for scientific studies in the upper atmosphere.

THE FUTURE

Concorde is the only supersonic passenger aircraft. Profits from fares are very small compared to the costs of designing and building this aircraft. There are plans for hypersonic passenger planes that could fly to the edge of space and travel across the Atlantic Ocean in under an hour.

HOTOL (*horizontal* takeoff and *landing*) aircraft could be hypersonic transports in the future. Flying outside the atmosphere, HOTOL aircraft would use liquid oxygen to burn fuel in their engines.

The Thrust supersonic car (SSC) is propelled by two jet-fighter engines. In 1998, it became the first land vehicle to break the sound barrier. Shock waves build up in front of the car as it approaches the speed of sound, in the same way it happens in supersonic flight. A supersonic vehicle cannot be heard as it approaches because all the sound it makes is concentrated in the shock wave left behind it.

SEE ALSO

70–1 Birds, 160 Jet engines and gas turbines, 196–7 Forces, 204–5 Relativity and gravity

FACTS AND FIGURES

MOVEMENT

Speed
Speed is how quickly an object moves. An object that moves one metre (1 m) in one second (1 s) has a speed of one metre per second (1 m/s):

$$\text{Speed (m/s)} = \frac{\text{distance moved (m)}}{\text{time taken (s)}}$$

Velocity
Velocity is a measurement that describes both the speed of an object and its direction of travel. A car that travels round a bend at a speed of 10 m/s does not have a fixed velocity because its direction changes at every instant.

Acceleration
Acceleration measures how quickly the velocity of an object changes. An object whose velocity increases by one metre per second in each second has an acceleration of one metre per second per second, or one metre per second squared (1 m/s²):

$$\text{Acceleration (m/s}^2) = \frac{\text{change in velocity (m/s)}}{\text{time taken (s)}}$$

PROPERTIES OF MATTER

Inertia is a fundamental property of matter that causes it to resist efforts to change its velocity.

Momentum is calculated by multiplying the mass in kilograms of an object by its velocity in metres per second.

Momentum (kg m/s) = mass(kg) x velocity (m/s)

FORCE, ENERGY, WORK AND POWER

Force
A force changes the momentum of an object. The unit of force is the newton (N). One newton (1 N) of force will accelerate a mass of one kilogram (1 kg) by one metre per second squared (1 m/s²).

Force (N) = mass (kg) x acceleration (m/s²)

Energy
The unit of energy is the joule (J). Ten joules (10 J) of energy will raise one kilogram (1 kg) of mass slightly more than one metre (1 m) vertically upwards against the force of gravity.

Work
The unit of work is the joule (J). Work is done when energy converts from one form into another. One joule (1 J) of work is

done when a force of one newton (1 N) moves a distance of one metre (1 m)

Work (J) = force (N) x distance (m)

Power
The unit of power is the watt (W). Power is the rate at which work is done or energy changes into another form. A machine that converts one joule (1 J) of energy in one second (1 s) has a power of one watt (1 W).

NEWTON'S LAWS OF MOTION

1st Law: An object will stay still or continue to move at a constant velocity unless a force pushes or pulls it.
2nd Law: The rate of change of the momentum of an object is proportional to the force that acts on it.
3rd Law: If one object exerts a force on another object, then the second object exerts an equal and opposite force (the reaction force) on the first.

SPEED

metres per second (m/s)	kilometres per hour (km/h)	miles per hour (mph)
10	36	22
20	72	45
30	108	67
50	180	112
100	360	224
200	720	447
500	1,800	1,119
1,000	3,600	2,237

kilometres per hour	metres per second	miles per hour
10	2.8	6.2
20	5.6	12
30	8.3	19
50	14	31
100	28	62
500	139	311
1,000	278	621

AMOUNTS OF ENERGY

5 J = work done to shut a door
20 J = work done to throw a ball
1,000 J (1 kJ) = work done to climb a flight of stairs
4,000 J (4 kJ) = electrical energy in a typical fully charged AA battery
40,000 J (40 kJ) = heat energy required to boil a cup of water
120,000 J (120 kJ) = chemical energy in a teaspoonful of sugar
300,000 J (300 kJ) = electrical energy in a car battery
34,000,000 J (34 MJ) = heat given out by burning a litre of petrol

AMOUNTS OF FORCE

5 N = force (push) required to turn on a light switch
8 N = force to accelerate a 4 kg mass at a rate of 2m/s²
9.8 N = downward force of gravity pulling on a 1kg mass
20 N = force (pull) to open a drinks can
2,000 N = force on a tennis ball struck hard by a racquet
5,000 N = force from car engine
200,000 N = force from a jet engine

AMOUNTS OF POWER

1 W = torch bulb
60 W = electric light bulb
250 W = washing-machine motor
300 W = running up stairs
1,000 W (1 kW) = cheetah running flat out
35,000 W (35 kW) = small car engine
250,000 W (250 kW) = large wind turbine
500,000,000 W (500 MW) = small power station
2,000,000,000 W (2,000 MW) = typical coal-fired power station
20,000,000,000 W (20,000 MW) = largest power station in the world (Lower Tungusta River, Russia).

KEY DATES

1586 Different weights fall at the same speed in a vacuum demonstrated by Dutch mathematician Stevinus.
1647 French scientist Blaise Pascal discovers that pressure at a depth underwater acts in all directions.
1668 Principle of conservation of momentum discovered by English mathematician John Wallace.
1684 Law of gravity described by English scientist Isaac Newton.
1842 The principle of conservation of energy – the basis for the first law of thermodynamics – discovered by German physicist Julius von Meyer (also independently by the British physicist James Joule).
1850 Second law of thermodynamics formulated by German physicist Rudolph Clausius.
1856 The term 'kinetic energy' invented by British physicist William Thomson.
1905 Third law of thermodynamics formulated by German physical chemist Hermann Nernst.
1926 First flight of liquid-fuelled rocket by US engineer Robert Goddard.
1931 Gyrostabilisers for ships invented by US engineer Elmer Sperry.
1947 Sound barrier broken by US pilot Chuck Yeager.

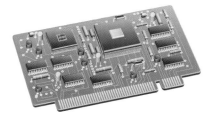

CHAPTER 8

ELECTRICITY
AND ELECTRONICS

E lectricity is all around us, present in the electrical charges of minute subatomic particles called electrons. Under certain conditions, these electrons move from atom to atom in an electrical current. Although Greek philosophers were aware of the power of static electricity as long ago as 600BC, it was only in the 18th and 19th centuries that scientists started to understand electricity. Through their studies, they learned how to generate and make use of electrical currents.

Such has been the success of the pioneers of electricity that it is now difficult to envisage a world without electricity. Much of the world's heating, lighting and power depends on electricity. Without electricity, there could be no computers, radio, television or spacecraft. Even the motor car, propelled by a non-electrical internal-combustion engine, relies on electricity to fire its spark plugs and power its lights and many of its controls. It is fair to say that electricity powers the modern world. It will continue to do so long into the future, even though some of the ways of generating electricity and transmitting it to where it is needed will be altered and improved.

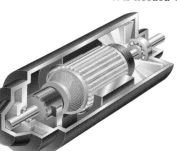

ELECTRICITY

Electricity is a form of energy created by the movement of minute charged particles called electrons. It is a vital source of power in the modern developed world.

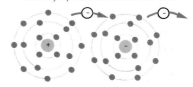

Electrons jump from atom to atom in a metal

When an electrical current flows through a metal, some electrons jump from atom to atom. The others orbit around the atomic nuclei, which are in fixed positions.

Italian scientist Alessandro Volta (1745–1827) was one of electricity's pioneers. By 1800, Volta had invented the first battery capable of holding an electric charge. It was called a voltaic pile.

US physicist Robert Millikan (1868–1953) won the Nobel Prize for physics in 1923 for being the first person to measure the charge of an electron.

All matter – from the paper this book is printed on to the air that we breathe – is made up of tiny particles called atoms. Each atom has a positively charged centre called a nucleus. Nuclei contain positively charged particles called protons and uncharged (neutral) particles called neutrons. Much smaller, negatively charged particles called electrons whizz around the nuclei at high speed.

In general, the number of electrons in an object exactly matches the number of protons. The negative charges of the electrons balance the positive charges of the protons and the object is neutral. An object that has fewer electrons than protons has a positive charge and will attract electrons. A negatively charged object has more electrons than protons, and it will readily pass its surplus electrons to a positively charged object.

STATIC ELECTRICITY

Static electricity has intrigued people for many centuries. Around 2,500 years ago, the Greek philosopher Thales (625–547BC) observed that rubbing a silk cloth over a piece of fossilized tree sap, called amber, caused the cloth, and other

lightweight objects such as feathers, to be attracted to the amber. It is now known that this attraction occurs because rubbing drags electrons from the surface of the silk to the surface of the amber. The negatively charged amber attracts objects as it tries to lose its surplus electrons to those objects.

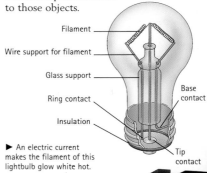

Filament

Wire support for filament

Glass support

Ring contact

Insulation

Base contact

Tip contact

▶ An electric current makes the filament of this lightbulb glow white hot.

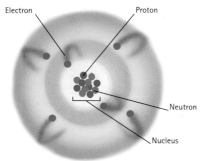

Electron

Proton

Neutron

Nucleus

Carbon atom

The nuclei of atoms consist of positively charged protons and uncharged neutrons. Electrons are negatively charged, travelling around a nucleus at great speeds. This carbon atom has six electrons, six neutrons and six protons.

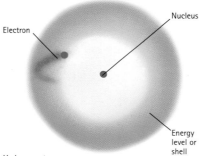

Electron

Nucleus

Energy level or shell

Hydrogen atom

The positive charge of the proton in this hydrogen atom attracts the negatively charged electron toward the nucleus. In all atoms, the electrons travel around the nucleus in energy levels, sometimes called shells.

A similar effect is produced by running a comb repeatedly through dry hair or by a person shuffling along a nylon carpet. This type of electricity is called static because the charge remains fixed (static) on the charged object – the amber, the comb or the person – until it finds some way to escape or be discharged.

Lightning is a spectacular example of a natural discharge. As they rise and fall, currents of air produce friction inside a thundercloud. The friction causes charges to build up within the cloud. From time to time, the charge becomes large enough to cause a discharge to the ground in an enormous spark – lightning. Much smaller sparks cause the crackling sound that is sometimes heard when folding woollen or nylon clothing.

CURRENT ELECTRICITY

An electric current is a flow of electrons from a point with too many electrons to a point with too few, just as a current of water flows from higher ground to lower levels. Whereas lightning is a catastrophic

VAN DE GRAAFF GENERATOR

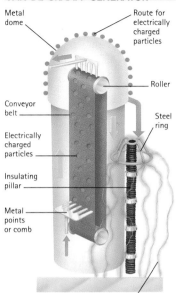

Metal dome

Route for electrically charged particles

Roller

Conveyor belt

Steel ring

Electrically charged particles

Insulating pillar

Metal points or comb

Discharged particles

The rubber conveyor belt in a Van de Graaff generator carries electrons from a high-voltage supply to a metal dome. As more electrons collect in the dome, the voltage of the dome increases. Large machines can produce 13 million volts of charge before the electrons in the dome jump to a steel ring and flash to the ground like lightning. Small machines collect enough charge to make a person's hair stand on end if they touch the dome, since electrons travel to the tips of the hair and repel each other.

burst of charge, the current electricity that provides heat, light and power for homes and industry is a regular flow that passes through cables to where it is needed. The cables are conductors – materials whose electrons can jump from one atom to the next with ease.

Most current electricity is produced by generators in power stations. Some comes from chemical reactions in batteries, or from the action of light on photocells.

Electricity is useful because it can be converted into other forms of energy. Electric motors convert electrical energy into mechanical energy which can be used to create movement. Electric heaters produce heat energy when a current flows through their heating elements. Electric lamps produce light energy in a similar way to heaters – a current flows through a fine wire, which glows white hot.

The skyline of central Toronto in Canada gives an indication of just how much people rely on electricity to provide illumination at night.

SEE ALSO

34–5 Winds, storms and floods, 240–1 Storage of electricity, 244–5 Conductors

ELECTRICAL CIRCUITS

Electrical circuits are collections of electrical components, such as resistors and diodes, that are linked by conducting wires and cables.

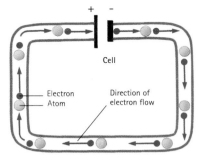

A cell drives the flow of electrons from atom to atom. The atoms stay in the same place; only some of the electrons move. Conventional current flows in the opposite direction to the electron flow.

André-Marie Ampère (1775–1836), a French physicist, studied electrical circuits. The unit of electric current is called the ampere, or amp, and is named after him.

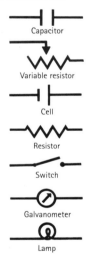

Capacitor

Variable resistor

Cell

Resistor

Switch

Galvanometer

Lamp

Electrical components each have their own unique symbol so that drawings of circuits, called circuit diagrams, can be understood worldwide.

Electrical circuits can be thought of as plumbing systems inside electrical devices such as toasters and computers. Current flows through the conducting wires between components, and each component has an effect on the current that flows through it.

A material that conducts well, such as a metal, contains a plentiful supply of free electrons. These are electrons that can move easily from atom to atom in the material. Plastics and rubber have no free electrons. Materials like these are good insulators – they cannot conduct a current.

In an unconnected piece of conducting wire, the free electrons move around randomly. No current flows because there is no overall movement of the electrons from one end of the wire to the other. However, this changes if the two ends of the wire are connected to a cell, which is commonly called a battery.

The chemical reaction that occurs in a cell produces an excess of electrons at one terminal of the cell – the negative terminal – and a shortage at the other – the positive terminal. When the two ends of a wire are connected to the cell, the negative terminal starts to feed electrons into the wire. Because like charges repel, free electrons in the wire move along the wire

Transistors can act as switches or amplifiers in circuits

towards the end that is connected to the positive terminal. There, the electrons drain back into the cell through the positive terminal. A current starts to flow.

Cells acts as pumps that create a high pressure of electrons at one terminal and a low pressure at the other. This is a potential difference and is measured in volts.

The ability of a power source to push electrons around a circuit is called its electromotive force (emf).

Before the discovery of the electron, scientists thought that positive charges moved around circuits. The conventional current of the positive charges flows

Bulb contacts

Battery terminals

Cell

▶ Lightbulbs wired in parallel each have the full voltage of the cell at their contacts. Each bulb shines as brightly as a single bulb in a simple circuit. Bulbs wired in series share the cell's voltage between them, and so shine less brightly.

In a parallel circuit, each bulb is connected to the terminals of the cell. All the bulbs shine brightly.

Lightbulbs in a series circuit share the same current. They glow less brightly than in a parallel circuit.

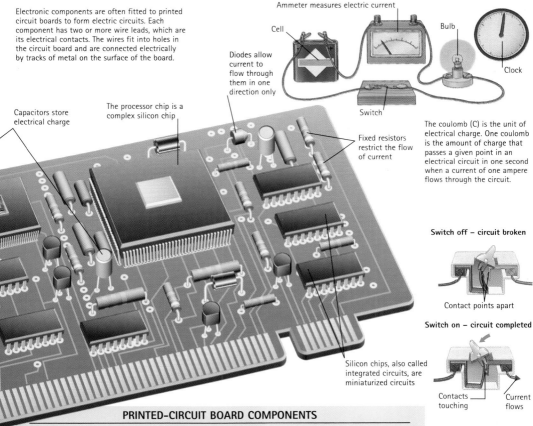

Electronic components are often fitted to printed circuit boards to form electric circuits. Each component has two or more wire leads, which are its electrical contacts. The wires fit into holes in the circuit board and are connected electrically by tracks of metal on the surface of the board.

Ammeter measures electric current

Cell

Bulb

Diodes allow current to flow through them in one direction only

Clock

Switch

Capacitors store electrical charge

The processor chip is a complex silicon chip

Fixed resistors restrict the flow of current

The coulomb (C) is the unit of electrical charge. One coulomb is the amount of charge that passes a given point in an electrical circuit in one second when a current of one ampere flows through the circuit.

Silicon chips, also called integrated circuits, are miniaturized circuits

Switch off – circuit broken

Contact points apart

Switch on – circuit completed

Contacts touching

Current flows

PRINTED-CIRCUIT BOARD COMPONENTS

Diodes	Resistors	Capacitors	Transistors	Silicon chips
Diodes are used to change alternating current into direct current in a process called rectification.	Resistors are used to add a fixed or variable amount of resistance to an electrical circuit.	Capacitors collect and store charge. They can be used to smooth the flow of a variable current.	Transistors can amplify current or switch a current on or off in response to a controlling signal.	Silicon chips are miniaturized electronic circuits. They are etched onto minute wafers of silicon.

Electrical switches operate by breaking a circuit and stopping the flow of current through a wire. When a switch is turned on, it completes the circuit and allows electrical current to flow around it.

from the positive terminal to the negative terminal. This is the opposite direction to the flow of electrons.

A cell drives a current in one direction only. This is a direct current, or d.c. The current from power stations changes direction many times per second and is called an alternating current, or a.c.

The amount of current that flows increases with the potential difference, or voltage, at the two ends of the circuit. If the voltage doubles, so does the current.

RESISTANCE

The amount of current that flows through a wire depends on the metal from which it is made. Some metals contain fewer free electrons than others; in some cases the 'free' electrons are less free to move.

The reluctance to allow electrons to flow freely is called resistance. Circuit components called resistors are designed to increase the resistance of a circuit. They can be connected in series – one after the other – or in parallel – side by side.

SEE ALSO

242 Electrochemistry,
244–5 Conductors, 252–3
Microprocessors, 254–5
Computers

MAGNETS AND MAGNETISM

Magnets can attract and repel each other over a distance through space because of their magnetic fields. The effect is called magnetism.

Physician William Gilbert (1544–1603), doctor to English monarchs James I and Elizabeth I, introduced the term 'magnetic pole'.

The north and south poles of a horseshoe magnet point in the same direction.

A bar magnet has a pole at each end. When it hangs on a string, it turns to point its north pole to Earth's magnetic north.

Some ring magnets have one pole on the outside of the ring and one on the inside. Others have one pole on either face.

The lines of force of a magnet's field can be 'seen' by sprinkling iron filings onto some paper around the magnet. The iron filings follow the lines of force, which are concentrated at the magnet's poles. The attractive and repulsive fields between two magnets can also be seen.

Magnetism is named after Magnesian stone, from Magnesia, northern Turkey. Over 2,000 years ago, the ancient Greeks found that pieces of Magnesian stone attracted some metals. The stone was magnetite, a form of iron ore. A magnet is an object that behaves in the same way as magnetite.

Some metals, such as chromium, become weakly magnetic when a magnet is placed nearby. This magnetism, which disappears when the magnet is removed, is called paramagnetism.

Only three metals – cobalt, iron and nickel – have the ability to become permanently magnetized when a magnet is placed near them and then removed. This property is called ferromagnetism.

A bar magnet can attract, pick up and support nails, paper clips and other small objects made of iron, nickel or steel. (Steel is a mixture of iron with small amounts of carbon and other materials.) Any piece of iron or steel can be turned into a magnet by stroking it several times in one direction with one end of a permanent bar magnet.

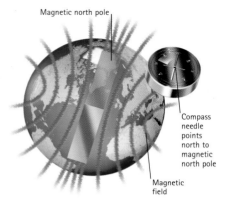

Magnetic north pole

Compass needle points north to magnetic north pole

Magnetic field

Earth acts like a huge magnet with a magnetic north and a magnetic south pole, both of which are close to the geographic north and south poles.

When an object is attracted to a bar magnet, it sticks to its ends, or poles, where the magnetic field is strongest. One pole of a magnet is north-seeking and the other is south-seeking. The north and south poles of two magnets will attract each other. Similar poles – north and north or south and south – will repel, forcing each other away.

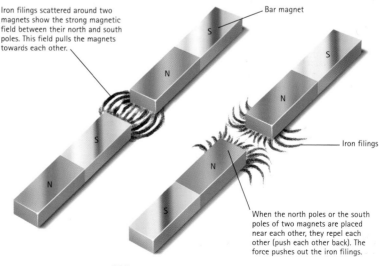

Iron filings scattered around two magnets show the strong magnetic field between their north and south poles. This field pulls the magnets towards each other.

Bar magnet

Iron filings

When the north poles or the south poles of two magnets are placed near each other, they repel each other (push each other back). The force pushes out the iron filings.

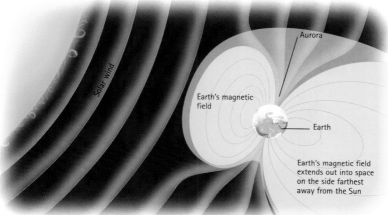

Apart from enormous amounts of heat and light energy, the Sun also pours subatomic particles into space, most of which are electrons. This invisible solar wind is electrically charged and affects the Earth's magnetic field. The solar wind makes the Earth's magnetic field lop-sided – it extends much farther into space on the side of the Earth farthest from the Sun.

Aurora

Earth's magnetic field

Earth

Earth's magnetic field extends out into space on the side farthest away from the Sun

The magnetic compass was first used around AD1100. Chinese sailors used it to navigate when clouds obscured the sky.

MAGNETIC FIELDS

A magnetic field is the region around a magnet in which other magnetic objects can be affected by its magnetism. A magnetic object will always try to align itself with another object's magnetic field. The stronger the magnet, the larger its magnetic field.

Earth has its own magnetic field, which is strongest at its magnetic north and south poles. The magnetized pointer of a magnetic compass aligns itself north to south and is a handy aid to navigation.

Magnetism is a force that, although it is not fully understood, is put to use in numerous ways. Many machines, from car ignition systems to electric motors, make use of the properties of magnets. Video and audio tapes are coated with a thin layer of magnetic material. It is this material that allows them to be used for recording and playback.

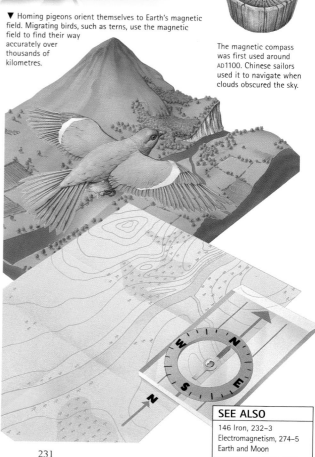

▼ Homing pigeons orient themselves to Earth's magnetic field. Migrating birds, such as terns, use the magnetic field to find their way accurately over thousands of kilometres.

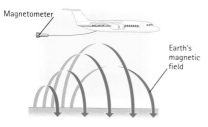

Magnetometer

Earth's magnetic field

Magnetometers are devices that measure the strength of a magnetic field. Towed behind aircraft, they can detect changes in the Earth's magnetic field.

SEE ALSO

146 Iron, 232–3
Electromagnetism, 274–5
Earth and Moon

ELECTROMAGNETISM

Electromagnetism connects the current that flows through a wire to the magnetic field around that wire. It is used to drive motors and to generate electricity.

Danish physicist Hans Christian Ørsted (1777–1851) was the first person to study the link between electric current and magnetism.

Horseshoe, bar and ring magnets are examples of permanent magnets. Their magnetism cannot be turned off and on. Electromagnets are not permanent magnets. Rather, they produce magnetism because of the flow of electrical current through a wire or a coil.

Danish physicist Hans Ørsted first noted the magnetic field produced by a current when, during one of his public lectures, he placed a compass near to a wire through which a current was flowing. The magnetic needle of the compass moved, indicating the presence of the magnetic field around the wire.

Electromagnets usually contain a piece of wire that is coiled many times to increase the magnetic field. A coil like this is called a solenoid. In most cases, the solenoid is wrapped around a core of a magnetic material such as iron. When a current flows through the coil, the iron core becomes temporarily magnetized. This increases the magnet's power by adding to the field of the coil.

It is possible to make a simple electromagnet by winding a coil of insulated wire many times around an iron nail and connecting the ends of the wire to the terminals of a battery.

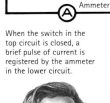

When the switch in the top circuit is closed, a brief pulse of current is registered by the ammeter in the lower circuit.

US physicist and Princeton professor Joseph Henry (1797–1878) discovered electromagnetic inductance. He built the electromagnetic motor in 1829.

Moving a magnet through a coil of wire induces an electric current in the wire. The direction of the current flow depends on the direction of movement of the magnet. Electrical generators produce an alternating current by spinning a coil in a magnetic field.

Electromagnets are useful because their magnetic field can be controlled by changing the current that passes through the solenoid. Sometimes the current is simply switched on or off, as in the case of a scrapyard crane's electromagnet.

In a loudspeaker, a variable current passes through a coil in the back of the loudspeaker cone. The varying magnetic field causes the cone to vibrate in the constant field of a permanent magnet. This produces air vibrations, or sound.

Maglev (magnetic levitation) trains and their tracks contain electromagnets. Magnetic repulsion makes the trains hover over the track, as well as driving them forwards and stopping them.

The ignition circuit of a car's engine uses electromagnetic induction to produce a high-voltage spark using the current from a low-voltage battery. While the switch is open, charge collects in the capacitor. When the switch closes, this charge flows through the primary coil as a pulse of current. This pulse induces a voltage in the secondary coil that is great enough to produce a spark at the spark gap. It is this spark that ignites the fuel and air mixture in the cylinder.

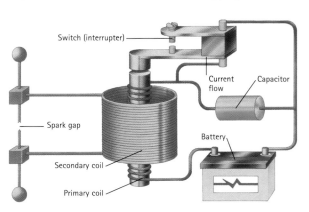

ELECTROMAGNETIC FIELD

When current passes through an electromagnet, a magnetic field is generated. The strength of this magnetic field depends on the number of turns in the coil of wire and the size of the electric current. Doubling either the number of turns or the size of the current produces a magnetic field that is twice as strong. The magnetic field drops to zero when the current is switched off.

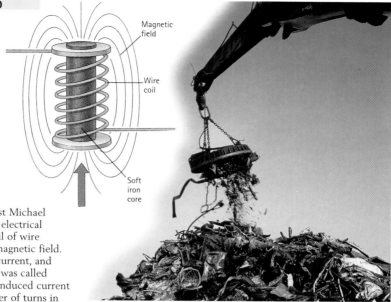

Magnetic field

Wire coil

Soft iron core

INDUCTION

In 1831, the British scientist Michael Faraday discovered that an electrical current flows through a coil of wire when it moves through a magnetic field. He called this an induced current, and the effect that produced it was called induction. The size of the induced current doubles if either the number of turns in the coil or the strength of the magnetic field doubles.

Electricity generators produce electrical current by making a coil rotate between two poles of a magnet. The movement in the field induces a current in the coil.

Induction coils use pulses of electric current in one coil to produce pulses in a second coil without either coil moving.

The pulse of magnetic field from the first coil affects the second coil as if it had been swept through a magnetic field. If the second coil has many more turns of wire than the first coil, a small current in the first coil will produce a much larger current in the second coil. Transformers use the same effect to change the voltage of an a.c. supply. This works with an alternating current because its direction of flow is constantly changing.

A large disc-shaped electromagnet hangs from the jib of this scrapyard crane. Steel and iron objects fly towards the magnet when the current is switched on. In this way, iron and steel can be separated for recycling.

Maglev trains use the repulsion of like magnetic poles to lift the train above the rail. A motor constructed out of many electromagnets uses magnetic fields to propel the train forwards. Maglev trains suffer much less friction and wear than conventional trains. Experimental maglev trains in Germany and Japan have reached speeds of over 400 km/h.

SEE ALSO

168–9 Electromagnetic spectrum, 230–1 Magnets and magnetism

GENERATORS AND MOTORS

Generators produce electricity by a magnetic field acting on a moving coil. Motors produce motion by a magnetic field acting on a current-carrying coil.

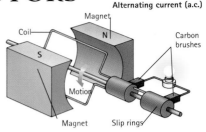

Alternating current (a.c.)

The current from the coil in this generator is almost zero and will start to reverse as the top arm of the coil gets closer to the south pole of the magnet.

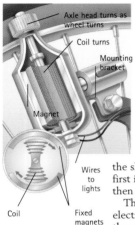

A bicycle dynamo is a simple electrical generator. As the bicycle wheel turns, it spins a coil between two fixed magnets.

A d.c. generator has several coils wound around the armature. As the armature rotates, carbon brushes tap current from whichever coil is in the strongest magnetic field. In this way, the generator produces an almost constant amount of electricity.

E lectricity generators use electromagnetic induction to convert mechanical energy into electrical energy. In an electric motor, the current that flows through a coil produces a magnetic field that makes the coil rotate between two fixed magnets.

GENERATORS

One type of generator has coils of wire that can be spun in a magnetic field by a turbine-driven shaft. As the shaft rotates, the magnetic field points first in one direction through the coils, and then in the other.

The motion of the coil creates an electrical current that is strongest when the wires of the coil are closest to the two magnets. As the shaft rotates further, the current drops to zero at the point where the wires of the coil are directly between the magnets. The current then starts to increase in the opposite direction as the magnetic field reverses.

The same generator can be used to produce either a.c. or d.c. electricity, depending on how it is connected. If the

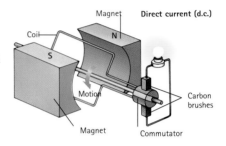

Direct current (d.c.)

The commutator in a d.c. generator reverses the connections of the coil at every half-turn. The output current flows in one direction only.

two ends of the coil are connected to the output cables through slip rings, the output cables are always connected to the same end of the coil and the current switches back and forth as the coil spins. Generators that produce an alternating current are sometimes called alternators.

The same coil can produce d.c. electricity by connecting its two ends to the output cables through a device called a commutator, which reverses the output connections every time the current starts to reverse. A d.c. generator is sometimes called a dynamo.

Power stations use steam, water or burning gas to drive the turbine that turns the generator. The turbine and generator are connected to the same shaft and the whole unit is called a turbogenerator.

In road vehicles and diesel locomotives, the main engine drives a generator that provides the electricity supply.

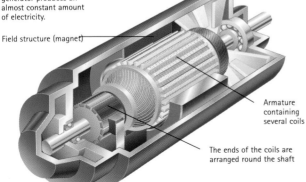

Armature containing several coils

The ends of the coils are arranged round the shaft

ELECTRIC MOTORS

An electric motor converts electrical energy into movement. Millions of electric motors, big and small, are used throughout the world. They power an enormous range of things, from cassette recorders and toys to ventilation fans and electric vehicles.

In most motors, the magnet stands still while the coil of wire carrying the electric current turns inside it. When a current flows through the coil, the coil becomes magnetized. Since unlike poles attract and like poles repel, the coil rotates between the two poles of the fixed magnet until the north pole of the coil is facing the south pole of the fixed magnet.

The direction of the current then reverses, which reverses the poles of the coil. The coil's north pole is now facing the north pole of the magnet. Since like poles repel, the coil swings round another half a turn to line up its poles again. As long as the current in the coil reverses at every half-turn, the coil keeps turning round.

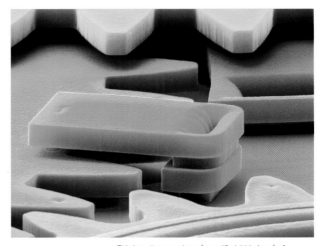

This is a close-up photo (magnified 200 times) of a micromotor's gear cogs. Micromotors have been developed for use in space missions and microsurgery.

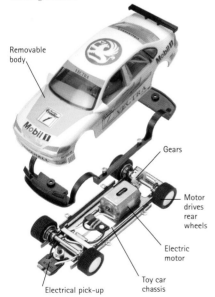

Some toy cars are driven by electric motors. The car connects to a direct-current (d.c.) supply through a pick-up that fits into slots in an electrified track.

A d.c. motor uses a commutator to reverse the current at each half turn. An a.c. motor is more complex, because the supply current reverses many times per second. Instead of a permanent magnet, some a.c. motors have an electromagnet that runs from the same supply as the coil. In this way, the magnetic field of the coil and the fixed magnet both reverse when the supply current reverses.

In this d.c. motor, current from a battery flows through a coil of wire. The magnetized coil then rotates in the magnetic field between the poles of a magnet.

FLEMING'S LEFT-HAND RULE

Fleming's left-hand rule predicts the direction in which a motor will turn by holding the thumb, forefinger and middle finger of the left hand at right angles to each other. By lining up the forefinger of the left hand with the magnetic field (south to north) and the middle finger with the current (positive to negative) the thumb shows which way the coil will move.

SEE ALSO

POWER STATIONS

Electricity is generated in power stations all over the world. Many burn fossil fuels to provide heat; others harness nuclear energy or renewable resources.

Natural petroleum can be refined to provide fuel oil for burning in power station furnaces.

Natural gas produces less pollution than fuel oil when it burns. It is often burned in gas turbines.

Coal is more plentiful than natural gas. However, burning coal produces more pollution than burning natural gas. Many modern power stations filter and remove this pollution from the furnace gases.

This generator shares a shaft with the turbine that drives it. A second turbogenerator is shown in the background.

Most electricity is generated in thermal power stations, which are factories that turn heat into electrical energy. In many power stations heat is provided by burning fossil fuels. Other power stations use the heat that is given off during a nuclear reaction to generate electricity.

FOSSIL FUELS
Fossil fuels are the remains of animals and plants that lived millions of years ago. The remains were covered by sediment and, over a period of millions of years, became underground fuel reserves.

Fossil fuels include oil, coal and natural gas. They are all compounds of carbon and hydrogen. Most power stations that use fossil fuels burn vast quantities of them to produce steam from water. The steam then drives turbines, which in turn provide mechanical power for generators.

Gas-turbine power stations use the hot gases from burning natural gas or fuel oil to drive turbines without making steam.

Fossil fuels are used in the vast majority of power stations around the world but they do have drawbacks. The gases formed

The hot water from the condensers of this coal-fired power station cools off in eight cooling towers. Some water is lost as steam, but most is re-used.

by burning fossil fuels include carbon dioxide, which causes global warming, and gases that form acid rain. Also, the fossil fuels will not last forever.

NUCLEAR POWER
Nuclear power stations use nuclear fission (the splitting of atoms) to release huge quantities of heat from small amounts of fuel. The nuclei of radioactive elements,

NUCLEAR POWER STATIONS

In principle, the only difference between a nuclear power station and a coal-fired or oil-fired power station is the heat source – the rest of the generating process is similar for both types. A coolant takes heat from the reactor to a boiler, where the heat turns water into steam. The steam drives the turbine that turns the generator. Then the steam turns back to water in the condenser before returning to the boiler. Cooling water from the condenser loses its extra heat through cooling towers.

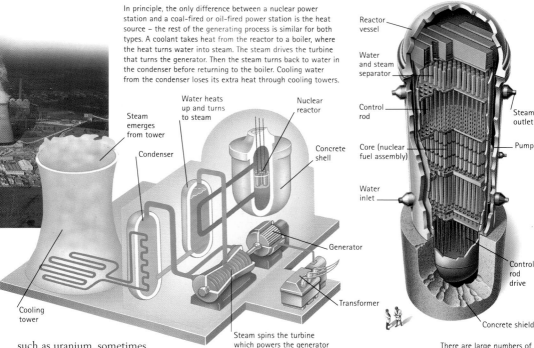

Reactor vessel

Water and steam separator

Control rod

Core (nuclear fuel assembly)

Water inlet

Steam outlet

Pump

Control rod drive

Concrete shield

Steam emerges from tower

Water heats up and turns to steam

Nuclear reactor

Condenser

Concrete shell

Generator

Transformer

Cooling tower

Steam spins the turbine which powers the generator

There are large numbers of fuel rods and control rods inside a nuclear reactor. In the reactor above, they are immersed in water, which is the moderator and also the coolant.

such as uranium, sometimes split. When they do, they release heat energy and tiny particles called neutrons. When these neutrons hit other radioactive nuclei they can make them split, starting what is called a chain reaction. Materials called moderators help the chain reaction to occur. They slow down the neutrons, which improves their chances of causing fission. Control rods slow or stop the chain reaction by absorbing neutrons and preventing them from causing fission.

In the 1950s, nuclear power was seen as the solution to the world's energy needs. One kilogram of uranium fuel can provide as much energy as more than 2,000 tonnes of coal without producing any carbon dioxide or acid rain gases.

Unfortunately, the high cost of

handling and using nuclear fuel safely makes nuclear electricity expensive. Unsafe handling can be disastrous, as was shown by the 1986 explosion at the nuclear power plant in Chernobyl in the Ukraine.

The neutrons released by fission reactions in the fuel rods are slowed by a moderator to improve their chances of causing other atoms to split. Control rods can stop or slow the chain reaction by absorbing neutrons.

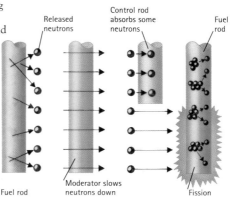

Released neutrons

Control rod absorbs some neutrons

Fuel rod

Fuel rod

Moderator slows neutrons down

Fission

In 1942, Italian physicist Enrico Fermi (1901–1954) built the first nuclear reactor in a squash court at the University of Chicago.

SEE ALSO

10–11 Fossils and geological time, 238–9 Renewable energy sources

RENEWABLE ENERGY SOURCES

Unlike fossil fuels, renewable energy sources will never run out. They include sunlight, wind, tides and rainwater in the form of hydroelectric power.

A vertical-axis wind turbine spins around an upright shaft. It can catch wind from any direction.

Three-quarters of the world's electricity is generated by thermal power stations that run on fossil fuels or nuclear power. Fossil fuels produce pollution and are in limited supply. Nuclear fuels are expensive to use safely, and their waste is dangerous and difficult to store.

There are several other ways of driving electricity generators without creating pollution or the risk of nuclear accidents. Many of these alternatives use natural resources that will not run out. They are known as renewable energy sources.

WIND TURBINES

For many centuries, windmills have been used to grind grain and to pump water. They convert the power of the wind into useful mechanical power. Nowadays, more sophisticated windmills, called wind turbines, use wind power to drive electricity generators.

The blades of a wind turbine can turn around a horizontal or a vertical shaft. Those that turn around a horizontal axis, similar to a windmill's, have a mechanism that points them into the wind if the wind direction changes. Vertical-axis turbines catch the wind from any direction.

A single wind turbine with 25-metre blades can provide enough electricity for a small community. They are most useful for remote villages and islands.

Most wind turbines are built in groups called wind farms. The largest wind farms are currently found in Denmark and the United States. A typical farm produces over 1,000 megawatts in windy weather.

Wind farms are often located on flat coasts or offshore, where the wind is stronger and less gusty than it is inland.

▶ Horizontal wind turbines look like propellers. The amount of power that they can generate depends on the speed and angle of the wind across their blades. An electronic control mechanism turns the turbine head to point into the wind direction at all times, thereby collecting the most power from the wind. A transmission system connects the turbine blades through a shaft to a generator in the turbine head.

Positioning gears

Transmission

Turbine head

Control electronics adjust position of wind turbine head

Drive shaft

Internal ladders allow access to wind turbine head

Electrical generator

Vents for cooling air

Turbine blade

▼ By 2030, more than 25 per cent of all Denmark's electricity is expected to be generated by offshore wind farms.

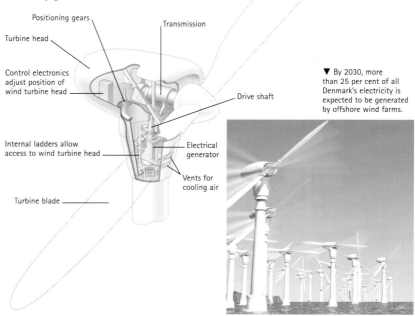

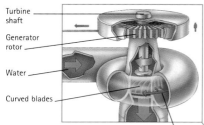

Turbine shaft
Generator rotor
Water
Curved blades
Water

▲ Water enters the turbine through a curved pipe around its edge. Water passes through the turbine blades and leaves through the middle.

WATER POWER

There are three forms of water power: hydroelectric power, tidal power and wave power. Hydroelectric power, or HEP, uses the force of water flowing out of a reservoir and through a dam to move turbines and generate electricity.

Tidal power harnesses the power of water as it flows into and out of a river estuary. A low-level dam, or barrage, is built across the estuary, and the tidal water turns turbines in the barrage as the tide rises and falls. The tidal power plant at La Rance, France, produces enough power for the needs of 300,000 people.

Wave-power devices use the up-and-down movement of waves to drive electricity generators. One such device is called a duck. Ducks are teardrop-shaped plastic floats (see page 459). They contain devices that pump water as they bob up and down in waves. Water from a string of ducks can pump enough water to drive a small turbine and generator.

SUNLIGHT

Heat from sunlight can be focused onto a boiler in the middle of thousands of mirrors in a field. Steam from the boiler drives turbogenerators similar to those found in a coal-fired power station. Devices called photovoltaic cells produce a current whenever light shines on them. They are used to provide current for spacecraft, and some calculators and radios.

HYDROELECTRIC POWER

Hydroelectric power plants (HEPs) are often built in mountainous areas. There, rainfall is plentiful, and natural valleys can be dammed to provide a store of water called a reservoir. The largest single scheme is at Itaipu, on the Brazil–Paraguay border. This plant can generate over 12,000 megawatts of power. A scheme in China will soon generate 20,000 megawatts. HEP generates almost 20 per cent of the world's electricity. Some hydroelectric power stations use a system called pumped storage. During the day, water flows from an upper reservoir down through the hydroelectric plant, turning turbines, before being collected in a lower reservoir. At night, when electricity demand is low, surplus power from other power stations pumps water up to the top reservoir.

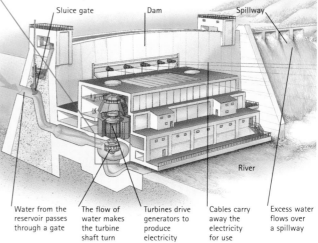

Sluice gate   Dam   Spillway

River

| Water from the reservoir passes through a gate | The flow of water makes the turbine shaft turn | Turbines drive generators to produce electricity | Cables carry away the electricity for use | Excess water flows over a spillway |

GEOTHERMAL

The rocks beneath Earth's surface are often hot. Geothermal power stations use the heat in these rocks to turn water into steam. The steam can be used to generate electricity or for heating.

▶ At this geothermal power station in Iceland, wells are drilled to tap into the heat lying beneath Earth's surface. Water is pumped into the hot rocks where it turns to steam. This steam can be used to power electricity generators or to heat homes and offices.

SEE ALSO
10–11 Fossils and geological time, 238–9 Renewable energy sources

STORAGE OF ELECTRICITY

Electrochemical cells and batteries are self-contained and often mobile sources of energy. Capacitors are used to store electrical charge.

Stopper
Brass rod

Outer foil
Inner foil

A Leyden jar is a type of capacitor. Sheets of foil cover half the jar inside and out. The electric charge, applied through the brass rod, collects on the inner foil.

This fruit cell consists of a steel paper clip and a brass drawing pin stuck into a lemon. The juice of the lemon starts the electrochemical reaction.

Fixed capacitor

Variable capacitor

▲ These symbols represent fixed and variable capacitors in circuit diagrams.

Electricity can be stored as charge, in devices called capacitors or as chemical energy in cells. Batteries are collections of cells linked together. A variety of chemical reactions can be used to store electricity. In primary cells the electricity runs out when the chemical reaction is complete, after which the cell must be discarded. In rechargeable cells, which are also called secondary cells or accumulators, the reaction can be reversed by pumping electricity into the cell. The recharged cell can then be used again.

CAPACITORS

A capacitor is a device that stores electrical charge. Capacitors consist of two charged metal plates that are separated by an insulator. Charges of opposite signs collect on each plate when a potential difference, or voltage, is applied across the plates. The amount of charge increases as the voltage increases. The ability of a capacitor to store charge is called its capacitance. Capacitors are used to help regulate the current in electrical circuits.

BATTERIES

The first cell was built by Italian physicist Alessandro Volta (1745–1827) in 1800. Volta's cell, also called a voltaic pile, was a stack of alternating plates of copper and zinc. Sheets of card kept the metal

plates apart, and the whole assembly was soaked in a solution of acid. The solution started the chemical reaction that produced electricity.

The chemical reactions that run cells occur in two parts, or half reactions. One half reaction produces electrons at the negative electrode, or anode. The other half reaction consumes electrons at the positive electrode, or cathode. A solution called an electrolyte is in contact with the electrodes and provides the materials for the reactions. In Volta's cell, the copper sheets were the cathode, and the zinc sheets the anode. The acid solution was the electrolyte of the cell.

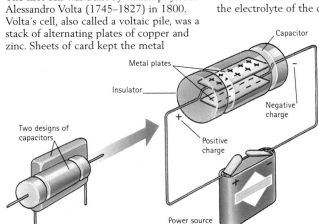

Capacitor
Metal plates
Insulator
Negative charge
Positive charge
Two designs of capacitors
Power source

A capacitor consists of two metal plates that are separated by an electrical insulator. Positive charge is stored on one plate, and negative charge on the other. The amount of charge stored at a given voltage depends on the size of the plates and the distance between them.

The rechargeable battery found in most motor vehicles is a collection of cells, in which lead plates are immersed in concentrated sulphuric acid and contained within a plastic casing. There are usually six cells, each consisting of one set of lead plates that is connected to the negative terminal and another set connected to the positive. A small dynamo, driven by the vehicle's engine, charges the battery whenever the engine is running.

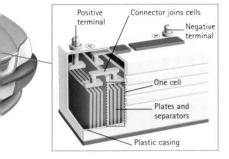

Positive terminal

Connector joins cells

Negative terminal

One cell

Plates and separators

Plastic casing

PRIMARY CELLS

The most commonly used type of primary cell is the zinc–carbon battery or dry cell, which was invented in the 1860s by French engineer Georges Leclanché (1839–1882). This type of battery is the standard 'battery' for use in equipment such as torches, toys and radios.

The steel casing of a dry cell covers a zinc cup, which is the anode. The cathode is a carbon rod that fits inside the zinc cup. The cathode is connected to a metal stud in the top of the battery. The space between the carbon rod and the zinc case contains a paste of ammonium chloride and zinc chloride.

Alkaline batteries are an improvement on zinc–carbon cells. They are made in the same shapes, sizes and voltages as zinc–carbon cells. When used in the same equipment, however, an alkaline cell will last about six or seven times longer than a regular zinc–carbon cell. This is because the combination of chemicals in an alkaline cell produces so much more energy before the chemicals are all used up by the electrochemical reaction.

SECONDARY CELLS

Although primary cells can be made to last longer by using more efficient chemical reactions, they can only be used once. In some cases, attempting to recharge a primary cell will make it explode. A secondary cell, on the other hand, can be recharged and reused a number of times before it stops working. Secondary cells are recharged by passing a small electrical current into the cell, often over a number of hours.

The lead–acid batteries used in many road vehicles contain secondary cells. The energy stored in the battery is used to power a vehicle's electrical systems. While the vehicle is running, its engine drives a small generator that charges the battery.

Nicads – cells with nickel and cadmium electrodes – are another type of secondary cell. Nicad cells can usually be recharged between 500 and 900 times before they have to be replaced. These rechargeable cells are often used to power portable electrical goods, including personal stereos and electric shavers.

BATTERY IMPROVEMENTS

Research work has already created new types of primary cells that produce more power for longer periods. Much work is being done to develop secondary cells that can store enough energy to power electric cars for long distances. Batteries of these cells must also be able to be recharged quickly at roadside recharging stations.

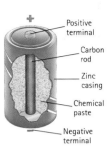

Positive terminal

Carbon rod

Zinc casing

Chemical paste

Negative terminal

This standard zinc–carbon cell has a carbon rod surrounded by a paste of ammonium chloride and zinc chloride. The paste causes a reaction between the carbon and the zinc. This generates a voltage.

This small disc-shaped cell is a zinc–mercuric-oxide cell. It is used to power hearing aids and watches, and also as a power source in digital cameras and portable electronic personal organizers.

The batteries of this French Citroën AX electric car may be charged overnight at home, then topped up during the day at rapid charge points in service stations, such as this one in La Rochelle, France. The batteries store enough charge to drive for about 75km.

SEE ALSO

120–1 Chemical reactions, 228–9 Electrical circuits, 242 Electrochemistry

ELECTROCHEMISTRY

Electrochemistry is the science of chemical reactions and electrical currents. It describes the conversion of energy between chemical and electrical forms.

Although Englishman Michael Faraday (1791–1867) had little formal education, he went on to become a renowned scientist and the pioneer of electrochemistry. Faraday invented the first electric generator and investigated electrolysis.

Some electrochemical reactions occur spontaneously when two electrodes, dipped in a liquid, are connected to an electrical circuit. This is what happens in a battery. Other reactions occur when an electrical current passes through a liquid.

One type of electrochemical process is electrolysis, which is the chemical breakdown of a compound by an electrical current. Some liquids, called electrolytes, contain particles called ions, which are atoms or fragments of molecules that carry a positive or negative charge.

If two electrodes are connected to the terminals of a battery and placed in the electrolyte, positive ions, called cations, will be attracted to the negative electrode, or cathode. At the same time, negative ions, called anions, will be attracted to the positive electrode, or anode.

When an ion reaches an electrode, it loses its charge and changes chemically. When acidified water is electrolyzed, for example, positive hydrogen ions are attracted to the cathode. There, they pick up electrons and form hydrogen gas.

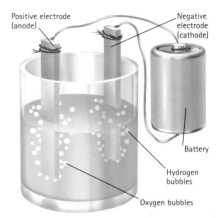

Positive electrode (anode)

Negative electrode (cathode)

Battery

Hydrogen bubbles

Oxygen bubbles

The current from a battery can split acidified water into hydrogen and oxygen. Bubbles of hydrogen form at the cathode, and bubbles of oxygen form at the anode.

Objects can be electroplated with a thin layer of certain metals in a cell where the object is the cathode, the plating metal is the anode and the electrolyte contains cations of the plating metal. The metal cations become atoms at the cathode and form a surface layer of pure metal. Copper, gold, silver and zinc are all used to plate objects in this way.

Many decorative objects, such as candlesticks, are made from a cheaper base metal and then electroplated with silver. The thin coating of silver provides an attractive finish and protection from corrosion.

These sheets have been coated with copper in an electroplating bath. They will be used to manufacture flexible printed-circuit boards.

SEE ALSO

124–5 Bonding and valency, 136 Oxidation and reduction, 240–1 Storage of electricity

POWER CELLS

Different types of power cells generate electricity from chemical reactions, sunlight and the action of pressure on crystals.

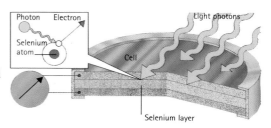

Panels of solar cells

Power cells are used to provide electrical power for a variety of applications. The type of cell used depends on where and how it will be used.

This Intelsat-V satellite orbiting Earth has two arrays of photovoltaic cells, commonly called solar cells, that are capable of generating electricity from sunlight.

One type of photoelectric cell uses selenium as its cathode. It releases electrons when photons of light shine on it. The flow of electrons forms an electric current.

FUEL CELLS

Like batteries, fuel cells use chemical reactions to produce electricity. Unlike batteries, they continue to operate just as long as they are supplied with fuel and oxygen. The fuel may be hydrogen or a substance, such as natural gas or petroleum, that contains hydrogen. The fuel and oxygen combine in the fuel cell to produce electricity and water.

Fuel cells are efficient and do not cause pollution. They are used in spacecraft, where the water they produce can be used for drinking, washing and cooking. Fuel cells are also used to power some types of electric road vehicles.

PHOTOELECTRIC CELLS

Photoelectric cells, or solar cells, produce electrical current from light. The cells contain a material that releases electrons when light shines on the cells. These electrons are then caught by a conductor, which is the negative terminal of the cell. The material itself is the positive terminal, since it becomes positively charged when the electrons jump out of it.

PIEZOELECTRICITY

Certain crystals, including quartz, develop a voltage when squashed or stretched slightly. This is the piezoelectric effect which may be used to power cells in the future. Some digital watches use the effect the other way round, passing an electric current through a crystal to obtain a very accurate vibration as a reference for time.

In a quartz watch, a voltage from the watch's battery makes a quartz crystal vibrate. The speed of these vibrations can be used to measure hours, minutes and seconds.

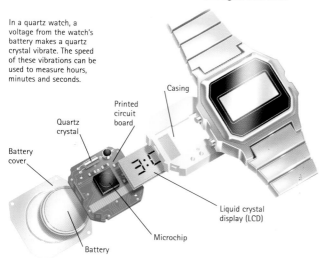

Casing

Printed circuit board

Quartz crystal

Battery cover

Microchip

Battery

Liquid crystal display (LCD)

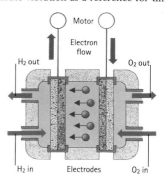

Motor

Electron flow

H_2 out

O_2 out

H_2 in

Electrodes

O_2 in

▲ Hydrogen and oxygen are pumped into a fuel cell, where a catalyst in the electrodes makes them react. The reaction produces electricity and water.

SEE ALSO

CONDUCTORS

Electric conductors will carry an electric current. Insulators will not normally conduct. Semiconductors can act either as conductors or insulators.

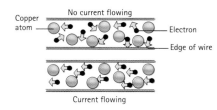

Current moves through materials that conduct electricity. When current flows through a copper wire, for example, the electrons all move in the same direction.

German physicist Gustav Kirchhoff (1824–1887) drew up a set of laws for calculating the amount of current flowing in a conductor. Working with German chemist Robert Bunsen (1811–1899), Kirchhoff also discovered the chemical elements, caesium and rubidium.

An electric arc produces intense heat where it contacts electrodes. This heat can be used to weld pieces of metal together.

Metals are good electrical conductors because they contain plenty of free electrons that can move from atom to atom, carrying the current. The journey is not completely smooth, however, and interference along the way can hinder the flow of current. This phenomenon, called resistance, also causes the material through which the current is flowing to become warm. Good conductors, such as silver and copper, have low resistance.

Insulators are substances, such as plastics and non-metals, whose electrons are not normally free to move between the atoms or molecules. Under most conditions, the resistance of an insulator is high. If a large enough voltage is used, however, some of the electrons can be ripped out of their atoms or molecules. When this breakdown happens, those electrons are free to move and the resistance of the material drops.

A great amount of heat is generated when current flows through an insulator at breakdown. The heat produced by a current, or arc, passing through air is used to melt metals in arc welding.

SUPERCONDUCTORS

Even good conductors, such as metals, offer some electrical resistance under normal conditions. There are, however, materials called superconductors. These offer almost no resistance. They are useful because they save a lot of energy when a strong current is necessary, as is the case in a powerful electromagnet.

Some metals, such as aluminium and lead, can superconduct when they are cooled to incredibly low temperatures – just a few degrees above absolute zero (–273°C). Liquid helium, which is costly and difficult to handle, is used for cooling.

▶ A magnet hovers over a ceramic superconductor made of yttrium, barium and copper oxides. This material only superconducts at very low temperatures.

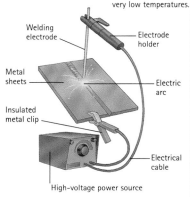

▲ In arc welding, an electrical current flows between a welding electrode and the metal objects that are to be welded. Heat from the electric arc melts the metal around the joint. A strongly welded joint forms once the metals have cooled and solidified.

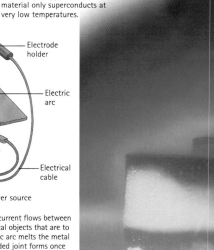

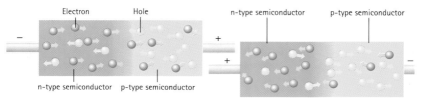

n-type semiconductor | p-type semiconductor

Electron | Hole

n-type semiconductor | p-type semiconductor

◄ Electrons move easily
from n-type to p-type
silicon, but much less
easily in the opposite
direction. In the left-hand
diagram, current flows in
the preferred direction.
In the right-hand diagram,
the voltage is reversed and
very little current flows.

Some superconductors function when they are cooled to –196°C using liquid nitrogen. This is cheaper and easier than using liquid helium. Scientists have already made materials that superconduct when cooled to –78.5°C using dry ice (solid carbon dioxide). In the future, there will be superconductors that function well at close to room temperature.

SEMICONDUCTORS

Semiconductors have a much higher resistance than a normal conductor, but much lower than an insulator.

Like other semiconductors, silicon conducts better when it contains traces of impurities than when it is pure. The addition of an impurity is called doping.

If silicon is doped with phosphorus, which has one more electron per atom than silicon, the extra electrons can carry negative charge through the silicon. This is called n-type silicon.

A boron atom has one less electron than a silicon atom, so boron-doped silicon is a few electrons short. These gaps, called holes, carry positive charges. This is called p-type silicon.

What makes silicon transistors work as switches is that electrons will not flow from p-type to n-type silicon.

An n-p-n transistor has a layer of p-type silicon between two n-type layers. Electrons will not flow between the two outer layers, the emitter and the collector, because they would have to pass from p-type to n-type silicon.

However, if electrons are fed into the middle layer, or base, they fill the holes in that layer and a current can flow from the emitter to the collector.

Nobel Prize-winning US physicist William Shockley (1910–1989) was a key member of the team that developed the first transistor in 1948.

Collector

Base

Emitter

This is the electrical symbol for a transistor, showing its three component layers: the emitter, base and collector.

Transistor

A small current flowing into the base of an n-p-n transistor allows a larger current to flow between its emitter and collector.

SEE ALSO

226–7 Electricity, 232–3 Electromagnetism

INSULATORS

Electrical insulators do not conduct electricity. They perform many valuable tasks in electrical circuits, as well as ensuring the safe transmission of electricity.

Single-core wire

Insulated multi-strand wire

Pair of wires with insulation

Electrical cable with three layers of insulation

▲ By placing a rubber or plastic insulator as a sheath around a good conductor, such as a copper wire, electricity can flow along the wire without causing harm.

A good insulator is an extremely poor conductor. It has no mobile electrons, or at least not enough to carry an electric current if a voltage is applied across a piece of the material. This means that insulators act as barriers to the flow of electricity.

Rubber is one of the most effective natural insulators. Glass and plastics are also good insulators, as are porcelain and many other ceramics. Air and other gases insulate well, as does a vacuum. Many liquids – water being an exception – also insulate very efficiently.

In an ordinary piece of electrical flex, the current-carrying cables are insulated in plastic to prevent them from forming a short circuit that could start a fire. Another insulating layer is wrapped around these wires to give extra protection against potentially harmful electric shocks.

This worker at a US hydroelectric power station is carrying out routine maintenance on a high-voltage insulator made from a ceramic material.

The wires used to form the coils for solenoids and transformers are coated with a thin layer of insulating varnish that allows them to be wound together without forming short circuits.

Heavy-duty glass and ceramic insulators are used to isolate high-voltage power-transmission lines from the structures that hold them in place.

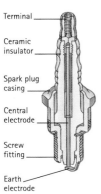

Terminal

Ceramic insulator

Spark plug casing

Central electrode

Screw fitting

Earth electrode

◄ Ceramic materials are excellent electrical insulators. They are used in many places. Here, a ceramic insulator prevents high-voltage charge from causing a short circuit inside a spark plug.

▶ The transformers at this electricity substation have heavy-duty ceramic insulators to keep the high-voltage cables from touching each other and prevent electricity from leaking to the ground.

SEE ALSO

226–7 Electricity, 156 Polymers

RESISTANCE

Resistance is the ability of a substance to resist the flow of an electrical current. Resistors are components in a circuit that provide a known amount of resistance.

German physicist Georg Ohm (1789–1854) developed Ohm's law and discovered how the resistance of a material depends on its length, thickness and the material from which it is made. Ohm's Law was published in 1827.

Ohm's triangle is used as a memory aid to calculate the current (I), voltage (V) or resistance (R) of a circuit using Ohm's law: V = I x R.

Resistance is a measure of how much a circuit, or a component of a circuit, restricts the flow of electrons. Atoms collide with the electrons, slowing them down and causing some electrical energy to be converted into heat energy and sometimes light.

The length and thickness of a conductor both have an effect on its electrical resistance. The thicker a wire, the easier the current flows. This is because a wider cross-section contains more free electrons to carry the current.

The longer a wire, the greater its resistance. This is because there are more atoms in the path of the free electrons. This is why circuit designers try to use as little wire as possible to cut down on energy losses caused by resistance.

Resistors are components that are designed to reduce the amount of current that travels through a particular part of a circuit. They are often used to protect delicate components from carrying too great a current.

Thermistors are resistors whose resistance depends on temperature. They are often used in circuits that measure temperature. Light-dependent resistors are used to measure light intensity.

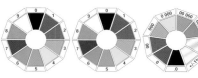

The value of a fixed resistor is encoded in three coloured bands. A fourth band shows the possible variation of resistance between individual resistors of the given type.

Band 1   Band 2   Band 3

55Ω   24,000,000Ω

The left-hand resistor has a value of 55 ohms, the right-hand resistor is 24 million ohms. The silver bands show that these values are correct to within 10 per cent.

Manually variable resistances can be adjusted by turning a knob or by moving a slider. They are used in the volume and tone controls of stereo equipment.

MEASURING RESISTANCE

The standard unit of resistance is the ohm (Ω), named after German physicist Georg Ohm. The resistance of a circuit is one ohm when a voltage of one volt makes a current of one amp flow.

Ohm's law states that the resistance in a circuit is the voltage divided by the current. A voltage of two volts would make a current of two amps flow in a circuit that has a resistance of one ohm. But two volts would produce a current of only one amp in a circuit of two ohms. Put in other ways: voltage is current multiplied by resistance; or current is equal to voltage divided by resistance.

The resistance of a thermistor (top) changes with temperature. Some variable resistors are adjusted by hand (middle). Others change when light strikes them (bottom).

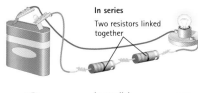

In series
Two resistors linked together

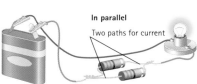

In parallel
Two paths for current

When resistors are wired in series, their resistances add together and the bulb glows dimly. When in parallel, there are two paths for the current to flow along, each with only one resistance, and the bulb glows brightly.

Fuses are safety devices that limit the size of the current that can pass through a circuit. If the circuit develops a fault, this sends a surge of current through the fuse. The resistance of the fuse wire makes it heat up and melt. The fuse then blows, breaking the circuit before further damage can occur.

Fuse wire

SEE ALSO

TELECOMMUNICATIONS

Telecommunication is the transmission of electronic or electromagnetic signals to carry words, sounds, images and other data over long distances.

Scottish-born inventor and teacher Alexander Graham Bell (1847–1922) built the first working telephone in 1875 and patented it one year later.

Bell's telephone used a coil and electromagnet to turn sounds into an electrical signal.

This telephone, dating from 1919, had a body with a mouthpiece and rotary number dial, and a separate earpiece.

People have needed to communicate over long distances for thousands of years. At first, the fastest way to send news and other information was by runners, then messengers on horseback or on boats.

Later, visible signals were passed through chains of observers and signallers who used smoke, reflected sunlight and devices such as flags. These methods conveyed simple messages much quicker than a messenger could.

The telegraph offered the first electrical communication system that could carry detailed messages over long distances in the form of coded electrical pulses.

In 1876, while working to improve the telegraph, Alexander Graham Bell invented the telephone. He did this by discovering a way to send the human voice along wires as electrical signals. Bell had no idea how successful the telephone would become – he believed that just a handful of his devices would ever be needed. Today, there are over a billion telephones in use worldwide.

Telephone systems advanced rapidly following Bell's invention. At first, human operators connected callers' lines manually by plugging the two ends of a cable into a socket for each caller. The first mechanical switchboard, or exchange, was patented in 1891. Most modern exchanges are now computerized, which allows callers to dial direct from their telephones to almost every country in the world.

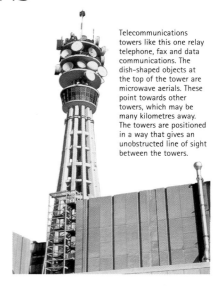

Telecommunications towers like this one relay telephone, fax and data communications. The dish-shaped objects at the top of the tower are microwave aerials. These point towards other towers, which may be many kilometres away. The towers are positioned in a way that gives an unobstructed line of sight between the towers.

CABLES AND WIRES

Telephones rely on a network of cables and wires to carry electrical signals for at least part of the journey from callers to receivers. Telephone lines can be strung between poles, buried underground or laid on the beds of oceans. The electrical signal in a telephone wire travels thousands of times faster than sound travels through the air. Electrical signals also travel much farther than the human voice, and amplifiers along the line help to keep

SATELLITE COMMUNICATIONS

Geostationary satellites orbit Earth at a height of approximately 35,900 kilometres. At this altitude, they orbit at the same speed as the planet rotates, so they stay above a fixed location on Earth's surface. In 1945, science-fiction writer Arthur C. Clarke suggested the use of geostationary satellites to relay phone calls, television images and other signals between ground stations separated by thousands of kilometres. The first geostationary communications satellite, called *Syncom 2*, was launched in 1963. Since then, hundreds of communications satellites have been put into stationary orbits. These satellites receive signals from transmitter antennas on Earth's surface, amplify them and then transmit them to receivers on the ground.

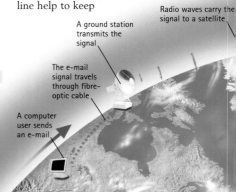

Radio waves carry the signal to a satellite.

A ground station transmits the signal

The e-mail signal travels through fibre-optic cable

A computer user sends an e-mail

THE TELEPHONE

A telephone handset has a mouthpiece and an earpiece. When a caller speaks, the sound waves hit a thin metal diaphragm and cause it to vibrate. This vibration repeatedly compresses carbon granules in a cylinder behind the diaphragm. This makes their electrical resistance vary so that the current that flows between two electrodes in the mouthpiece mimics the sound waves of the speaker's voice. The incoming signal makes an electromagnet and diaphragm in the earpiece vibrate to reproduce the sound of the other caller's voice.

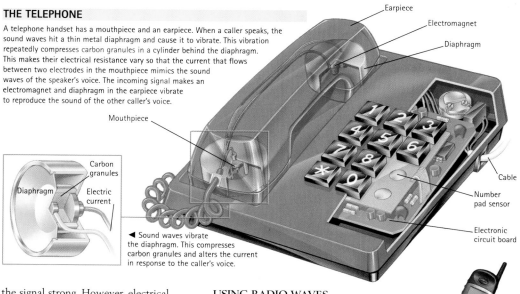

Earpiece

Electromagnet

Diaphragm

Mouthpiece

Cable

Number pad sensor

Electronic circuit board

Carbon granules

Diaphragm

Electric current

◀ Sound waves vibrate the diaphragm. This compresses carbon granules and alters the current in response to the caller's voice.

the signal strong. However, electrical interference reduces the sound quality over very long distances.

Many telephone networks use optical fibres to connect exchanges. Optical fibres are thin threads of glass. Exchanges turn electrical signals into light signals, which travel through optical fibres without interference. A pair of optical fibres can carry up to 6,000 calls at one time.

USING RADIO WAVES

Many telecommunications systems use high-frequency radio waves called microwaves. Relays and satellites pass the signals around the world.

Mobile telephone networks use numerous antennas to transmit and receive signals within small areas called cells. When a caller uses a mobile phone, a microwave signal is sent to the nearest antenna, which connects the caller to the network. Regular signals from each mobile telephone let a central computer know where to direct incoming calls for that telephone.

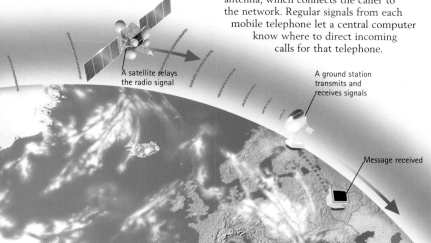

A satellite relays the radio signal

A ground station transmits and receives signals

Message received

Mobile telephones provide portable communication for people on the move.

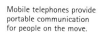

Pagers are small radio receivers that monitor a specific radio frequency. When a message is sent, a signal that is unique to the pager switches the pager on. The pager then collects and displays the message.

SEE ALSO

226–7 Electricity, 256–7 Information technology

TELEVISION AND VIDEO

Television captures and transmits images and sounds to millions of television receivers all over the world. Video records images and their soundtracks.

Scottish electrical engineer and inventor John Logie Baird (1888–1946) was the first person to transmit a television picture by radio waves. He applied for patents as early as 1923, and worked almost until his death on new methods of transmission.

Television broadcasts began in Britain in 1936 and in the United States in 1939.

At first, few people owned television sets, and the cameras, transmitters and receivers could process images only in black and white. Since then, the popularity of television has grown enormously, especially since the introduction of colour television in 1953.

One of the pioneers of television was John Logie Baird. In 1926, Baird demonstrated a system called the Baird Televisor. It used a rotating disc with a spiral of lenses, called a Nipkow disc, to divide an image into horizontal lines. Each lens would focus a horizontal line from the image onto a photoelectric cell. The cell produced an electrical current that varied with the amount of light that fell on it. This process, called mechanical scanning, converted the image into a series of signals that could be transmitted as radio waves. The receiver turned the signal into a light of varying intensity. When this light passed through another Nipkow disk, it reproduced the image on a screen.

Modern television sets use electron guns to produce beams of electrons, which 'paint' television images in lines on a screen. The beams are deflected by electrically charged plates that steer the beams from side to side and from top to bottom. Chemicals called phosphors glow when the beam hits them.

Standard European television systems use 625 lines per frame, or image, and 25 frames per second. Japanese and US systems use 525 lines and 30 frames per second. High-definition television, or HDTV, uses 1,125 lines to give sharper, more detailed images.

HOW TELEVISION WORKS

Light detectors in a TV camera convert light into electric signals, which are processed before being transmitted as radio waves. The screen of a television is the front of a funnel-shaped glass tube. The inside of the screen is coated with strips of phosphors – chemicals that glow three different colours when electrons hit them. Three electron guns at the back of the television tube fire beams of electrons at the three sets of phosphors behind the screen, which illuminate to create the picture.

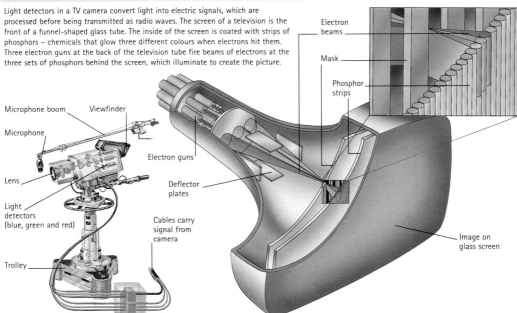

Electron beams

Mask

Phosphor strips

Microphone boom

Viewfinder

Microphone

Lens

Light detectors (blue, green and red)

Electron guns

Deflector plates

Cables carry signal from camera

Trolley

Image on glass screen

VIDEO RECORDING

The tape in a video cassette has one broad track on which the picture information is recorded, a narrow control track and a track on which sound is recorded. The recording head magnetizes the tape in a way similar to an audio cassette recorder. The difference with a video recorder is that the recording head rotates while the tape passes over it at an angle. This leaves diagonal stripes of recorded information on the tape.

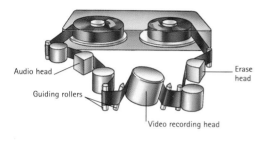

Audio head

Guiding rollers

Erase head

Video recording head

TRANSMISSION AND RECEPTION

Television cameras and sound-recording equipment collect pictures and sounds in television studios and from outside-broadcast locations.

The lens in a camera focuses incoming light onto filters that divide the image into its blue, green and red components. These components shine onto light sensors called charge-coupled devices, or CCDs. There are three CCDs, one for each colour. The CCDs convert light into electrical signals.

Often, several television cameras film at the same time. Editors select the best image at any one time and sometimes process the images with special effects.

Television transmitters broadcast television signals as radio waves. Relay stations receive the signals, amplify them and retransmit them to cover audiences in large geographical areas. Relay stations can be on Earth or on geostationary satellites.

Television receivers normally use aerials to collect signals from the transmitter or from a relay on Earth. Some use small satellite dishes to receive signals from transmitters on satellites. Others receive signals through underground cables, which are supplied with signals by local TV stations or from a community antenna.

VIDEO RECORDING

The first video cassette recorder, or VCR, was made by Sony in 1969. VCRs record TV and sound signals onto magnetic tape.

A video camera-recorder, or camcorder, combines a video camera and microphone with a VCR machine. The VCR records the image and sound signals as they are produced. The recording can be replayed on the camera's liquid-crystal display or through a normal television set.

Camcorders store video footage on tape. Some are small enough to fit in the palm of a hand.

Outside-broadcast television cameras are linked by radio signals to a transmitter that sends signals representing images and sound to the television studio.

SEE ALSO
184–5 Colour, 186–7
Colour mixing, 248–9
Telecommunications

MICROPROCESSORS

Microprocessors consist of thousands of electronic circuits etched on silicon chips. They are used to control electrical and electronic devices.

Photoengraving and chemical processes etch circuits on a silicon wafer.

Silicon wafers, each about 0.25 mm thick, are cut from a cylinder of extremely pure silicon.

▲ The manufacture of silicon chips starts with a rod of pure silicon that may be several centimetres in diameter. The rod is cut into round slices, called wafers, usually no more than 0.25 mm thick. Hundreds of integrated circuits are etched into the surface of each wafer. The wafers are then broken up into hundreds of chips, each of which carries a complete integrated circuit.

Modern electronic devices use vast arrays of circuits and switches to perform logic tests, modifications and calculations on electrical signals. Input signals come from sensors within the device, such as the temperature sensor in a washing machine. Output signals are instructions that control devices such as motors, displays and audible alarms.

Until the 1940s, vacuum tubes were the only electronic switches available. They were large and often unreliable. In 1948, however, a team at Bell Telephone Laboratories developed semiconducting transistors that could replace much larger vacuum tubes in many applications.

By the 1960s, transistors were replacing vacuum tubes in many electrical goods, most notably in portable radios.

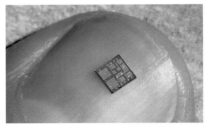

An integrated circuit on a tiny silicon chip may contain thousands of switches and other components.

SILICON CHIPS
In the 1960s, the miniaturization of circuits, which had started with the change from vacuum tubes to transistors, took another great step forwards with the

The control unit instructs other parts of the execution unit to collect data, perform calculations and store results. It also gives instructions to send information through the bus interface unit to the random-access memory (RAM).

The bus interface unit controls the links between the microprocessor and other components. In a computer, it also manages the movement of information between the components of the microprocessor and memory-storage devices such as RAM chips. The RAM is not part of the microprocessor.

Circuitry in the arithmetic logic unit (ALU) performs the microprocessor's calculations. Many processors are built with integral maths coprocessors, which can handle complex mathematics speedily.

The paging and segment units help the bus interface unit to locate information.

Paging unit

Segment unit

Control unit

Bus interface unit

Copper or gold connectors provide electrical links through which signals can flow to other devices.

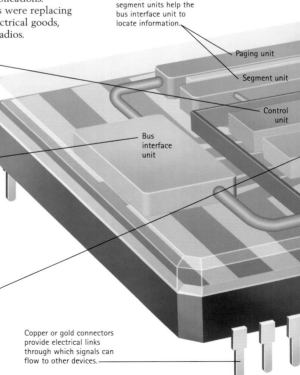

invention of integrated circuits. Integrated circuits contain all the conductors and transistors on the surface of a silicon chip.

Integrated circuits are produced by creating conducting paths of n-type and p-type silicon. The conducting paths are surrounded by channels of insulating silicon dioxide, which prevent short circuits between conductors.

With advances in technology, integrated circuits can now be made with thousands of transistors in little more than a square centimetre of silicon.

IMPACT OF MICROPROCESSORS

Microprocessors are collections of silicon chips that perform calculations and make decisions in electronic devices.

Simple microprocessors control the functions of digital watches, as well as washing machines and other appliances.

More complicated microprocessors are at the hearts of laptop computers and control systems for satellites and aircraft. If these technologies used vacuum tubes rather than microprocessors, a laptop computer would fill a swimming-pool and fly-by-wire aircraft would be too heavy to get off the ground.

Microprocessors are the controlling devices of computer games. These two players are engaged in a two-person boxing simulation in an amusement arcade.

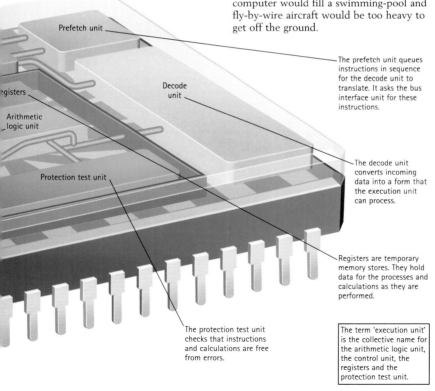

Prefetch unit

Decode unit

gisters

Arithmetic logic unit

Protection test unit

The prefetch unit queues instructions in sequence for the decode unit to translate. It asks the bus interface unit for these instructions.

The decode unit converts incoming data into a form that the execution unit can process.

Registers are temporary memory stores. They hold data for the processes and calculations as they are performed.

The protection test unit checks that instructions and calculations are free from errors.

The term 'execution unit' is the collective name for the arithmetic logic unit, the control unit, the registers and the protection test unit.

The physical layout of a microprocessor is called its architecture. Even though many different types of architecture exist, all microprocessors work in a similar way. Microprocessors take instructions and data from memory. Units in the microprocessor handle, queue and control the incoming information in sequence. An execution unit checks what it should do, performs its tasks and then passes on the results and instructions. Microprocessors are often linked to RAM chips. These store data. Microprocessors are also linked to interface chips, which control other parts of a machine, such as keyboards, monitors, motors or robot arms.

SEE ALSO
162–3 Automation, 228–9 Electrical circuits, 244–5 Conductors, 254–5 Computers

COMPUTERS

Computers use microprocessors to process information according to sets of instructions. They are used for many tasks in education, leisure and work.

The Electronic Numeric Integrator And Calculator (ENIAC), completed in 1946 at the University of Pennsylvania, used 18,000 vacuum tubes as switches.

In the 1960s, computers that were the size of a large room used circuit boards of transistors to perform operations.

A modern computer can perform millions of calculations every second. Instructions from a user tell the computer the operations it should perform. The computer then carries out these instructions according to the rules of software programs. Computers process information in the form of electrical signals. The results are often presented on the screen or as a printed document.

The first electronic computer was built in Britain in 1943. It was the first of ten Colossus machines, which used vacuum tubes as electronic switches. A machine called ENIAC was completed in 1946 in the USA. These early computers were used to automate scientific calculations.

Since the 1970s, computers have used integrated circuits to perform calculations and to store information. Vastly powerful microprocessors and compact memory devices have been made possible by advances in integrated circuits. Due to these improvements, a modern home computer can handle animation and 3D design tasks that were once only possible using enormous computers.

Devices such as image scanners, colour printers, digital cameras and sound cards have created new uses for computers. These include desk-top publishing, image editing and speech synthesis.

Early home computers, such as the Sinclair ZX81, were very basic computers by modern standards. The ZX81 stored programs slowly and unreliably on audio cassette tape and had just one kilobyte of internal memory.

INSIDE A COMPUTER

The part of a computer that performs calculations is called the central processing unit (CPU). Programs and information are stored in one or more hard drives and in random-access memory, or RAM. Data can be read from compact discs (CDs) and digital-versatile discs (DVDs). Data can also be stored on removable disks.

Input devices such as a keyboard, mouse and joystick are used to enter instructions. Output devices, such as a monitor and a printer, are used to display information on a screen or in printed form (hard copy).

Network devices, modems and ISDN cards allow computers to communicate with each other through network cables or through the telephone. Telephone cables are used to gain access to the Internet through the computer of an internet service provider, or ISP.

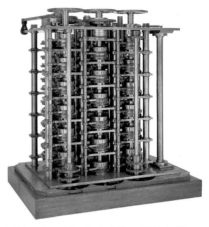

British mathematician Charles Babbage (1792–1871) completed his prototype Difference Engine No.1 in the 1830s. It was designed to calculate mathematical tables mechanically.

COMPUTER COMPONENTS

A computer's base unit contains a number of printed-circuit boards. The main board, called the motherboard, holds the central processing unit, the computer's clock and the system memory. The random-access memory, or RAM, holds data that is being processed. Switching off the computer clears the data from the RAM.

KEYBOARD
Pressing a key on the keyboard sends an electrical signal to the computer.

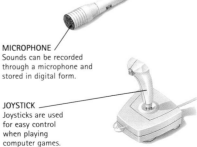

MICROPHONE
Sounds can be recorded through a microphone and stored in digital form.

JOYSTICK
Joysticks are used for easy control when playing computer games.

Read-only memory, or ROM, does not clear when the computer is switched off. The ROM includes instructions for the computer to start up when the power is switched on. Most of the information in a computer is stored on magnetic disks in the computer's hard drive. The base unit has slots where circuit boards called cards are fitted. These cards control the sound, graphics and other functions of the computer. Spare slots, called expansion slots, allow further cards to be added to extend the capabilities of a computer. Modem cards allow computers to communicate through telephone lines. They can be fitted in the base unit or in their own separate housing.

The monitor screen, keyboard and mouse plug into the base unit, as do printers, scanners and other devices. These devices are called peripherals.

Laptop computers cram much of the computing power of a desktop machine into a small, portable case.

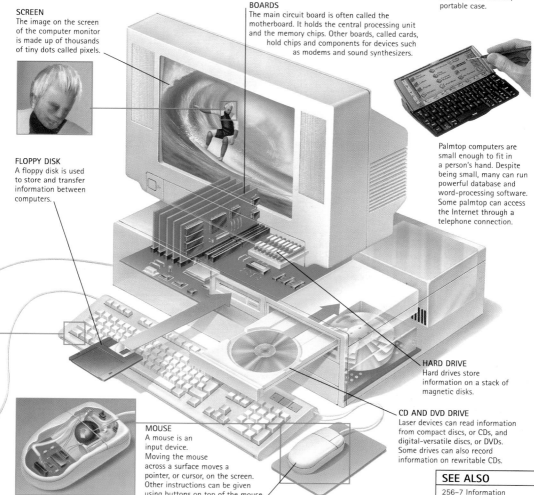

SCREEN
The image on the screen of the computer monitor is made up of thousands of tiny dots called pixels.

BOARDS
The main circuit board is often called the motherboard. It holds the central processing unit and the memory chips. Other boards, called cards, hold chips and components for devices such as modems and sound synthesizers.

Palmtop computers are small enough to fit in a person's hand. Despite being small, many can run powerful database and word-processing software. Some palmtop can access the Internet through a telephone connection.

FLOPPY DISK
A floppy disk is used to store and transfer information between computers.

HARD DRIVE
Hard drives store information on a stack of magnetic disks.

CD AND DVD DRIVE
Laser devices can read information from compact discs, or CDs, and digital-versatile discs, or DVDs. Some drives can also record information on rewritable CDs.

MOUSE
A mouse is an input device. Moving the mouse across a surface moves a pointer, or cursor, on the screen. Other instructions can be given using buttons on top of the mouse.

SEE ALSO

256-7 Information technology

INFORMATION TECHNOLOGY

Information technology is the storage, processing and transmission of information by computers or computerized systems.

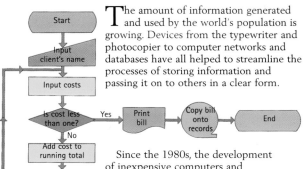

This flow diagram shows the sequence of steps performed by a typical computerized billing system. Programs like this are designed to reduce the work of a human operator by asking questions and making decisions based on the information the programs hold.

The amount of information generated and used by the world's population is growing. Devices from the typewriter and photocopier to computer networks and databases have all helped to streamline the processes of storing information and passing it on to others in a clear form.

Before the invention of computers and word processing, large organizations used large 'pools' of typists to produce documents. This is a 1930s typing pool.

Since the 1980s, the development of inexpensive computers and telecommunications systems has made access to information faster and easier. Whereas information used to be stored as written, typed or printed documents, a large proportion of information is now stored in digital form on computers.

Database programs store vast amounts of information in an organized manner. A database program can use this information to do calculations and produce reports in much less time than a person would require to do the same job.

Many different kinds of information are kept in databases. A library might have a database that records the title, author and publication date of each of its books. A more detailed database could be used to see how often each book is borrowed.

Companies use databases to manage lists of their customers and contacts. Databases are also used to keep track of a company's stock levels so that supplies can be ordered before stocks run out.

Keeping information in databases helps organizations to function efficiently. Information must often be kept securely, however, since rivals could cause damage to an organization if they had access to its databases. For this reason, most databases can only be entered using a password.

Photocopiers make quick, clear copies of documents. This photocopier works by shining light onto the original document. The reflected image is focused onto a drum charged with static electricity. The static charge remains on the drum where there are dark areas corresponding to a replica of the original document. Toner powder is attracted to the static charge on the drum, transferred to a sheet of copy paper and fixed by heat.

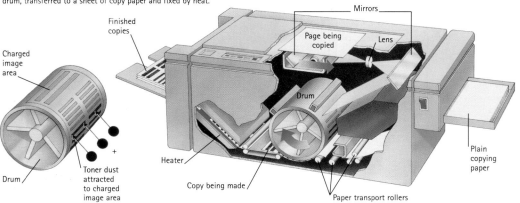

NETWORKS AND THE INTERNET

The information stored in a database is often used by many people on different machines. Users can share this information by connecting their computers to a network. A network might serve a room or a building. Telephone links can extend networks to cover several separate locations, sometimes in different countries.

The Internet is a special type of network. The development of the Internet started in the 1960s, when the United States government set up a network called ARPANET. This network was designed to withstand a nuclear attack by being able to carry information even if one part of the network failed. A growing number of universities and institutions connected to the Internet during the 1970s and 1980s.

Since 1989, when the World Wide Web started, Internet access has grown enormously. Hundreds of millions of people worldwide now access the Internet. Many people use the Internet to send electronic messages, called e-mails, which can be sent rapidly between any two Internet connections. Other people search for information using Internet programs called search engines. Search engines look for collections of information, called websites, that are related to the search topic keyed in by the user.

Information such as images from a camera, computer files or music can be sent by e-mail or viewed at websites.

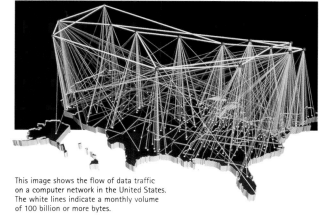

This image shows the flow of data traffic on a computer network in the United States. The white lines indicate a monthly volume of 100 billion or more bytes.

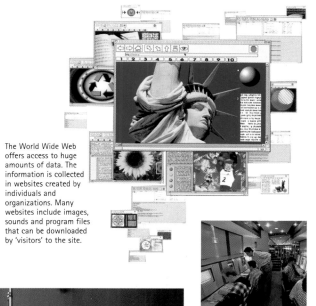

The World Wide Web offers access to huge amounts of data. The information is collected in websites created by individuals and organizations. Many websites include images, sounds and program files that can be downloaded by 'visitors' to the site.

Video conferences save the time and expense of travelling to meetings by passing sounds and images between the two ends of a telephone connection.

▲ Internet cafés such as this one in Bangalore, India, allow customers to hire a computer and surf the Internet.

SEE ALSO

248–9 Telecommunications, 254–5 Computers

FACTS AND FIGURES

LAWS AND RULES OF ELECTRICITY

Electrical charge (measured in coulomb)

= **current (amp) x time (second)**

Potential difference (volts)

$$= \frac{\text{energy transferred (joule)}}{\text{charge (coulomb)}}$$

Resistance (ohms)

$$= \frac{\text{potential difference (volt)}}{\text{current (amp)}}$$

Domestic electrical energy consumption (kilowatt hours; kWh)

= **power (kilowatt) x time (hour)**

Current drawn by an electrical appliance in normal operation (amp)

$$= \frac{\text{power rating of appliance (watt)}}{\text{voltage (volt)}}$$

COMPUTER STORAGE TERMS

Bit	smallest unit of computer data
Byte	eight bits
Kilobyte	1,024 bytes
Megabyte	1,024 kilobytes
Gigabyte	1,024 megabytes

The word 'bit' is shortened from *binary digit*. A digit of binary code can have one of two values: 0 or 1. These values can be stored as different magnetizations of specific locations on a computer hard drive, for example. Bits can be transmitted as pulses of current through cables, or as pulses of light through optical fibres.

ELECTRICAL CIRCUIT SYMBOLS

Electrical and electronic engineers and physicists use a number of symbols to enable them to draw complex circuit diagrams in limited spaces. Some of these symbols are shown in the table below.

KEY DATES

BC

c.600 Greek philosopher Thales of Melitus discovers that amber resin attracts light objects when rubbed.

AD

1600 British physician William Gilbert publishes the results of his experiments with electricity and magnetism.

1672 German physicist Otto von Guericke constructs the first electrostatic generator. It is a hand-cranked sulphur sphere.

1729–32 British physicist Stephen Gray discovers the principles of electrical conduction.

1745 Dutch physicist Petrus van Musschenbroek invents the Leyden jar, a simple capacitor.

1752 US diplomat and scientist Benjamin Franklin invents the lightning conductor after flying a kite in a storm.

1784-89 French physicist Charles Coulomb establishes laws of electrostatics.

1800 Italian physicist Alessandro Volta invents the voltaic pile, the first electric battery.

1827 German physicist Georg Simon Ohm devises a law that describes the relationship between current, resistance and voltage in a circuit.

1831 British chemist and physicist Michael Faraday produces a current by turning a copper loop in a magnetic field.

1859 French physicist Gaston Planté invents the lead–acid storage battery.

1871 Belgian electrical engineer Zénobe Gramme begins manufacturing dynamos.

1873 British physicist James Clerk Maxwell publishes equations that describe the properties of electromagnetic waves.

1876 US inventor Alexander Graham Bell invents the telephone.

1877 French engineer Georges Leclanché invents the zinc–carbon dry cell.

1882 French engineer Marcel Deprez achieves the first transmission of high-voltage electricity through cables over a distance of around 55 km.

1895 Italian electrical engineer Guglielmo Marconi makes first radio transmission over a distance of 2.4 km.

1910 French chemist and engineer Georges Claude invents the neon lighting tube.

1911 Dutch physicist Heike Kamerlingh Onnes discovers superconductivity of mercury cooled to 4.2 K (−268.7°C).

1914 US inventor Thomas Alva Edison develops an alkaline storage battery.

1928 Swiss-born US physicist Felix Bloch produces a complete theory of semiconductors.

1929 The BBC makes the world's first transmission of television programmes.

1942 Italian-born US physicist Enrico Fermi initiates the first controlled nuclear fission chain reaction.

1948 US physicists John Bardeen, Walter Brattain and William Shockley invent the transistor.

1951 The Experimental Breeder Reactor, EBR-1, at Idaho Falls, USA, becomes the first nuclear reactor to produce electrical power.

1960 United States launch Echo 1, the first telecommunications satellite to orbit Earth.

1961 Unimate becomes the first robot in service in a factory.

1969 US Department of Defense establishes ARPANET, a data communications network that spawned the Internet.

1971 Intel Corporation launches the first microprocessor.

1981 IBM starts production of the first personal computer.

1985 Philips launch the CD-Rom as a computer data medium.

1989 British computer scientist Timothy Berners-Lee devises the World Wide Web to simplify Internet usage.

COMMON CIRCUIT SYMBOLS

Fixed resistor Fixed capacitor Triac SCR

Transistors
NPN   PNP

Potentiometer or Adjustable resistor   Polarized capacitor   LED   Iron core transformer

Darlingtons

Variable resistor   Variable capacitor   Variable coil   Adjustable coil

MOSFETs

Variable resistor   Variable resistor   Variable resistor   Variable resistor

JFETs

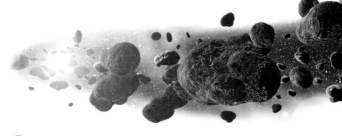

CHAPTER 9
SPACE AND TIME

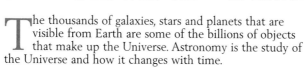

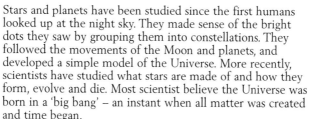

T he thousands of galaxies, stars and planets that are visible from Earth are some of the billions of objects that make up the Universe. Astronomy is the study of the Universe and how it changes with time.

Stars and planets have been studied since the first humans looked up at the night sky. They made sense of the bright dots they saw by grouping them into constellations. They followed the movements of the Moon and planets, and developed a simple model of the Universe. More recently, scientists have studied what stars are made of and how they form, evolve and die. Most scientist believe the Universe was born in a 'big bang' – an instant when all matter was created and time began.

Almost all the objects in the Universe are too far away for missions from Earth to visit. Astronomers use telescopes on Earth and in orbit to gather information from the light, X-rays, radio and infrared radiation given out by such distant objects. Space probes have visited some of the planets, comets and asteroids, and twelve men have walked on the Moon. Each year, astronomers and space scientists discover more objects in space and learn new details about those objects that have already been known for some time.

THE UNIVERSE

The Universe consists of everything in space, and space itself. It is unimaginably large, constantly changing and getting larger all the time.

The Universe is everything that exists – from the smallest creatures on Earth to the largest, distant structures in space. The Universe is also a dynamic place. Everything in it follows a life cycle, whether it is a human who lives for 70-80 years, or a star for 10 billion years. The process is continual as each object in the Universe is born, lives and then dies.

CONTENTS OF THE UNIVERSE

Our planet seems important and large compared to a human that lives upon it. Yet Earth is a tiny speck when compared to the rest of the Universe. Earth is one of nine planets orbiting the star called Sun. There are more stars than anything else in the Universe. It is impossible to count them but we can estimate there are at least 100 billion billion.

The Sun, like all other stars, is a glowing sphere of hot gas. It is a fairly average, medium hot, yellow star which, along with billions more, make up a galaxy. This vast spiral-shaped grouping of stars is

Quasars are a type of active galaxy that produce huge amounts of energy. They are some of the most powerful objects in the Universe. In this false colour image of quasar 3C 272, the first quasar to be identified, the intensity of X-rays emitted is shown by the different colours. The most intense are pale blue, green and yellow.

our home galaxy, the Milky Way. With about 30 others it forms a cluster of galaxies in space. Together they are called the Local Group. All over the Universe are many more clusters of galaxies which in turn are arranged in superclusters.

We see many galaxies from Earth but many more are beyond our vision. In all there are about 100 billion galaxies in the Universe. We can see them in every direction. Some are so close, or so large, we can make out their shapes and their individual stars. We see them through telescopes, either on land or in orbit around Earth.

▼ The time light takes to travel is used for measuring distances in space. The light from the closest night-time star takes 4.2 years to reach Earth. The star is therefore 4.2 light years away. The time for light to travel distances from more local places are shown as a comparison.

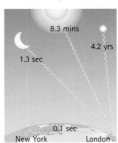

1 Everything in the Universe – the stars, the planets, and human life – is made of the same elements.
2 Humans are just one of billions of life forms on planet Earth. More than 6 billion people now live on its land.
3 Our planet, Earth, is the only place in the Universe known to support life.
4 Earth is one of nine planets orbiting the star called Sun. Together they form the Solar System.
5 The Sun, and its system of planets, is just one of billions of stars in the Milky Way Galaxy.
6 Countless galaxies we have never seen fill the Universe. They are grouped in clusters.
7 Clusters of galaxies are strung together in superclusters which stretch across the Universe. In between are huge, empty voids.

1 Swirling clouds of gas and dust, called nebulae, are the birthplace of stars. Often, they are themselves the remains of earlier generations, material blown out from short-lived, massive stars.

2 Spheres form in the gas as gravity pulls it together. As the gas compresses it begins to heat up.

3 The young star – a protostar – continues to spin. Unused material is expelled.

4 Nuclear reaction begins in the hot core. Light and heat are produced.

5 The star shines steadily for billions of years.

MEASURING DISTANCES

It is easy to make measurements on Earth. We use familiar systems such as the metric (kilometre, metre, centimetre) or imperial (mile, yard, foot, inch) to measure how big, or how long something is. These measuring systems are used to measure some of the smaller things in space such as a crater on the Moon, or even the distance of Pluto from the Sun.

Some things are so big, or just so far away, these measuring systems will not work as the numbers become too long to handle. Astronomers use different units. Within the Solar System they use the Astronomical Unit (AU). One AU is the distance between Earth and the Sun and this unit is used to measure the distance to other objects; the Sun to Mars is 1.5 AU, to Jupiter 5.2 AU and so on.

When measuring outside the Solar System, to stars and galaxies, or the size of a galaxy from side to side, astronomers use light years (ly). Light travels faster than anything else, at 299,792 km per sec. A light year is the distance that light travels in a year. That is 9,460,700,000,000 km, or put another way, 9.46 billion km, or better still, put simply as 1 ly. The Milky Way Galaxy is 100,000 ly across.

▲ Stars are born all the time. Spinning spheres of gas form inside large clouds of gas and dust. Once hot enough, the sphere's hydrogen is converted to helium. Light and heat are produced and the star is born. Through its life, then death, it will shed material which will in turn be used to make new stars.

◀ New stars are born in the Eagle Nebula, 7,000 light years away. At the tips of the huge columns of gas and dust are oval-shaped clumps. They look tiny compared to the columns, but they are new stars being born.

EXPANDING UNIVERSE

Everything in the Universe is moving. Earth spins on its axis once each day and we experience, alternately, day and night as we bathe in sunlight, then face dark space. Every other planet and moon, each tiny piece of space rock, and each of the countless stars spins on its own axis.

Such objects also move through space as they spin. Earth, for example travels round the Sun once every year, each complete trip is known as an orbit. At the same time, the Sun, Earth and all the other bodies that make up the Solar System, move as a whole. They orbit round the centre of the Milky Way Galaxy. From our place in the Milky Way, we look deep into space and see galaxies. The galaxies live in clusters which are rushing away both from our galaxy and each other. The further away the cluster, the faster it recedes. The Universe itself is expanding all the time, and has been doing so ever since it was created.

American astronomer, Edwin Hubble (1889–1953) sits in front of the telescope he used to observe the Universe. In 1924, he presented the first evidence to show that other galaxies existed outside our own. In 1929, he proved that the Universe is expanding.

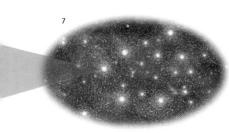

SEE ALSO

166–7 Heat and light from the Sun, 264–5 Galaxies, 266–7 Stars, 268–9 The Sun, 272–3 The Solar System

THE UNIVERSE: ORIGINS AND FUTURE

The Universe was born in an explosion called the Big Bang about 13 billion years ago, and has been changing and growing ever since. It will exist for trillions of years.

Almost all astronomers believe the Universe was created about 13 billion years ago in a huge explosion called the Big Bang. In the tiniest fraction of time, too short to be measured, the Universe was born. It was an enormous amount of energy packed into an unimaginably small space, but in a fraction of a second the Universe inflated. It grew from smaller than a pinprick to larger than a galaxy. The Universe has been expanding ever since.

The energy created in the Big Bang was transformed into atomic particles. Within three minutes, the temperature had dropped from 10^{28} °C to 1 billion°C and continued to fall. By this time the Universe was made up of 77 per cent hydrogen and 23 per cent helium. All the other elements and compounds that now exist were created from these two.

When it was around 300,000 years old, the Universe, which until this time had been like an opaque soup, became transparent. Its temperature dropped to 3,000°C. About one billion years after the Big Bang, gravity was pulling the hydrogen and helium into clouds. Spinning spheres of gas formed and the first stars and galaxies were born.

▼ This cool, dense cloud of hydrogen in the constellation of Orion is producing stars. The colours indicate the different temperatures of the cloud and stars embedded in it. The dust of the cloud is yellow. A bright, newborn star (red) is at centre right.

▲▶ Matter has been cooling, changing, and expanding out from the point of explosion ever since the Big Bang. For the first quarter of a million or so years it was a 'soup' of hydrogen and helium particles, slowly coming together to form stars and galaxies.

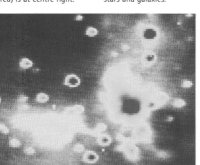

Big Bang

10 yrs

100 yrs

1000 yrs

10,000 yrs

1 Nothing existed before the Big Bang. All we know is that within the tiniest fraction of a second, enough energy to make all the material in the Universe had come into existence. As the Universe cooled down, this energy was transformed into atomic particles.

2 Just after the Big Bang, the Universe was unimaginably small. It then began to expand suddenly, inflating hugely in every direction. The Universe was full of energy and its temperatures soared.

3 The Universe had to cool down from a temperature of about 10^{28} °C to about 3,000°C before atoms could form. Atoms are the minute units of matter. The atoms were mainly of hydrogen, which is the simplest and most plentiful substance found in the Universe. The rest were more complex atoms of helium.

4 Hydrogen and helium filled the Universe with a thin dark fog. The gas atoms in denser parts were pulled into separate, much smaller, clouds by gravity. (Gravity is the force by which objects attract one another.) The centres of the clouds, where the gas atoms were packed together, heated up, giving birth to stars as the galaxies formed.

EVIDENCE FOR THE BIG BANG

Light from distant galaxies takes billions of years to reach us and in this way we can see what galaxies were like years ago. Using very powerful telescopes we can look back to a time when galaxies were young and the Universe was in its infancy. The most distant objects we can see are galaxies as they were ten billion years ago.

The galaxies are moving apart, which suggests that everything was once concentrated in a single place. Additional evidence for the Big Bang came in 1965 when scientists found heatwaves left over from the vast explosion coming from every direction in space. In 1992, the COBE satellite detected the ripples in the heat caused by cooling after the Big Bang. Yet, astronomers realize they still haven't found the majority of the Universe's matter – the material of which it is made. They call this dark matter, and it makes up about 90 per cent of the Universe. Once found, it may help to fill some of the gaps in the Universe's life story.

THE FUTURE OF THE UNIVERSE

Astronomers who study the origin of the Universe are called cosmologists. They are also interested in the future of the Universe. Some believe the Universe will continue to expand, growing larger and cooler. Eventually all the stars will die, and the Universe will be cold and dark. We know that as the galaxies move apart the gravity of one pulls on the other, slowing this expansion. Some cosmologists think that, in trillions of years' time, gravity will have slowed the galaxies completely, until they are stationary.

Gravity will then pull the galaxies back toward each other. The Universe will contract until it is all together in one point. As its material packs ever closer together it will heat up. Eventually, the Universe will collapse violently inwards in an implosion called the Big Crunch. Everything will be destroyed and it will be the end of the Universe. But, this may be followed by another Big Bang and the creation of a new Universe.

In 1992, it was announced that the satellite Cosmic Background Explorer (COBE) had traced the background radiation and 'ripples' left over from the Big Bang, 13 billion years before.

We do not know exactly when the Universe was created, but now believe it was 13 billion years ago. All the matter in the Universe, time and space were created together in the Big Bang. The matter was at first atomic particles, then atoms and compounds. All the complex materials we know today, including our own planet, our homes and our own bodies, started in this way.

First stars form within large clouds of gas and dust

Solar system forms

Lightning storms on the young Earth

3

100,000 yrs

1,000,000 yrs

13,000,000,000 yrs

10,000,000 yrs

4

5

6

7

5 About 4.6 billion years ago, our local star, the Sun, was formed. Around it was a cloud of gas and dust containing substances such as carbon and oxygen. These had been formed in older stars and blasted out into space when the stars died. This material came together to form the planets.

6 The first living cells appeared on Earth 3.5 billion years ago. How life began is uncertain. Maybe lightning storms provided the energy to start chemical reactions in the 'soup' of elements on the young planet.

100,000,000 yrs

1,000,000,000 yrs

7 It has taken one ten-thousandth of the time since the Big Bang for recognizable humans to develop from apes. Today, scientists try to work out the story of the Universe by sending out space satellites which look back across time.

10,000,000,000 yrs

SEE ALSO

260–1 The Universe,
264–5 Galaxies

GALAXIES

Galaxies exist throughout the Universe. They are enormous groups of stars held together by gravity. We live in a spiral arm of the Milky Way Galaxy.

▼ Galaxies begin as huge masses of gas. As they shrink under the pressure of gravity, the gas at the centre becomes dense enough to start forming stars. Some galaxies start spinning, forming a spiral or barred spiral galaxy.

1 A slowly spinning mass of gas collapses under the pressure of gravity, and the first stars are formed at the centre. As the cloud shrinks, its turning speed increases.

2 Gas clouds meet in the swirling disc, and attract more clouds because of their extra gravity. Stars start to form here too.

3 There is no gas left at the centre to make new stars, but the arms are rich in raw star material. The galaxy is now in the prime of life.

Wherever you look in the Universe there are galaxies. There are billions of them – vast collections of stars, gas and dust. Each one can contain hundreds of thousands of stars or even many billions.

Galaxies are classified according to their shape. The three main types are elliptical, spiral and barred. The fourth type, the irregular, are galaxies that have no distinctive shape. More than half of all the galaxies we can see are ellipticals. They range in size from the smallest to the largest galaxies and contain little gas or dust and so no star formation takes place. About a third are either spirals or barred spirals. Their centres contain old stars, and their arms hold young stars and new ones being born from clouds of gas and dust. The young ones are very bright and the outshine the stars between the spirals.

THE MILKY WAY

The galaxy we live in is called the Milky Way. Viewed from the outside, we would see it is a barred spiral galaxy made up of over 200 billion stars. Our star, the Sun, is located in one of the spiral arms. The arms are made up of stars, and nebulae of gas and dust where new stars are born. The centre contains older stars. The Sun takes about 220 million years to orbit the centre. The Galaxy is 100,000 ly across and 2,000 ly thick.

▶ THE SUN'S NEIGHBOURS IN THE MILKY WAY
1 Cone Nebula   2 Rosette Nebula   3 Orion Nebula
4 Lagoon Nebula   5 Solar System   6 California Nebula
7 Trifid Nebula   8 Vela Supernova Remnant
9 North America Nebula

Astronomers do not know exactly why the galaxies have these shapes but it could be due to the way that they are formed. The amount of material in a galaxy, the speed of its spin and the rate its stars form all help to determine the shape of a galaxy. Astronomers have also studied colliding galaxies and seen that two can join to form one new large galaxy. This can be as a result of a direct hit or a glancing blow, and leads to a period of intense star formation in the new galaxy. It is thought that galaxies will continue to join in this way so that in the future there will be fewer but larger galaxies than now.

ACTIVE GALAXIES

There are some galaxies known as active galaxies, and these are different from the others in their characteristics and appearance. A typical active galaxy has a very luminous core, which may vary in brightness, and from which two huge jets of material are emitted. In the very centre,

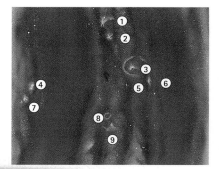

The Milky Way

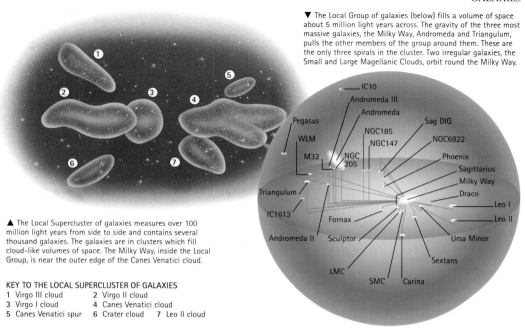

▼ The Local Group of galaxies (below) fills a volume of space about 5 million light years across. The gravity of the three most massive galaxies, the Milky Way, Andromeda and Triangulum, pulls the other members of the group around them. These are the only three spirals in the cluster. Two irregular galaxies, the Small and Large Magellanic Clouds, orbit round the Milky Way.

▲ The Local Supercluster of galaxies measures over 100 million light years from side to side and contains several thousand galaxies. The galaxies are in clusters which fill cloud-like volumes of space. The Milky Way, inside the Local Group, is near the outer edge of the Canes Venatici cloud.

KEY TO THE LOCAL SUPERCLUSTER OF GALAXIES
1 Virgo III cloud   2 Virgo II cloud
3 Virgo I cloud   4 Canes Venatici cloud
5 Canes Venatici spur 6 Crater cloud   7 Leo II cloud

a glowing ring of dust and gas surrounds a supermassive black hole. Quasars, blazars, Seyfert galaxies and radio galaxies are all examples of active galaxies. However, astronomers think that one model can account for the different types, and that they have been classified into groups because of the angle at which they appear when we see them in our sky.

CLUSTERS AND SUPERCLUSTERS

Galaxies gather in groups called clusters. Our galaxy, the Milky Way, is one of about 30 galaxies that make a small cluster called the Local Group. Most of the group are small and faint. About half of them are elliptical, and around a third irregular

in shape. Three are more massive and much brighter spiral galaxies. Several thousand galaxies make up the larger clusters. The Virgo Cluster, about 50 million light years away, contains more than 2,000 galaxies with three giant ellipticals and many bright spirals.

Clusters are arranged into even larger groups called superclusters. These are the largest structures in the Universe, and measure in excess of 100 million light years across. A supercluster can contain dozens of clusters. The Local Group belongs to a supercluster which is centred on the Virgo Cluster. The supercluster consists of 11 main cloud-shaped clusters in a roughly oblong volume of space.

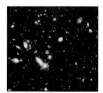

This view into space took ten days to make. All but one of the objects are galaxies, seen for the first time in this image. The faintest are about four billion times dimmer than we can see with our eyes. Because of the vast distances their light had to travel to reach Earth, the galaxies are seen as they were around ten billion years ago.

Those galaxies without any particular shape are classified as irregular galaxies (Irr). They contain much gas and dust.

Elliptical galaxies are ball-shaped galaxies, ranging from an almost spherical shape (E0) to an elongated squashed-ball shape (E7).

Spiral galaxies are disc-shaped. Arms radiate from a central nucleus. They range from tight arms (Sa) to loose arms (Sc).

Barred spirals have a bar-shaped nucleus from which the arms radiate. They range from tight arms (SBa) to loose arms (SBc).

SEE ALSO

266–7 Stars, 268–9
The Sun, 270–1
Constellations

STARS

There are billions of stars in the Universe. Each one is a ball of hot, glowing gas that changes size, temperature and colour during its long lifetime.

Danish astronomer Ejnar Hertzsprung (1873–1967) studied stars. In 1906, he observed that the temperature and the luminosity of a star were linked. He found that stars could be arranged in a family, from hot bright stars to cool dim ones.

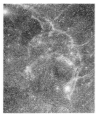

These strands of gas are the remnant of the Vela Nebula, a supernova that exploded about 12,000 years ago. The gas will eventually be used to form new stars.

▶ Stars like the Sun follow a life cycle. Like all other stars they are born in a nebula of gas and dust. Once nuclear reaction starts, the star shines. The Sun is a main sequence star, and will shine steadily for several billion years until the hydrogen is used up and it matures into a red giant. Next, it will slowly shed its outer layers into space, until only the shrunken core is left. The star will then die. Its small, hot body will cool into a white dwarf, a cold, dark cinder in space.

There are more stars than anything else in the Universe. Our Sun is the closest, the rest only appear as bright dots of light in the night sky. Astronomers study their light to find out what they are made of, and how big and how hot they are.

Stars come in different sizes and colours and, although all are hot, they come in a range of temperatures. Each one is made of gas familiar to us on Earth and held together by gravity. Inside the core, hydrogen is converted into helium. In this nuclear reaction, enormous amounts of energy are produced and then given off. We feel some of the Sun's energy in the form of heat, and see it as light. But other forms of energy such as ultraviolet and X-rays are also emitted by the Sun and stars.

THE LIFE OF A STAR

The stars that we see are at different stages of their lives. All stars start by converting hydrogen into helium in their core. After the hydrogen is used up, the helium is converted into new elements such as carbon and oxygen. During this process the stars change size, throw off material, and change temperature and colour. The hottest are the bluest, the yellow ones cooler, and the red cooler still,

The Pleiades star cluster was born about 78 million years ago from a nebula of gas and dust. About 100 stars are surrounded by remains of the nebula.

although the red dwarves still have a surface temperature of 3,000°C. Some stars convert their gas more quickly than others. The amount of material a star is made of, the star's mass, influences how long a star will live, the development stages it will go through, and how it will eventually die. Many stars, such as the Sun, shine for thousands of millions of years before they run out of fuel. Others, those that have a higher mass than the Sun, use up their fuel more quickly and so have shorter lives. A star's mass also influences its brightness. The more massive the star, the greater its luminosity.

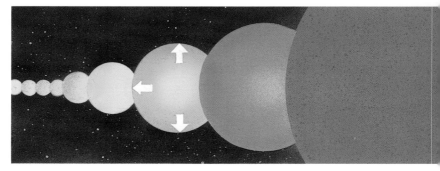

▲ At first the star shines steadily as it converts hydrogen into helium.

▲ The core's hydrogen is nearly used up. The core collapses and becomes hotter.

▲ The helium starts to convert to carbon. The outer parts of the star move outwards.

▲ The star gets larger and larger. The outer layers cool and turn from orange to red.

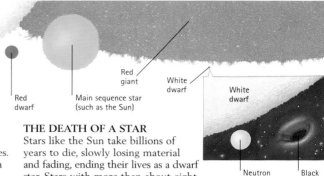

Blue giant

Red dwarf

Main sequence star (such as the Sun)

Red giant

White dwarf

White dwarf

Neutron star

Black hole

All stars are born from clouds of gas and dust called nebulae. Small clouds form within a nebula and condense, spinning and pulling the material into the centre by gravity. As the material in the centre becomes compressed, its temperature rises. At about 10 million°C, a nuclear reaction starts and the star is born.

Most stars are born in pairs, like twins. These pairs and other stars are born within a group of stars called a cluster. Some stars spend the rest of their lives as part of a cluster. The stars in other clusters drift apart. This is what happened to the Sun and the brother and sister stars it was born with. There are two sorts of star clusters in the night sky. The open clusters are loose collections of young stars which will eventually disperse. Globular clusters are tightly knit groupings of older stars. These stars were born together, have spent their lives as part of a cluster, and will die in the cluster. Once a star has started shining it is called a main sequence star. The Sun is a main sequence star, and will remain so for billions of years yet. As the Sun uses up its helium it will mature into a red giant, before starting the final stage of its life-cycle, its death.

THE DEATH OF A STAR

Stars like the Sun take billions of years to die, slowly losing material and fading, ending their lives as a dwarf star. Stars with more than about eight times the mass of the Sun end their lives in a much more dramatic way. Once nuclear reaction ceases, the star collapses and blows itself apart in a supernova explosion, leaving only a core behind.

The name supernova comes from the star's sudden change of appearance. The explosion looks like a super, new (*nova* is Latin for new), bright star in the sky. The future of the core-remains depends on the mass. Those stars with three times the mass of the Sun continue collapsing until they produce a black hole. Those with less mass leave a tiny neutron star. They are so densely packed that a handful weighs billions of tonnes. A rapidly spinning neutron star is called a pulsar, sending out beams of energy, like the rotating beam of a lighthouse. It spins many times a second, and if Earth is in the path of the beam, the star's pulses of energy can be detected.

Stars vary enormously in size. The largest are the supergiants which can be up to several hundred times larger than the diameter of the Sun. Next in size are the giants, either blue or red, which measure up to 100 times the Sun's diameter. Blue stars are the hottest and brightest. The smallest stars are the dwarves which are much smaller than the size of the Sun.

▼ After the collapse of an ordinary star it becomes a white dwarf, finally cooling to become a dead black dwarf.

▲ The star is a red giant. It is up to 100 times larger than the original star.

▲ The fuel in the core is used up and nuclear reaction stops. The core collapses.

▲ The outer layers are pushed off into space. The core shrinks to the size of a small planet.

SEE ALSO

166–7 Heat and light from the Sun, 264–5 Galaxies, 268–9 The Sun

THE SUN

Without the Sun there would be no life on Earth. It also helps us to understand other stars in the Universe. Like them, it is a changing ball of hot gas.

The hot gas on the Sun's surface is constantly moving about. It gives the Sun a mottled appearance. By looking at its edge we can see the jets, flares and prominences which appear all over its surface.

Short-lived flares of hot gas burst from the Sun's surface. They usually last for only minutes at a time.

The Sun is the nearest star to Earth. It is the only one that we can see in close-up. The Sun is a globe of hot gas, mainly hydrogen, which is so large that 1,300,000 Earths would fit inside it, and 109 would fit across its face.

Inside its core it is very hot, around 15 million°C. Here, nuclear reactions are converting the hydrogen into helium. In the process, huge amounts of energy are produced which, after tens of thousands of years, eventually reach the Sun's surface. Once there, the energy escapes into space as heat, light and other types of radiation. The light and heat are essential to us on Earth, but the other radiations such as ultraviolet rays can be harmful. Earth's atmosphere shields us from much of it.

The surface of the Sun is called the photosphere. It is cooler here, around 5,500°C. The photosphere is not solid,

but is the Sun's visible outer edge. The Sun spins, and its equatorial regions take about 25 days to go round once, while the polar regions take about ten days longer. We can also observe how the surface and the atmosphere change from week to week and year to year.

SURFACE FEATURES

Sunspots are darker, cooler areas on the Sun's photosphere, ranging in size from a few thousand to tens of thousands of kilometres across. The spots are caused by strong magnetic fields within the Sun that slow down the flow of heat from inside. They are not permanent features, following a regular 11-year cycle of appearances, and only last for weeks at a time. Sunspots form and disappear progressively closer and closer to the equator before disappearing altogether as a new cycle starts at higher latitudes.

The Sun's surface is a violent place and regularly releases short-lived bursts of energy called flares into space. Larger eruptions, prominences, stretch for many thousands of kilometres into the Sun's atmosphere and can last for months.

Flare

Spicule

Sunspots are cooler, darker areas on the Sun's surface. Nearby are faculae (singular, facula), clouds of glowing hydrogen just above the surface.

Prominence

Sunspot

Facula

The corona, the outer atmosphere of the Sun, is not normally visible – it is outshone by the Sun's disc. However, the corona can be seen during a total eclipse.

A huge stream of glowing gas flows out from the Sun's surface. The prominence has arched over to form a loop. It will last for several hours before collapsing back.

The photosphere is the Sun's outer surface. It is a very active, 500 km-thick layer of gas. All over the surface are circular granules formed by the ever-moving hot gas, and spicules, jets of gas stretching upwards. Immediately above the surface is the inner atmosphere, called the chromosphere.

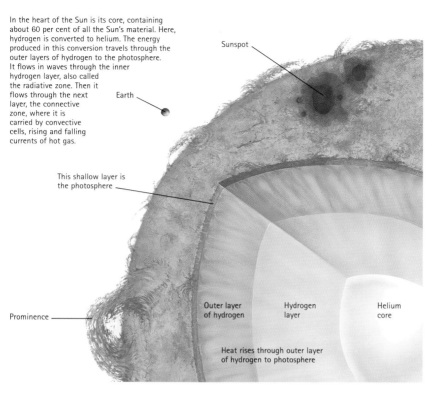

In the heart of the Sun is its core, containing about 60 per cent of all the Sun's material. Here, hydrogen is converted to helium. The energy produced in this conversion travels through the outer layers of hydrogen to the photosphere. It flows in waves through the inner hydrogen layer, also called the radiative zone. Then it flows through the next layer, the connective zone, where it is carried by convective cells, rising and falling currents of hot gas.

Sunspot

Earth

This shallow layer is the photosphere

Prominence

Outer layer of hydrogen

Hydrogen layer

Helium core

Heat rises through outer layer of hydrogen to photosphere

The American astronomer George Ellery Hale (1868–1938) studied the Sun and its spots. His work led him to discover magnetic fields within sunspots.

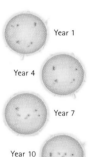

Year 1

Year 4

Year 7

Year 10

Year 12

At the start of each 11-year cycle, sunspots appear well to the north and south of the Sun's equator. As time continues they appear closer until, at the end of the cycle, they appear along the equator. Finally the spots in this cycle disappear, and a new cycle starts by producing spots well to the north and south.

THE SUN'S ATMOSPHERE
Immediately above the Sun's surface is the chromosphere, a layer of hydrogen and helium, about 5,000 km thick. At its furthest edge, where it merges into the outer layer of atmosphere, the corona, its temperature is about 500,000°C. The corona stretches for millions of kilometres into space and is very thin and hot, around 3 million °C. It gives off a constant stream of particles, called the solar wind, which travel through the Solar System.

THE LIFE OF THE SUN
The Sun was formed from a nebula, a cloud of gas and dust, about five billion years ago in the Milky Way Galaxy. It has been shining steadily ever since, using up about seven million tonnes of material every second. It is presently about halfway through its life. In about five billion years' time, when the hydrogen in its core has been converted to helium, the core will collapse. The Sun's outer layers will swell out, cool and change colour. The Sun will have moved from the main sequence stage of its life to become a red giant.

When the helium starts to form carbon, the Sun will change colour from red to yellow and enter another stage of its life, losing its outer layers of material as it continually shrinks and expands. The inner layers will collapse and what remains of the original Sun will be packed together into a small star, a white dwarf, which will slowly fade and die as a black dwarf.

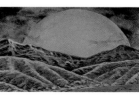

The Sun will continue to shine for billions of years. But in about five billion years it will swell into a red giant. Earth will grow hotter, water will boil away, and life will die out. The Sun will become so large it will engulf Mercury and Venus, and Earth will be inside its atmosphere.

SEE ALSO
128 Carbon, 166–7 Heat and light from the Sun

CONSTELLATIONS

A constellation is an area of the night sky. Its stars make a dot-to-dot picture. Astronomers use 88 of these pictures to help them find their way about the sky.

The Greek astronomer and geographer Ptolemy (AD100–170), who lived in Alexandria, Egypt, listed the 48 constellations that had been used by ancient astronomers from around 2000BC. This list grew into the 88 constellations we use today.

The Plough

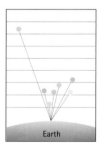

Earth

The seven stars in the Plough appear to be the same distance away (top) when seen from Earth. In fact they are different distances from us. The star at left is up to three times as far away as the others.

▶ Part of the figure of Orion, the hunter, has been drawn around the stars in the night sky. Two bright stars mark his shoulders. The three that mark his belt are the key stars to look for when identifying Orion. The line bottom left is his sword, hanging from the belt.

From anywhere on Earth you can look into the sky and see thousands of pinpoints of light. They are all stars, and belong to our Galaxy, the Milky Way. There are so many stars that, at first, it is difficult to tell them apart. But there are distinctive star patterns in the sky, and astronomers use these to find their way around. They draw imaginary pictures around the star patterns to help them remember them. The very first pictures were created over four thousand years ago. We still use these, and some more recently devised ones, today.

CELESTIAL SPHERE
Astronomers looking out from Earth imagine a great star-studded sphere surrounding them. They look at the inside edge of this imaginary celestial sphere. The sphere is divided into 88 pieces like an enormous jigsaw. Each piece is a constellation and has a star pattern with an imaginary picture. Astronomers around the world use this system of 88 constellations with their star patterns.

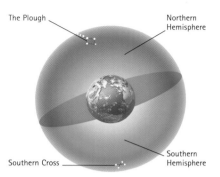

The celestial sphere surrounds Earth. It is divided into two domes, the Northern and Southern hemispheres, which are made into flat maps (see right). The celestial equator is the dividing line between the two maps.

SKY PICTURES
The pictures in the sky include real and imaginary humans, animals, sea creatures, and tools of the artist and scientist. We do not know exactly who created the first pictures, or when, but a system of 48 was in use by around AD150. This is when the Greek geographer and astronomer Ptolemy listed them in his book *Almagest*. The 48 constellations include figures from Greek mythology, such as the hunter, Orion, the flying horse, Pegasus, and the Centaur who is half-man and half-horse. The other 40 constellations were created in more recent times, and include some of the great inventions, such as the telescope and the clock.

THE ZODIAC
A group of constellations known as the zodiac covers a special area of the sky. They form a band of sky 20° wide with the ecliptic passing along the centre. This stretches right round the celestial sphere. As we look to the sphere from Earth we see the Sun, Moon and all the planets move across the sky against the background of this band of stars. The ancient astronomers had 12 constellations in this band, today it is 13. The Sun seems to spend approximately one month in each constellation.

LOOKING AT THE CONSTELLATIONS

It is impossible to see all 88 constellations from one point on Earth. People living in Earth's Northern Hemisphere see those in the Northern Hemisphere of the celestial sphere, as well as some from its southern half. People living in Earth's Southern Hemisphere see the constellations in the celestial sphere's Southern Hemisphere, and some from its northern half. But even these stars are not seen at one time. The Earth rotates on its axis once a day, and follows a yearly path around the Sun, within the celestial sphere. These movements mean that the portion of available sky seen from any one spot changes during the course of the year, and that portion of sky seems to rotate during the course of a night.

NORTHERN HEMISPHERE SKY

The stars in the northern sky seem to move around the star Polaris, the bright star in the very centre of the map. It is in the constellation of the Little Bear, close to its companion the Great Bear. The seven stars in the tail and lower back of the Great Bear are called the Plough. This is one of the easiest patterns to find and is a good starting point for finding your way around the northern sky.

Pegasus, the flying horse

Ursa Major, the Great Bear

Leo, the Lion

SOUTHERN HEMISPHERE SKY

The Southern Cross is the smallest of all the 88 constellations, but it is bright and easy to find. It is a good place to start when exploring the southern sky. It is in the thick, starry band of the Milky Way. There are so many stars here that, as we look into the disc of our Galaxy, they make a river of milky light across the sky.

Phoenix, the firebird

Hercules, the giant

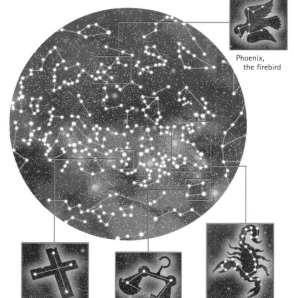

THE STAR MAPS

The maps show the major constellations of the sky. The stars in the centre of the maps can usually be seen all year round from the corresponding hemisphere on Earth. It is those near the edge that are seen only during particular seasons, or at certain times of night. The brightest stars are those with the largest dots. Some appear bright only because they are close to Earth. Others are bright because they really are bright stars. All the stars are so far away they appear to be the same, equal distance from Earth. In fact, the stars in any constellation are unrelated and are vast distances from each other.

Crux, the Southern Cross

Libra, the Scales

Scorpio, the Scorpion

SEE ALSO

264–5 Galaxies,
266–7 Stars, 272–3
The solar system

THE SOLAR SYSTEM

The Solar System is the Sun and its family of objects
that orbit it. They were made from the same cloud,
and stay together because of the Sun's strong gravity.

The Sun is the largest
object in the Solar System.
Next comes Jupiter, the
largest planet, followed by
the other three gas giants,
Saturn, Uranus and
Neptune. All four have
rings and large families of
moons. Earth and the
other planets are smaller.

The Sun, our local star, dominates the
Solar System. It is the largest object in
the system and the most massive. It is
made of over 95 per cent of all the Solar
System material. The rest is used up in
the objects that orbit the Sun. These are
9 planets, more than 60 moons, billions of
asteroids and billions of comets. Because
of the Sun's great size, it has a
powerful gravitational pull, and this
pull keeps the Solar System
together and controls the
movements of the planets.

THE BIRTH OF THE SOLAR SYSTEM

About five billion years ago, the material
that now makes up the Sun and planets
was part of a great cloud of gas and dust
called the solar nebula. The cloud, made
up mainly of hydrogen and helium but
including a tiny percentage of other
elements, spun round and its material was
pulled toward the centre. The solar nebula
was now a ball of gas surrounded by a disc
of gas and dust. The central ball became
the Sun and the disc material produced
the planets and other bodies. Much
more unused material was blown
away into space.

Sun

Pluto

Neptune

Uranus

Jupiter

Saturn

Moon

Venus

Earth

Mars

Mercury

▼ Measurements in
the Solar System are
measured in astronomical
units (AU). One AU is
149.6 million km, the
average distance of the
Earth from the Sun.

Mercury 0.39 AU
Venus 0.72 AU
Earth 1 AU
Mars 1.52 AU
Jupiter 5.20 AU
Saturn 9.54 AU
Uranus 19.19 AU

0 1 2 3 4 5 6 7 8 9 10 11 12 13 14 15 16 17 18 19 20 21 22 23 24 25 26

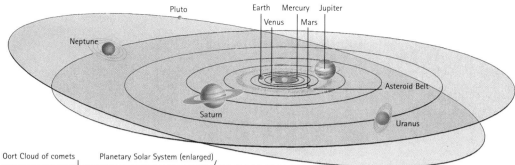

Pluto Neptune Earth Venus Mercury Mars Jupiter Asteroid Belt Saturn Uranus

Oort Cloud of comets   Planetary Solar System (enlarged)

◀ A spherical-shaped region of comets surrounds the planetary Solar System. It contains billions of comets each following its own orbit around the Sun. The sphere is called the Oort Cloud and measures 100,000 AU from side to side. Beyond it is interstellar space.

▲ The planetary Solar System is disc-shaped. The planets go round the Sun in elliptical orbits, and in an anticlockwise direction. The length of orbit and the time taken to complete an orbit increases with successively distant planets.

PLANETS
The planets started to form about 4.6 billion years ago. The material closest to the Sun made the four rocky worlds, Mercury, Venus, Earth and Mars. Further away in the outer disc where it is much colder, rocky bodies formed and then pulled large amounts of gas toward them. These were the large gas planets, Jupiter, Saturn, Uranus and Neptune. Pluto was created from different material.

ASTEROIDS AND COMETS
Rocky material between Mars and Jupiter failed to form into a planet because Jupiter's strong gravity kept pulling it apart. This material became the Asteroid Belt, a doughnut-shaped ring of rocks. At the edge of the Solar System, way beyond the planets were billions of chunks of dust, rock and snow. Many were thrown out of the system, but many remained and formed the Oort Cloud of comets. Each asteroid and comet follows its own path around the Sun.

OTHER SOLAR SYSTEMS
There are many more stars like the Sun in the Universe. Astronomers have always thought it was possible that one of these had a system of planets orbiting around it. The stars, however, are so distant and so bright that detecting a dull planet close to a star was for a long time impossible.

During the 1980s, astronomers discovered the first stars surrounded by discs of gas and dust. A decade later the first planets around stars other than the Sun were discovered. The first was in 1995 around the star 51 Pegasi. The planet is at least 150 times more massive than Earth, 20 times closer to its star than Earth to the Sun, and orbits every 4.2 days. About 15 other stars with a planet have now been found. Astronomers cannot see the planets directly but they can detect the small wobble in the star that is produced by the pull of the planet's gravity as it orbits round it. In the not too distant future new telescopes will give us our first views of these new worlds.

There are more than 60 moons in the Solar System. The smallest moons are around 30km across. The third largest, Callisto, whose surface is shown here, is the same size as the planets Pluto and Mercury.

Most of the Solar System material is found in the Sun, planets and moons. The small amount that remains is found in the billions of minor members of the system, such as in this asteroid.

AU closest	Neptune 30.10 AU																	Pluto 49.50 AU (at its furthest)	
29	30	31	32	33	34	35	36	37	38	39	40	41	42	43	44	45	46	47	48	49	50

SEE ALSO
268–9 The Sun, 270–1 Constellations, 274–5 Earth and the Moon

EARTH AND THE MOON

Earth is the largest of the rock planets, and the third planet from the Sun. The Moon, another ball of rock, about a quarter of Earth's size, accompanies it in space.

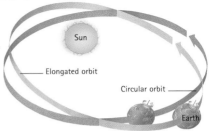

The gravitational pull of the planets in our Solar System changes the orbit of Earth around the Sun from circular to elongated over a period of about 100,000 years.

Life is found everywhere on Earth – in its oceans and on its land. It is home for millions of different species. Humans evolved on it about three million years ago.

From space our planet appears blue and white. The blue is from the water that covers two-thirds of its surface, and the white is from the clouds in the atmosphere surrounding it.

The Earth and the Moon move together through space. They are both spheres of rock but are very different worlds. Earth is unique in the Solar System as it is a very wet place and full of life. It has also changed a lot since it was formed 4.6 billion years ago, and continues to do so. In contrast, the Moon is a dry and dead world where very little has happened in the past three billion years.

PLANET EARTH

About two-thirds of Earth is covered in water. Most of it is in liquid form, but in the polar regions, where it is much colder, it is solid ice. The rest of the surface is made of land masses, the continents. The water and the land sit on top of Earth's outer shell of rock, its crust. This is split into plates of rock; eight giant ones and several smaller ones. In turn, they sit on top of a mantle of hot, molten rock.

The plates move about on the molten rock and produce changes in Earth's landscape. When two plates push into each other, the crust can be forced up and mountains result. Under the oceans when plates move apart, molten rock wells up between them and forms ridges. Molten rock also erupts from volcanoes and flows across the land.

Surrounding Earth is an atmosphere of about 78 per cent nitrogen and 21 per cent oxygen. It provides the air we breathe and protects us from harmful radiation from the Sun and space. It lets sunlight through and helps control temperature. Water is absorbed in the lower atmosphere, condenses into clouds and then falls back to Earth as rain or snow. The water and winds also have a changing effect on the landscape as they erode and shape it.

ORBITS AND SPIN

The Earth takes 365.25 days to make one complete orbit around the Sun. As it moves along its orbit it also spins on its axis, one spin takes 23.9 hours. The Moon accompanies Earth as it travels round the Sun, orbiting Earth at an average distance of 384,400 km. This is an average figure because the Moon's orbit is elliptical, and so the Moon's distance from Earth varies. The Moon takes 27.3 days to make one orbit round Earth. As it travels, it spins on its axis. The time for one spin is also 27.3 days, the same amount of time as that taken for one orbit. As a result the Moon always has the same side facing Earth.

Earth's atmosphere is divided into layers. Near the planet's surface is the troposphere which contains more than 75 per cent of the atmosphere. The highest layer, the exosphere, around 500 km above the surface, contains very little air, and fades away into space.

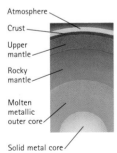

Atmosphere
Crust
Upper mantle
Rocky mantle
Molten metallic outer core
Solid metal core

▲ When Earth was younger and hotter its heavy iron sank to its centre to form a core. The lighter rocks floated above the metal. The Earth has cooled since then, but the mantle and part of the core material is still molten.

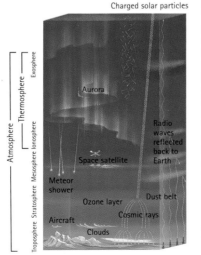

Charged solar particles

Aurora

Radio waves reflected back to Earth

Space satellite

Meteor shower

Ozone layer

Dust belt

Aircraft

Cosmic rays

Clouds

Exosphere
Thermosphere
Mesosphere
Ionosphere
Stratosphere
Troposphere

Atmosphere

EARTH'S MOON

The same side of the Moon always faces Earth. The dark maria, several hundred kilometres across are easily seen. Surrounding them are mountain ranges formed when the huge crater bowls were created. Smaller craters cover the surface.

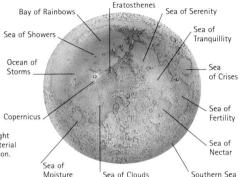

Bay of Rainbows
Eratosthenes
Sea of Serenity
Sea of Showers
Sea of Tranquillity
Ocean of Storms
Sea of Crises
Copernicus
Sea of Fertility
Sea of Nectar
Sea of Moisture
Sea of Clouds
Southern Sea

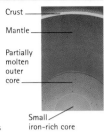

Crust
Mantle
Partially molten outer core
Small iron-rich core

▲ The Moon's crust is about 60–100 kilometres thick. Above it is the regolith, a surface of dust and rock. Due to the Moon's very weak gravity there is no atmosphere, wind or weather.

No one is certain how the Moon formed. It is thought that a Mars-sized ball of rock struck the Earth. Material from Earth combined with the rock to form the Moon.

THE MOON

The Moon does not have any light of its own, but shines because sunlight is reflected off it. One half of the Moon is always lit by the Sun and the other half is in darkness. In the same way, Earth always has one half in daylight and one in the darkness of night. As the Moon travels around us we see differing amounts of the sunlit half. In this way, the Moon appears to change shape.

These different shapes are the phases of the Moon. A complete cycle of the Moon's phases takes 29.5 days. It goes from new Moon, when the side facing us is completely dark, to full Moon, when this side is completely lit, and back to new Moon, to start all over again. It is possible to see some of the Moon's surface features by looking at it with the naked eye. The lighter areas are older, higher land, and the darker ones, younger, lower and flatter land. Binoculars or a telescope will reveal that the whole of the surface is covered in impact craters. Most of these were formed when the Moon was bombarded by space rocks between three and four billion years ago. Over the next billion years lava seeped through cracks in the Moon's crust and flooded the largest craters to form maria (*mare*, singular, is the Latin for sea). These are the dark regions visible with the naked eye. The first astronomers to observe them called them seas because that is what they thought they were.

Large areas of Earth's surface have been changed by humans. As population increases, their buildings, farming and communication systems will cover more of it.

There are about 850 active volcanoes on Earth. Lava flows from them and changes the surrounding landscape. Occasionally it forms a new island.

Half of the Moon is always in sunlight. The phases of the Moon depend on how much of the lit half we can see from Earth. At New Moon, when the Earth, Moon and Sun are roughly in line, we cannot see any of the lit half. About a week later, at the first quarter, we can see half of the part of the Moon that is in sunlight, and at Full Moon we can see all of it. By the last quarter we can again see only half of the lit part.

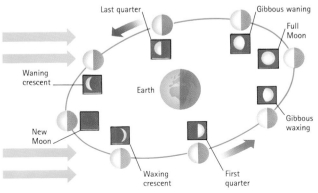

Last quarter
Gibbous waning
Full Moon
Waning crescent
Earth
New Moon
Gibbous waxing
Waxing crescent
First quarter

SEE ALSO

8 Earth and the Solar System, 12–13 Earth's structure, 14–15 Earth's atmosphere, 272–3 The Solar System

ECLIPSES

When the Moon is in Earth's shadow the Moon is eclipsed. When the Moon covers the Sun and stops its light reaching Earth, the Sun is eclipsed.

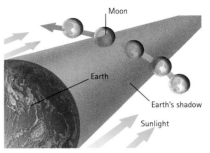

The Moon is eclipsed as it passes through Earth's shadow. It does not become invisible because a little sunlight is diffused into the shadow by the Earth's atmosphere.

In a total lunar eclipse the Moon is completely covered by Earth's shadow. Unlike a solar eclipse, no special equipment is needed to observe a lunar eclipse. As long as the Moon is in the sky and there are no clouds covering it, the eclipse should be visible.

A s Earth orbits the Sun, the Moon orbits around Earth. From Earth, the Moon is sometimes in the sky at the same time as the Sun. Although the Sun is four hundred times larger than the Moon, they appear roughly the same size in our sky. This is because the Moon is four hundred times closer to Earth than the Sun. If the Moon is directly in line with the Sun it will eclipse the Sun. When the Moon is on the opposite side of the Earth to the Sun, it might pass through Earth's shadow and be eclipsed itself.

LUNAR ECLIPSE

As the Moon travels around Earth it can move out of the Sun's light and into the shadow cast by Earth. But the Moon's path does not take it through Earth's shadow on each orbit. So the Moon is not eclipsed every time it travels round the Earth, only up to three times each year.

When the Moon is completely in Earth's shadow it is a total lunar eclipse. When the Moon is only partially covered by shadow it is a partial lunar eclipse.

SOLAR ECLIPSE

The Moon passes between Earth and the Sun each time it orbits round Earth. As it does so the Moon's phase is a New Moon. The Moon facing away from Earth is lit, and the side facing Earth is dark. When the Moon lies directly between the Sun and the Earth it blocks out the Sun's light and casts a shadow onto Earth. The Sun is eclipsed. If the Sun's disc has been completely covered by the Moon's, the Sun is totally eclipsed. When the Moon partially covers the Sun it is a partial eclipse. Eclipses do not happen at every New Moon, as usually the Earth, Moon and Sun are not directly aligned and the Moon does not cover the Sun's disc. Solar eclipses happen only once or twice each year, and then are only visible from the parts of Earth covered by the Moon's shadow. During a total eclipse it is possible to see the Sun's faint outer atmosphere, the corona, normally lost in the glare of the Sun's disc.

Just before and after the Sun's disc is totally eclipsed by the Moon, a diamond ring effect can be seen. For a few seconds, the Sun's light is shining between mountains on the lunar surface to create an enormous sparkling diamond in the sky.

▼ The Moon casts a shadow into space. When the shadow reaches Earth, anyone standing in the darkest part, the umbra, will see a total solar eclipse. The Sun will be eclipsed for up to about 7½ minutes.

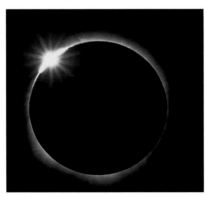

An eclipse of the Sun

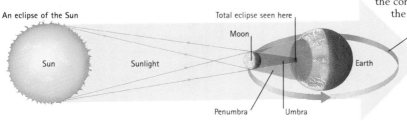

Sun

Sunlight

Moon

Total eclipse seen here

Earth

Moon's orbit

Penumbra

Umbra

SEE ALSO

166–7 Heat and light from the Sun, 268–9 The Sun, 274–5 Earth and the Moon

MERCURY

The closest planet to the Sun is a small, rocky, airless world. Mercury's crater-covered surface faces extreme temperatures but has not changed for billions of years.

Wherever you look on Mercury the view is of a crater-covered, rocky landscape. Above it is a black sky because of the lack of atmosphere to reflect sunlight.

Mercury is difficult to see and study from Earth because it is never far from the Sun's glare in our sky. Much of what we know about Mercury comes from Mariner 10, the only spaceprobe to visit it. It took photographs of nearly all of one side which, once radioed back to Earth, were combined to give an overall picture of much of the planet's surface.

SURFACE

Mercury is covered with craters which were formed when it was bombarded by meteorites about four billion years ago. The craters range from a few metres to hundreds of kilometres across. The largest, the Caloris Basin, is about 1,300 km wide. Inside the basin are flat plains of solidified lava which once oozed from the planet's interior. Mercury's surface is also crossed by wrinkles and ridges a few kilometres high. These were formed as hot, young Mercury cooled and shrank. The planet's outer mantle and crust wrinkled like the skin on a dried apple.

Mercury is the second smallest of the planets. From Earth we can only see faint markings on its rocky surface.

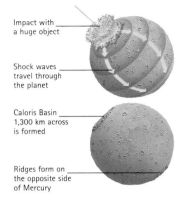

Impact with a huge object

Shock waves travel through the planet

Caloris Basin 1,300 km across is formed

Ridges form on the opposite side of Mercury

The dominant feature on Mercury, the Caloris Basin, was formed when a 100 km-wide body crashed into the planet. Shock waves from the impact travelled through the planet and threw up hills and mountains.

ORBIT AND SPIN

Mercury's orbit is elongated. Its distance from the Sun varies from 70 million km when farthest away, to only 46 million km when nearest. So the Sun in Mercury's sky sometimes appears one and a half times the diameter, and feels twice as hot, as at others. Mercury orbits at 48 km per sec making it the fastest moving planet, and takes only 88 days to complete one orbit. It spins as it travels, taking 59 days to make one spin. The combination of its fast orbit and slow spin means anyone standing on Mercury would have to wait 176 days between one sunrise and the next. For them a year (88 days) would be shorter than a day (176 days).

Extremely thin atmosphere

Crust

Rocky mantle

Huge iron-nickel core

Mercury is a rocky planet with a particularly large metal core. Surrounding this is rock, and above it an atmosphere so thin it barely exists. During sunrise to sunset the temperature on the surface reaches 450°C, but during the long night the temperature drops to about −170°C.

The Mariner 10 spaceprobe flew by Mercury three times during 1974 to 1975. Its cameras took sequences of overlapping photographs that were combined to give us a global view of this planet.

SEE ALSO

268–9 The Sun, 272–3 The Solar System, 290–1 Exploring space

VENUS

The hottest planet in the Solar System is a hostile world. A thick choking atmosphere surrounds a rock planet with a surface shaped by volcanic activity.

The Sun is barely visible in the dull, orange sky. It is hidden by the thick clouds of Venus' atmosphere. At night-time the stars are totally hidden from view.

Venus is a rock planet, second from the Sun and similar in size to Earth. It appears as a bright star in Earth's sky, either after sunset or before sunrise, and is so called the 'morning star' and 'evening star'. It shines brightly because sunlight is reflected off its cloud tops. But the rock world below is hidden from view. Venus spins slowly, taking 243 days to turn once, and is the only planet to turn clockwise when viewed from its North Pole.

Maat Mons, at 9 km high, is one of the largest volcanoes on Venus. Lava flows, hundreds of kilometres wide stretch across its base. Radar information recorded by Magellan was computer-processed to produce this image.

The surface of Venus is completely hidden by dense clouds. The clouds circle the planet every four or so days, forming Y- or V-shaped patterns.

VENUS' ATMOSPHERE

Venus' dense atmosphere consists mainly of carbon dioxide. It contains so much material that it would feel denser than water and the pressure is 90 times greater than that on Earth. The atmosphere also contains sulphur dust, and droplets of sulphuric acid from erupting volcanoes from when the planet was young. Sunlight penetrates the atmosphere and warms the surface of the planet. The ground radiates heat but the atmosphere traps the heat, warming the planet still more. The average surface temperature reaches 464°C.

SURFACE FEATURES

Spaceprobes sent by the former Soviet Union have travelled through Venus' atmosphere to land on the surface. However, the planet's physical features have been mapped by probes using radar and working from above the thick clouds. The most recent and successful was the American probe Magellan in 1990–1994. Its probes revealed a surface about half a billion years old and formed by volcanic activity. Venus is a smooth planet; 85 per cent of its surface is volcanic plain, dotted with hundreds of volcanic craters and lava flows. There are also over 900 impact craters that were formed when rocks collided with the planet.

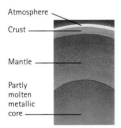

Atmosphere
Crust
Mantle
Partly molten metallic core

Venus is a typical rock planet, with a metal core, rocky mantle and crust. But the atmosphere surrounding it makes it a hostile place for humans.

▶ Venus is not the closest planet to the Sun but it is the hottest because its thick atmosphere is very efficient at holding in the Sun's heat. This is called the greenhouse effect, as it works like glass trapping heat in a greenhouse.

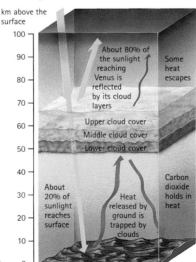

km above the surface

100
90 — About 80% of the sunlight reaching Venus is reflected by its cloud layers
80
70 — Upper cloud cover
60 — Middle cloud cover
50 — Lower cloud cover
40
30 — About 20% of sunlight reaches surface
20 — Heat released by ground is trapped by clouds
10
0

Some heat escapes

Carbon dioxide holds in heat

▲ The Magellan spacecraft mapped 98 per cent of the Venusian surface between 1990–1994. It recorded features of the surface using radar to 'see' through the cloud layers.

SEE ALSO

268–9 The Sun, 272–3 The Solar System, 290–1 Exploring space

MARS

Mars is the planet most similar to Earth. It is a dry, rocky, red world about half the size and much colder. Robot spacecraft have landed on it to search for life.

Mars has two tiny, dark, irregular-shaped moons. They are made of carbon-rich rock and are thought to be two asteroids caught by Mars' gravity.

Deimos

Phobos is about 27 km long; the larger moon and closer to Mars. Deimos is about 15 km long.

Phobos

Mars is known as the Red Planet. It not only has a red surface but also a red sky. The colour comes from wind-borne dust blown from the reddish, iron-rich surface.

Mars has cold regions at its north and south poles. These polar regions are covered with water ice, and carbon dioxide ice.

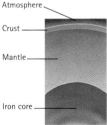

Atmosphere

Crust

Mantle

Iron core

Mars is made of dense rock with a metal core. An atmosphere, about 100 times thinner than Earth's, comprising mainly of carbon dioxide gas, surrounds the planet.

▶ The Pathfinder probe landed on Mars on 4 July 1997. It opened up to allow the Sojourner robotic vehicle to move down its ramp (at left) and explore the surface.

Mars is the fourth planet from the Sun, about one and a half times further from the Sun than Earth. Its distance makes it a much colder planet than Earth and its average temperature is well below freezing. It does, however, appear fiery red and was named after the Roman God of War due to its blood-like appearance. The red colouring comes from the rock and soil strewn across its surface. The rock is rich in iron-oxide, better known as rust.

SURFACE FEATURES

Volcanic activity, meteorite bombardment and running water, all in Mars' distant past have shaped the surface we see today. Much of the surface is rock strewn desert plain, with dusty dunes, and craters formed by meteorites. But giant features rise above it and cut into it. The volcano Olympus Mons, the highest mountain in the Solar System, is 600 km across and rises 24 km above the surrounding plain. Equally spectacular is the Valles Marineris, an enormous canyon system that runs for around 4,500 km across Mars, and is, in parts, 8 km deep. There are also winding valleys that look like dried-up river beds formed when Mars had running water over 3 billion years ago.

SEARCH FOR LIFE

Less than a century ago people believed that Martians, a form of intelligent life, lived on Mars. Today we know this is false, but primitive life may exist there. In the past Mars was warmer and wetter and under these conditions life may have developed. Two American spaceprobes both called Viking landed on Mars in 1976. Experiments on board tested for signs of life but none were found.

More spacecraft were sent to investigate Mars in the 1990s. They orbited the planet, mapping it and studying its weather, and landed on it. Pathfinder landed after a 7-month journey in 1997. It carried a six-wheel microwave-oven-sized robotic rover, Sojourner, which explored the landing site. More craft are being prepared for missions to Mars.

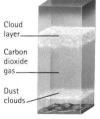

Cloud layer

Carbon dioxide gas

Dust clouds

Above the surface, winds raise huge dust clouds. At night the carbon dioxide freezes to form hoar frost.

SEE ALSO

272–3 The Solar System, 274–5 Earth and the Moon, 290–1 Exploring space

JUPITER

Jupiter is the largest and most massive of all the Solar System planets. It is a gas giant with a surface of colourful clouds, 16 moons and a thin ring system.

There is no solid surface on Jupiter. The outer edge that we can see is clouds at the top of layers of gas. Its 16 moons move across its sky as they orbit round the planet.

The Galileo probe arrived at Jupiter in 1995, circling the planet for two years before sending a probe into the atmosphere and investigating the moons.

Jupiter is so huge that 11 Earths would fit across its diameter, and 1,300 would fill it. It is made up of around 90 per cent hydrogen, mostly in the gaseous state, with about 10 per cent helium and traces of other elements. The outer part of Jupiter is gas, below this liquid, and in the centre is a solid core. It is the fifth planet from the Sun, and takes twelve years for one orbit. Jupiter rotates fast on its axis, taking less than 10 hours for one turn.

JUPITER'S WEATHER
Jupiter's rapid spin, combined with heat rising from inside, help create and drive the planet's weather. Different pressure systems, and winds of up to 650 km/h result. The planet's stripy appearance is due to dark zones of falling gas, and light zones of rising gas, creating storms where these zones meet. The largest of these are visible from Earth, and spaceprobes have detected giant bolts of lightning. A huge hurricane storm, the Great Red Spot, is an atmospheric whirlpool which rotates anticlockwise every six days.

Io is one of Jupiter's Galilean moons, named after the astronomer Galileo Galilei who studied them. Its rocky surface is coloured red and yellow because of the sulphur that erupts from its many volcanoes.

JUPITER'S MOONS
Surrounding Jupiter is a thin ring system and a large, varied family of moons. There are sixteen moons, four large and twelve small. The four largest ones, Io, Europa, Ganymede and Callisto, are known as the Galilean moons. Ganymede is the largest moon in the Solar System and is bigger than the planets Pluto and Mercury. Io is the most volcanically active body known.

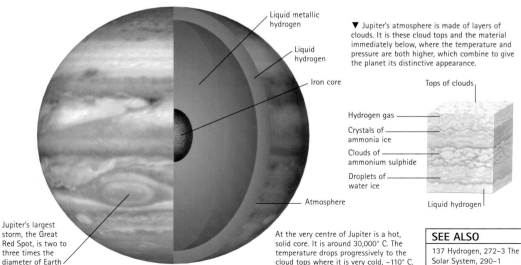

Jupiter's largest storm, the Great Red Spot, is two to three times the diameter of Earth

Liquid metallic hydrogen

Liquid hydrogen

Iron core

Atmosphere

At the very centre of Jupiter is a hot, solid core. It is around 30,000° C. The temperature drops progressively to the cloud tops where it is very cold, –110° C.

▼ Jupiter's atmosphere is made of layers of clouds. It is these cloud tops and the material immediately below, where the temperature and pressure are both higher, which combine to give the planet its distinctive appearance.

Tops of clouds

Hydrogen gas

Crystals of ammonia ice

Clouds of ammonium sulphide

Droplets of water ice

Liquid hydrogen

SEE ALSO
137 Hydrogen, 272–3 The Solar System, 290–1 Exploring space

SATURN

Saturn appears to be a bright yellow star from Earth. Spaceprobes have revealed a colourful planet with an amazing system of rings and a large family of moons.

Saturn, the second largest planet and the sixth from the Sun, is a gas giant, made up mainly of hydrogen. Its outer surface is not solid but made of bands of clouds, of ammonia, water and methane, coloured by phosphorus and other elements. The bands are surrounded by a haze which hides the stormy weather below. Some weather disturbances can be seen from Earth. About three times every century violent storms disturb the surface. They are seen as large white spots near the planet's equator.

Three spaceprobes sent to Saturn have taught us much about the planet. Pioneer 11 was the first in 1979, then Voyagers 1 and 2 in 1980–1981. They not only revealed details of the planet but observed its ring system, and discovered 12 of its moons. Another craft, Cassini, is heading toward it now. On arrival in 2004, it will orbit and study Saturn for four years, and will drop a small probe into Titan's atmosphere.

Saturn does not have a solid surface. Its outer layer, a haze of ammonia crystals is difficult to see through. The ring system dominates the sky.

Due to its fast spin Saturn is squashed in shape with a bulging middle. A gap in the rings, known as the Cassini Division, contains ring pieces.

Atmosphere

Liquid hydrogen

Liquid metallic hydrogen

Iron core

Saturn is made up mainly of hydrogen. This is in a gaseous form furthest from the surface, turning to liquid further in. Close to the iron core the liquid hydrogen becomes metallic in form.

▶ The small probe Huygens, carried to Saturn by the Cassini spaceprobe, will use a parachute to slow its descent to Titan.

SATURN'S RINGS

All four gas giants are surrounded by a ring system but Saturn's is the most extensive and spectacular. The broad but thin system appears to be divided into several wide rings but each of these is made up of thousands of separate narrow ringlets. Each ringlet is made of icy rock pieces ranging from tiny specks to house-sized chunks. The system is only metres thick in parts but stretches out to almost 500,000 km – further than the distance of our Moon from Earth.

The rings are not as old as the planet. They are thought to be only a few hundred million years old and are the remains of a moon or comet that was torn apart. Saturn is tilted on its axis as it orbits the Sun, so we see the rings from different angles. Twice in each orbit they are full face on to us, and twice they are edge on and disappear from our view.

SATURN'S MOONS

Saturn has at least 18 moons and probably over 20. It is expected that Cassini will find more smaller ones. Saturn's largest, and the second largest in the Solar System, Titan is the only moon known to have an atmosphere – mainly made up of nitrogen. The smallest, Pan, is only about 20 km across. It is one of a handful that orbits Saturn within the outer ring system.

The Huygens probe will test Titan's thick, nitrogen atmosphere as it descends. No one yet knows what kind of surface it will find underneath.

Saturn's rings are made up of billions of particles in orbit round the planet, ranging in size from tiny ice-coated rock grains, to large building-sized icy rocks. The largest fragments are found in the inner rings, and fine particles in the outer rings. The rings could be the remains of a small moon 100 km across.

SEE ALSO

137 Hydrogen, 272–3
The Solar System, 290–1
Exploring space

281

URANUS

Uranus is the seventh planet from the Sun, and only just visible with the naked eye. It is a gas giant with a ring system and a large family of moons.

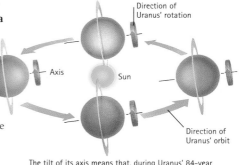

Axis — Sun

Direction of Uranus' orbit

The tilt of its axis means that, during Uranus' 84-year orbit each pole has a 40-year summer when the Sun shines constantly, followed by a 40-year winter period.

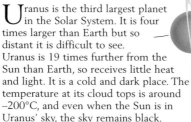

The sky above Uranus is black. The Sun is only a distant, bright star. A thin system of rings surrounds Uranus. Each is made of dark chunks of rock.

Uranus' surface is almost featureless. It consists mainly of hydrogen and helium. Another gas, methane, gives the planet its blue-green colouring.

Atmosphere of hydrogen, helium and methane

Mantle of ammonia, water and methane ice

Iron silicate core

The only solid part of Uranus is its core. Above this is a cold layer of liquid and ice, and above that a layer of gas.

▶ An artist's view of Uranus from the smallest of its five main moons, Miranda. It is thought that Miranda was once smashed apart by an enormous impact, and then re-formed.

Uranus is the third largest planet in the Solar System. It is four times larger than Earth but so distant it is difficult to see. Uranus is 19 times further from the Sun than Earth, so receives little heat and light. It is a cold and dark place. The temperature at its cloud tops is around −200°C, and even when the Sun is in Uranus' sky, the sky remains black.

Uranus travels round the Sun on its side. Its axis is tilted from the vertical by 98°, so its north pole is pointing slightly south. Its rings and moons orbit round its middle.

DISCOVERY

Its remote position in the Solar System meant that Uranus was not known in ancient times. The planet was only discovered in 1781 when the astronomer William Herschel (1738–1822) observed it through his telescope. From that time telescopes have been used to investigate the distant planet, but even the most powerful reveal little. It was only in 1986, when the spaceprobe Voyager 2 reached Uranus, that we got our first good look. Voyager 2 took images of the planet, its rings and moons, and transformed our knowledge of the Uranian system.

MOONS AND RINGS

The five largest of Uranus' moons were known before Voyager 2 reached Uranus. Ten more, each less than 100km across, were discovered by the probe. Since then, more have been found. By the start of the year 2000, 18 were known and three more potential moons were being watched. Most of the moons are named after Shakespearean characters, for example the largest moon is Titania, and others are Miranda, Oberon and Puck.

Uranus' ring system, made of 11 major rings or more, seems to be almost upright around the planet. This is because Uranus is tilted on its side. The rings consist of pieces of rock roughly one metre across.

In 1781, William Herschel discovered Uranus through his homemade telescope set up in his garden in Bath, England.

SEE ALSO

272–3 The Solar System, 288–9 Astronomical telescopes, 290–1 Exploring space

NEPTUNE

Neptune is the most distant of the four giant gas planets. It is a blue, windy and cold world, with a ring system and a family of eight moons.

Neptune, and the distant Sun, are seen in Triton's sky. The planet's cloud tops give it a blue, banded appearance. Triton is a frozen, rock world.

The Great Dark Spot (centre left) turns round every 16 days, and is large enough to contain Earth. The white clouds are methane ice.

Atmosphere of hydrogen helium and methane

Icy mantle

Iron core

▲ Neptune's solid core is surrounded by a mantle of icy water, methane and ammonia. Above is the gas layer made of about 80 per cent hydrogen.

▶ Neptune is very large in Triton's sky. The moon's surface is frozen nitrogen and methane, and at −235°C, is the coldest surface in the Solar System.

This bright blue gas world is the eighth planet from the Sun and the fourth largest in the Solar System. Yet for about 20 in every 248 years it becomes the most distant planet as Pluto comes closer to the Sun than Neptune. Neptune is difficult to see from Earth, and like Uranus, the first clear view of it was from the Voyager 2 spaceprobe. It reached Neptune in 1989, after its successful trip to Uranus in 1986.

NEPTUNE'S APPEARANCE
Neptune's atmosphere is mainly hydrogen and helium. Like Uranus it also contains methane, which gives the planet its brilliant blue colour. Neptune is bluer than Uranus because there is more methane in Neptune's upper clouds.

White and dark features appear on Neptune's surface. The Great Dark Spot was discovered by Voyager 2 but had disappeared by 1994 when the Hubble Space Telescope looked at the planet. The dark spots and white clouds are forced round Neptune by high winds. Neptune spins in an anticlockwise direction (when viewed from the north) but the winds blow in the opposite direction, from east to west. Speeds of 2,200 km/h make them the fiercest in the Solar System.

DISCOVERY
Neptune was not known to ancient astronomers. It is not visible to the naked eye and was discovered after astronomers had been observing Uranus. They noticed that Uranus' path was affected by the gravitational pull of an unknown body. In 1845, John Couch Adams (1819–1892) in England, and Urbain Le Verrier in France, worked out the position of an unknown planet whose gravity pulled on Uranus. In 1846, Neptune was discovered, in the predicted position, by German astronomer Johann Galle (1812–1910).

MOONS AND RINGS
Neptune's largest moon, Triton, was also discovered in 1846, and a second moon, Nereid in 1949. A further six moons were discovered by Voyager 2 in 1989. Triton is a rock body, larger than Pluto, which orbits in the opposite direction to the other moons. It may not have started life as a moon but was captured by Neptune's gravity. Voyager 2 also found four faint and narrow rings around Neptune.

▶ Urbain Le Verrier (1811–1877), a French astronomer, calculated the position of an unknown planet that affected the orbit of Uranus. He gave his predictions to Johann Galle who, in 1846, found the planet Neptune.

SEE ALSO
272–3 The Solar System,
288–289 Astronomical
telescopes

PLUTO AND THE MINOR PLANETS

Pluto is the smallest, coldest and most distant planet. We know little about it, but what we do know suggests that Pluto is only a minor member of the Solar System.

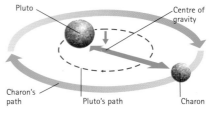

Charon is large compared to the planet it orbits, being about 25% of Pluto's mass. This affects how they both move. They both revolve around their centre of gravity, so that Pluto, as well as Charon, moves in a circle.

Charon and the Sun are seen in Pluto's sky. Pluto is so far from the Sun that it is a very cold and dark world. The Sun appears a thousand times fainter than seen from Earth.

Pluto is the ninth and final planet in the Solar System. Some astronomers think that it is not a planet at all, but is one of the minor members of the Solar System. There are vast numbers of these, and they form three main groups.

The asteroids are rocky bodies in the planetary part of the Solar System. Most are in the Asteroid Belt between the planets Mars and Jupiter. Further out are the Kuiper Belt objects which form a belt stretching from Neptune's orbit to the outer Solar System. The outer edge is occupied by the spherical Oort Cloud made up of the third group, the comets.

PLUTO

Pluto is a small ball of icy rock which probably became captured into a planetary-like orbit. Its orbit is different from those of the other planets – it is much less circular, and is also tilted by 17° to Earth's orbit. This means Pluto's distance from the Sun varies between 4.45 billion to 7.38 billion kilometres. When Pluto is at its closest, it is closer to the Sun than Neptune. This happens once during each of Pluto's 248-year orbits when it remains the second most distant planet for around 20 years until it moves back beyond Neptune.

Pluto is a ball of rock with a frozen nitrogen and methane surface. It is thought to have impact craters on it.

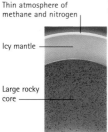

Thin atmosphere of methane and nitrogen

Icy mantle

Large rocky core

Pluto is always cold, about –220°C, but when its orbit takes it closest to the Sun it warms up, and some of its ice turns to gas and forms an atmosphere.

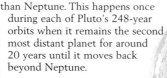

No one has seen either Pluto or Charon in detail, and so this artist's impression shows what astronomers think they are like. Pluto is shown in Charon's sky, above a frozen sea of methane.

Pluto was discovered in 1930 by US astronomer Clyde Tombaugh (1906–1997). Its moon, Charon, was discovered in 1978. It is about half the size of Pluto and also made of rock and ice. They follow paths around each other like a double planet. As they circle, they spin and keep the same sides facing towards each other.

As a small rocky body, Pluto seems to be out of place when compared with the gas giants, its four nearest planets. Pluto is so distant from Earth that it has not yet been investigated by a spaceprobe, but the Pluto-Kuiper Express will start its journey to Pluto in December 2004, and then go on to investigate the Kuiper Belt.

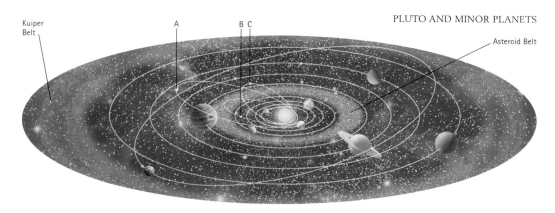

Kuiper Belt

A

B C

Asteroid Belt

ASTEROIDS

Billions of asteroids orbit the Sun. Over 90 per cent make up the Asteroid (or Main) Belt, a doughnut-shaped ring between Mars and Jupiter. They are sometimes called the minor planets, as they are each rocky bodies orbiting the Sun. They are the leftovers of a planet that did not form when the Solar System was young, 4.6 billion years ago.

The first to be discovered, Ceres in 1801, is also the largest. It is spherical and about 932 km wide. Most asteroids are much smaller, about one billion are more than one kilometre wide, but many more are only metres across. Asteroids are made of rock, metal or a combination of the two.

Over ten thousand individual asteroids have been discovered and had their orbits recorded. Each has been given a name. Our only close-up views have been provided by two spaceprobes. The first to be seen was Gaspra, by the Galileo probe in October 1991. The NEAR (Near Earth Asteroid Rendezvous) probe showed us Mathilde in 1997 and Eros in 1998.

KUIPER BELT

The Kuiper Belt is named after Dutch-born astronomer Gerard Kuiper (1905–1973). It is a ring of comet-like objects, thought to be made of ice, snow, rock and dust. It is estimated that there are at least 70,000 of them larger than 100 km across. They are so distant that at present we can only see the closest of them. The first were discovered in the last years of the 20th century.

The Asteroid Belt, between Mars and Jupiter, is made up of billions of asteroids. The largest, Ceres, has a face pitted with impact craters and a width of 932 km. Other, individual asteroids follow paths (A, B and C) within the inner Solar System. The Kuiper Belt, made up of tens of thousands of icy rock bodies, lies beyond Neptune's orbit.

▲ The asteroid Ida, imaged by the Galileo spaceprobe, is about 52 km long. Its surface is covered in craters where other asteroids have collided with it. A tiny asteroid, Dactyl, thought to have broken off Ida, orbits around it like a moon.

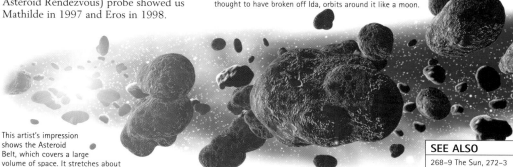

This artist's impression shows the Asteroid Belt, which covers a large volume of space. It stretches about 344 million km from its inner to outer edge.

SEE ALSO
268–9 The Sun, 272–3 The Solar System, 290–1 Exploring space

COMETS

Comets are dirty snowballs. Vast numbers of them live at the very edge of the Solar System. When one travels close to the Sun it grows a giant head and tails.

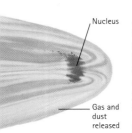

Nucleus

Gas and dust released

A cometary nucleus is heated up when close to the Sun. The gas and dust released forms a coma, many times the diameter of Earth, and tails that can be 100 million kilometres in length.

A comet forms a new coma and tails on each return to the Sun. The tails are longest after the closest approach, and they always point away from the Sun. A comet will average about 100 returns before its gas and dust is used up.

Astronomers have estimated that about 10 trillion comets make up the Oort Cloud, the enormous spherical cloud that surrounds our Solar System. The cloud is incredibly large, about 7.6 million million kilometres across. Each comet follows its own orbit around the Sun. They are all potato-shaped balls of snow and rocky dust, only kilometres across.

HEAD AND TAILS

Occasionally a comet leaves the Oort Cloud and travels into the inner Solar System toward the Sun. Close to the Sun the comet grows in size and brightness. The Sun's heat turns the snow on the surface of the snowball nucleus to gas, and some dust is released. The gas and dust form a bright head around the nucleus, the coma. More gas and dust is blown away and forms two tails. The dust tail is yellowish-white, the gas tail, bluish.

PERIODIC COMETS

When a comet is close to the Sun and has a head and tail it can be seen in Earth's sky. Some comets return regularly to our sky. The time between returns is a comet's period. Encke's Comet has the shortest period of all, just 3.3 years. Halley's Comet returns about every 76 years. These are just two of the 135 or so comets with periods of less than 200 years. They are known as short-period comets. Other comets that return, but not for thousands of years, are known as long-period comets.

The regular return of Halley's Comet has meant it has been seen by thousands of people through history. Spaceprobes were sent to it on its last return in 1986. The most successful, Giotto, travelled inside its coma, and people saw a cometary nucleus for the first time. Another probe, Stardust, will visit Comet Wild 2 in 2004. This probe will return to Earth, bringing some dust and gas from Wild 2 with it.

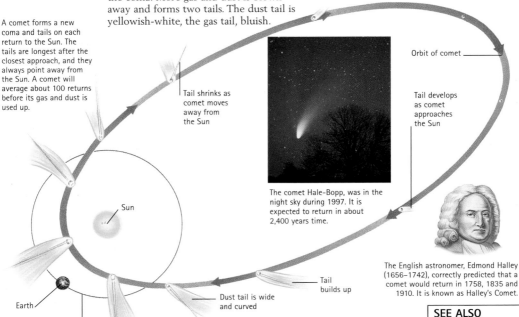

Orbit of comet

Tail shrinks as comet moves away from the Sun

Tail develops as comet approaches the Sun

The comet Hale-Bopp, was in the night sky during 1997. It is expected to return in about 2,400 years time.

Sun

Tail builds up

Earth

Dust tail is wide and curved

Earth's orbit

Gas tail is straight and narrow

The English astronomer, Edmond Halley (1656–1742), correctly predicted that a comet would return in 1758, 1835 and 1910. It is known as Halley's Comet.

SEE ALSO

268–9 The Sun, 272–3
The Solar System, 290–1
Exploring space

METEORS AND METEORITES

Tiny pieces of Solar System space rock burn up in
Earth's atmosphere and create meteors. Larger chunks
that land on Earth's surface are called meteorites.

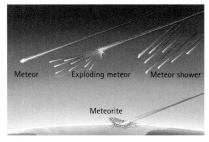

Meteor   Exploding meteor   Meteor shower

Meteorite

Meteors occur at between 120 and 80 km above Earth's
surface. Meteors in a shower all seem to start at the
same point in the sky. A space rock that reaches the
surface, known as a meteorite, can produce a crater.

Pieces of space rock and dust, called
meteoroids exist throughout the Solar
System. As Earth moves through space it
collides with many of these pieces. Over
200,000 tonnes of space rock enters
Earth's atmosphere each year. Most pieces
are tiny and burn up in the atmosphere.
Larger ones survive and land on Earth.

METEORS AND METEOR SHOWERS

Tiny dust pieces, no bigger than a grain
of sand, speed through the atmosphere
toward Earth faster than a bullet. The dust
is heated and parts of it burn off,
producing a column of light that is
known as a meteor. Meteors are
about one metre across and 20 km
long and last for less than a second.
Viewed from Earth a meteor looks like
a streak of starlight shooting across the
sky, and that is why they are also called
shooting stars. About ten an hour can be
detected in a moonless, and cloud-free sky.
A meteor shower consists of related
meteors coming from the same region of
sky, at the same time every year. Showers
are caused by the Earth travelling through
a stream of meteoroids. The stream forms
from dust that is shed by comets.

This iron and nickel
meteorite, discovered in
California in 1976, is the
second largest found in
America. It is nicknamed
'The Old Woman'.

A meteor is a column of
light produced by a tiny
piece of space dust burning
up in Earth's atmosphere.

METEORITES AND CRATERS

Large space rocks that do not burn up
in the atmosphere crash on the Earth's
surface and are known as meteorites. They
start life as part of a comet or an asteroid.
More than 3,000 of them, weighing
around a kilogram or more, land each year.
Most are made of rock, some are iron,
others are a mixture of the two. Craters
are formed when large meteorites collide
with the land. They range in size from a
few metres to 140 kilometres across. Most
were formed more than 50 million years
ago. The 200 km-wide Chicxulub Crater,
now under the Gulf of Mexico, was
formed when a huge rock impacted with
Earth 65 million years ago.

The Barringer Crater in
Arizona, USA, was formed
about 52,000 years ago
when a large meteorite
struck Earth. The iron
meteorite, thought to be
about 30 m wide blasted
the surface material away
on impact and formed this
1.2 km-wide crater.

SEE ALSO

14–15 Earth's atmosphere,
272–3 The Solar System,
274–5 Earth and the Moon

ASTRONOMICAL TELESCOPES

Telescopes are astronomers' most important tool. They provide information by collecting light, radio and other energy waves from space.

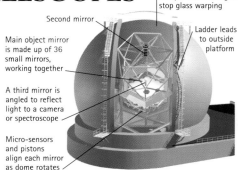

Temperature of dome is kept near freezing to stop glass warping

Second mirror

Main object mirror is made up of 36 small mirrors, working together

A third mirror is angled to reflect light to a camera or spectroscope

Micro-sensors and pistons align each mirror as dome rotates

Ladder leads to outside platform

Keck I, the first telescope to use a segmented mirror, was completed in 1992. It is housed next to an identical telescope, Keck II, 4,200 m above sea level, at the top of Mauna Kea, an extinct volcano in Hawaii.

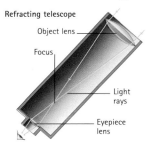

Refracting telescope

Object lens

Focus

Light rays

Eyepiece lens

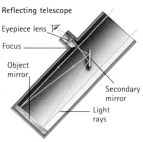

Reflecting telescope

Eyepiece lens

Focus

Object mirror

Secondary mirror

Light rays

These are the two simplest designs of optical (light-gathering) telescopes used by astronomers. In a refractor the light enters and is bent to form an image at the focus. This is then magnified by the eyepiece lenses. Reflectors use a main object mirror, and then a smaller secondary mirror to direct the light to the eyepiece.

Telescopes have been used by astronomers since the early 17th century. One of the first astronomers to use one was the Italian, Galileo Galilei. His telescope enabled him to see craters on the Moon's surface, observe the stars in the Milky Way, and study four of Jupiter's moons. The telescope was then a new invention, mainly used on land. Galileo published his discoveries, and so encouraged others to turn the telescope to the night sky.

HOW A TELESCOPE WORKS

The first telescopes used lenses to collect light from a distant object. This was then focused to form an image of the object. Another lens then magnified the image for the astronomer. The result was that a faint, distant object became larger, brighter and clearer.

A telescope using a lens for collecting light is called a refractor. Most telescopes today use a mirror and are known as reflectors. The first reflector was designed in 1668 by the English scientist, Isaac Newton. Both types were equally used until the 20th century when the reflector rose in popularity. By the end of the 20th century astronomers wanted telescopes with light-gathering power beyond the capabilities of a single mirror. So, many of today's most powerful telescopes use more than one mirror to collect light. The two Keck telescopes in Hawaii, for example, each have 36 six-sided mirrors joined together to make one large mirror, 10 metres wide.

USING THE TELESCOPE

In modern astronomy the astronomers' eyes have been replaced at the telescope's eyepiece by sensitive electronic cameras using CCDs (charged-coupled devices). These build up an image over minutes or hours and pick up information the human eye could never see. A second instrument, a spectroscope, which splits the light into its spectrum of colours, may also be attached to the telescope. The study of lines in the spectrum can reveal the elements inside a star, and its temperature. The computer is vital to the astronomer. He uses it to control the telescope, record the information it collects and help him analyse that information.

▶ Earth's spin means that stars, caught on a long-exposure photograph, appear as trails of light in the sky. Telescopes inside these domes on Hawaii are fixed on mounts with a counter movement so they stay fixed on a star as Earth moves round.

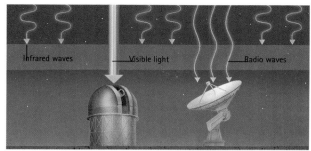

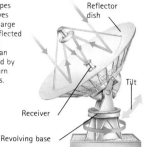

► Radio telescopes collect radio waves from space in a large dish. They are reflected to a receiver and collected. They can then be processed by a computer to turn them into images.

Reflector dish

Tilt

Receiver

Revolving base

Only some of the radiation emitted by stars reaches Earth's surface. Much of it is blocked by the Earth's atmosphere. Light and short-wave radio, as well as some infrared radiation gets through and can be collected by telescopes on the surface.

TELESCOPES ON EARTH

Earth's atmosphere blocks our best view of the Universe. It makes the stars twinkle and blurs the images of galaxies. Sites on mountain-tops are chosen for telescopes because the air is clear and thin. These telescopes are so powerful that the data collected in just a few nights can keep astronomers busy for months or years.

Modern telescopes are large but sensitive instruments, and they need protection from the environment. They are housed inside domes which open up to reveal the night sky above. The dome revolves, thus allowing the telescope to point in any direction. The astronomers and technicians using the telescope are in a separate control room.

TELESCOPES IN SPACE

Some observations of objects in space are totally impossible from Earth. The atmosphere absorbs some forms of energy such as X-rays, and reflects others back into space. The only place to collect these forms of radiation is above the Earth's atmosphere. Telescopes have been used in space for about thirty years and are launched by rocket or space shuttle. Once in orbit around Earth they can make continuous, uninterrupted observations of the Universe. They collect a range of energy waves, including light, in the same way as conventional telescopes. Data collected are sent to a ground control station on Earth where they are stored in a computer. The satellites work for a few

years and are then replaced by newer instruments. At present, for example, the Chandra X-ray Observatory is looking at the Universe for X-rays, SOHO (Solar and Heliospheric Observatory), an ultraviolet telescope, has been looking at the Sun since 1995, and the Hubble Space Telescope which collects lightwaves, can see objects fainter than those seen by ground-based telescopes.

▼ This view into deep space, made by the Hubble Space Telescope, combines 342 separate images. The telescope looked at a tiny area of the sky, the size of a grain of sand at arm's length. It kept looking for 10 days, all the time collecting light from distant space. This view contains 1,500 galaxies never seen before.

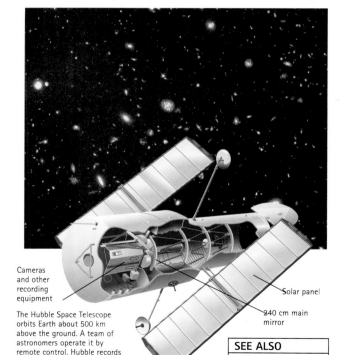

Cameras and other recording equipment

The Hubble Space Telescope orbits Earth about 500 km above the ground. A team of astronomers operate it by remote control. Hubble records views from space and sends them, as TV signals, to Earth.

Solar panel

240 cm main mirror

SEE ALSO

182–3 Lenses and curved mirrors, 290–291 Exploring space

EXPLORING SPACE

Astronomers send robot spacecraft to explore space. These send back to Earth stunning images and data about the planets, moons, asteroids and comets.

Robot space explorers have been working in the Solar System for over thirty years. They are called spaceprobes, and are about the size of a family car. They have shown us the rocky deserts of Mars, an active volcano on Io and the snowball heart of a comet, returned rock from the Moon and dropped through poisonous atmospheres.

They have been incredibly successful at making investigations and discoveries on our behalf. We have learnt much about the Solar System by using them; and will continue to do so from new missions already journeying to their targets, and others being planned.

Most spaceprobes are launched into space by a rocket but a few are taken there aboard a space shuttle. Once released into space, a probe starts its journey toward its target. It may not arrive

The Soviet Luna 9 was the first spacecraft to make a soft (controlled) landing on another world. It landed on the Moon in February 1966, and transmitted the first close-up photographs of the lunar surface.

Two identical Soviet robot vehicles, both called Lunokhod, explored the Moon, one in 1970, the second in 1973. They were driven by remote control from Earth. Together they covered nearly 48 km on the Moon's surface.

for years, but on arrival its real work begins. Scientific equipment on board is switched on and the probe starts its investigations for the astronomers back on Earth. When its work is over the probe is switched off and left in space.

SPACEPROBE FIRSTS

Luna 2 was the first successful spaceprobe to reach another world. It crash-landed on the Moon in 1959 and was the first of a series of Soviet Luna probes that studied the Moon for almost 20 years.

Mariner 10 successfully visited Mercury and Venus in 1973, and was the first probe to visit two planets. Venus, Mars, Jupiter and Saturn have each been investigated by a number of probes. Uranus (1986) and Neptune (1989) were the last planets to be visited. They were studied by the same probe – Voyager 2. Pluto remains the only planet still not visited.

The Sun and some of the Solar System's smaller members have also been the subject of spaceprobe investigation. Craft visiting planets such as Jupiter and Saturn

LANDING ON COMET WIRTANEN

The Rosetta spaceprobe is now being prepared for an encounter with Comet Wirtanen in 2012. The probe is in two parts, a lander craft will land on the comet's surface, as an orbiter circles at a distance. Together they will investigate the comet's material which dates from the time of the formation of the Solar System.

► The spaceprobe Cassini, launched in 1997 will reach Saturn in 2004. On arrival it will orbit the planet and its moons for four years. Cassini will release the mini-probe Huygens to investigate Titan, Saturn's largest moon, and the only one in the Solar System that has an atmosphere.

▼ Before launch, technicians fitted the heat shield to the mini-probe Huygens. Its 160 insulating silica tiles will protect the craft against temperatures of up to 2,000°C as it drops through Titan's upper atmosphere. The heat shield will then be ejected and Huygens' instruments will test Titan's lower atmosphere.

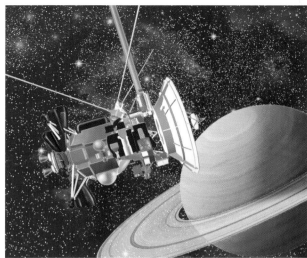

have looked closely at their moons, many of them far too small to be seen from Earth. The Giotto spaceprobe gave us our first close-up of a comet when it visited Halley's Comet in 1986. Our first view of an asteroid was obtained in 1991 by the Galileo spaceprobe on its way to Jupiter.

WORKING ROBOT

A spaceprobe carries on board everything it will need for its mission. This not only includes scientific experiments to carry out its work, but power sources, small thruster rockets for changing direction, and equipment for recording and transmitting the data it has collected.

The task a probe has to carry out, as well as the way it will do it, influences its design. There are three basic ways that probes complete their tasks. A fly-by probe studies its target as it flys past it. The probes Voyagers 1 and 2 have been the most successful of this type so far. Between 1979 and 1989, they flew past Jupiter, Saturn, Uranus and Neptune.

Alternatively, a probe can be an orbiter; this type reaches its target and then goes into orbit around it. The probe Magellan orbited Venus between 1990 and 1994 and gathered data to produce maps of 98 per cent of Venus' surface. The third way is for a probe to land on its target.

Landers work alone, or journey to their target with an orbiter. They then split to carry out their individual tasks. Landers have already successfully worked on the Moon, Venus and Mars.

A similar sort of probe to the lander is the mini-probe which splits from a larger probe once at the target. The Galileo spaceprobe released a mini-probe into Jupiter's atmosphere in this way in 1996. The probe Cassini will release a mini-probe called Huygens when it reaches Saturn in 2004. It will descend through the thick atmosphere of Titan, Saturn's largest moon.

▼ Planning, building and testing of spacecraft takes many years. Here, four identical Cluster satellites are being tested by technicians in Munich in 1994. But their launch a year later was not a success. Four new satellites with identical experiment packages were built and launched in 2000. Orbiting Earth in formation, they study the interaction between the solar wind and the Earth's atmosphere.

SEE ALSO

272–3 The Solar System, 292–3 Rockets and the space shuttle

ROCKETS AND THE SPACE SHUTTLE

Rockets are the only vehicles that travel at 11 km per second and so escape Earth's gravity and reach space. They carry space probes, satellites and astronauts.

The American rocket pioneer Robert Goddard (1882–1945) carried out rocket-based experiments and launched the first liquid-fuel rocket in 1926.

The knowledge to travel into space was developed in the late 19th century, when the first rockets were designed. A significant step in space-rocket technology came in 1926 with the launch of the first liquid-fuel rocket. Designed and launched by the rocket scientist, Robert Goddard, its flight lasted just two and a half seconds and it reached a height of 12.5 metres. His work, and that of other rocket scientists, led to the launch of the first space rockets in the 1940s and 1950s.

An Ariane 44LP rocket lifts off from the European Space Agency's launch site at Kourou in French Guiana. The main rocket has four boosters – two solid and two liquid fuel.

ROCKET TECHNOLOGY
Most of a rocket is taken up by the fuel needed to power it into space. Only a small part of it is set aside for the cargo which travels in the nose. The rocket's fuel needs an oxidizer, a chemical that contains oxygen. At lift-off the fuel and oxidizer are mixed and burned. This expands the fuel and converts it into hot gas. The gas is forced out of nozzles at the bottom of the rocket, powering the rocket and its cargo off the ground and into space. The heavier the cargo, the

more fuel is needed, but the addition of fuel also increases the weight of the rocket. Discardable booster rockets, attached to the outside of the lower body of large rockets, help give extra thrust at the launch. The main rocket is also designed to work in stages. Each has its own fuel supply and engines, and is used in turn and then discarded as the rocket climbs further from Earth. Only the final stage of the rocket will make the journey

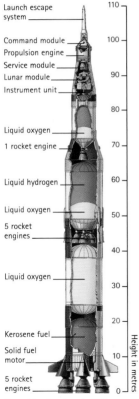

Launch escape system

Command module

Propulsion engine

Service module

Lunar module

Instrument unit

Liquid oxygen

1 rocket engine

Liquid hydrogen

Liquid oxygen

5 rocket engines

Liquid oxygen

Kerosene fuel

Solid fuel motor

5 rocket engines

Height in metres

110
100
90
80
70
60
50
40
30
20
10
0

Saturn V rockets launched US astronauts from Cape Canaveral to the Moon between 1969 and 1972.

Titan III launched Viking spaceprobes to Mars in 1974.

The European rocket, Ariane V, first flew in 1997.

Chinese Long March rockets launch satellites.

Since 1967 Soyuz has launched Soviet cosmonauts.

The Japanese H-IIA rocket launches satellites into orbit.

into space where it will release its cargo. Hundreds of rockets have lifted off for space in the past 50 years, mostly carrying satellites and spaceprobes, but also astronauts. However, rockets are very short-lived and expensive as each one can be used only once. Within about ten minutes of a launch a rocket has lifted its cargo into space and its work is over.

SPACE SHUTTLE
A new type of space vehicle, the space shuttle, was launched in 1981. The shuttle's main difference from the conventional rocket is that it is a reusable system. It uses rocket technology to launch into space, but then returns to Earth like a glider plane. Two of its three main parts can be reused many times. The three main parts of the shuttle are the orbiter, the fuel tank, and the two booster rockets. The crew and cargo fly inside the orbiter, the space plane. This is the part of the shuttle that is launched into space. It acts as a workshop for the astronauts while in space, and then brings them safely home.

There are four different orbiters, *Columbia, Discovery, Atlantis* and *Endeavour*. Only one orbiter is in space at any given time. They are used as space laboratories where astronauts and scientists carry out experiments and test equipment. They are also used for launching, retrieving and repairing satellites, launching spaceprobes, and for building the International Space Station (ISS).

A Japanese N-II rocket lifts off from the launch centre on Tanegashima Island, a small island in southern Japan. The launch centre includes spacecraft assembly and testing sites, and a satellite-tracking site.

THE SPACE SHUTTLE
A typical shuttle trip into space lasts a week. The actual launch is operated by onboard computers in the charge of the launch control team. Once in space, another team, mission control, take over. The orbiter is about the size of a small aeroplane. The large orange-red fuel tank supplying the main engines contains about one-third liquid oxygen and two-thirds liquid hydrogen. It is the only part of the shuttle not reused. The booster rockets are later recovered by ship for reuse.

1 The shuttle is ready for lift off. This is the only time all its components are together. It is lifted off the ground by two boosters and the three engines on the orbiter.

Main fuel tank contains liquid fuel; tank burns up as it falls back to Earth

Nose-cap protects against heat of 1,260°C on re-entry

Heat insulation felt and tiles are fixed to the outside of the orbiter

Two side reusable rocket boosters of solid fuel provide a thrust of 1.5 million kg at take-off

Two engines either side of the tail move the orbiter into and during orbit

Three main rocket engines

2 About two minutes after leaving the ground, explosive bolts release the two booster rockets. The main fuel tank is ejected six minutes later.

3 The orbiter is in space travelling round Earth. Its payload (cargo) bay doors are open. Astronauts inside control the arm to place the cargo, a satellite, in space.

4 On its return to Earth, the orbiter descends through the atmosphere, returning to land like a glider plane. Its rudder, movable edges on the wings, and a drag chute bring it to a stop on the runway.

SEE ALSO
136 Oxidation and reduction, 290–1
Exploring space, 294–5
Humans in space

HUMANS IN SPACE

Nearly 400 men and women have travelled into space since 1961. Just 26 of them, all American men, have left Earth's orbit and visited the Moon.

On 12 April 1961, Soviet cosmonaut Yuri Gagarin, aboard Vostok 1, became the first human to travel into space.

On 16 June 1963, the Soviet cosmonaut Valentina Tereshkova, aboard Vostok 6, became the first woman to travel into space.

American astronaut, Neil Armstrong, became the first human to step onto another world, the Moon, on 20 July 1969.

In July 1975, an American craft, Apollo 18, and a Soviet craft, Soyuz 19, docked together in space for the first time.

The first travellers in space were not humans. Dogs, chimpanzees and monkeys were among the first living creatures to test spaceflight. Their trips led to the first flights by humans. The Soviet Union and the United States of America both prepared to launch the first man into space in 1961. The very first flight, lasting 108 minutes, was by Yuri Gagarin and started a race between the two nations. This continued for almost a decade, and culminated in man landing on the Moon.

Today, these two countries still offer the only way that humans can travel into space, either aboard the Soviet Soyuz rocket or the American space shuttle. They each take space travellers from other countries, so now people from many nations have been into space. These travellers are called astronauts, although those on Soviet spacecraft are called cosmonauts. Each undertakes about two years' training before a first flight.

DRESSED FOR SPACE

An EMU (Extravehicular Mobility Unit) spacesuit protects the astronaut from radiation and the temperature extremes of space. The suit is made of up to 15 separate layers of material. Heated or cooled water is pumped through tubes in the suit to keep the body's temperature constant. A backpack includes a radio and enough oxygen for several hours supply. Each astronaut carries an instrument to measure his or her exposure to radiation. This restricts the amount of time an astronaut can spend working in space.

American astronauts perform an experiment in the shuttle, *Atlantis*. They are observing the growth of ice crystals in the weightless conditions of space.

SURVIVING IN SPACE

The human body is not built for living in space. Not only do space travellers have to take everything needed for survival, including air to breathe, but once in space they have to cope with the weightless environment. Gravity is not felt in space, and so here neither the astronaut, nor anything in the spacecraft, has any weight. Normal activities such as eating, sleeping and working, have to continue, but all in weightlessness. Astronauts and anything else float effortlessly around the craft.

Inside a spacecraft there is air for breathing, bunks for sleeping, a small area for eating and relaxing in, and a toilet. Astronauts work in a separate indoor area, or can go outside the craft to carry out tasks. They may need to work outside for several hours if, for example, they are repairing a satellite which involves catching, mending and then releasing it back into space. Any trip outside requires a protective spacesuit. Once in space astronauts are secured to the craft to stop them floating away. Alternatively they wear a manned manoeuvring unit (MMU), a backpack with rocket thrusters to control movement and direction.

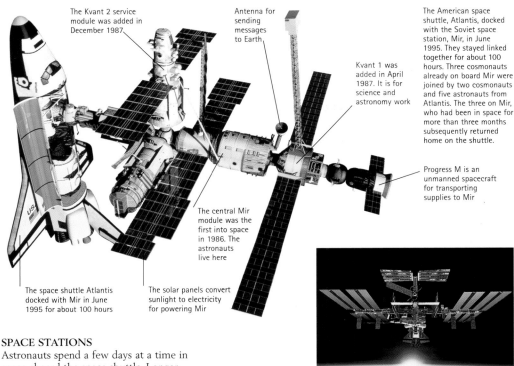

The Kvant 2 service module was added in December 1987

Antenna for sending messages to Earth

Kvant 1 was added in April 1987. It is for science and astronomy work

The American space shuttle, Atlantis, docked with the Soviet space station, Mir, in June 1995. They stayed linked together for about 100 hours. Three cosmonauts already on board Mir were joined by two cosmonauts and five astronauts from Atlantis. The three on Mir, who had been in space for more than three months subsequently returned home on the shuttle.

Progress M is an unmanned spacecraft for transporting supplies to Mir

The central Mir module was the first into space in 1986. The astronauts live here

The space shuttle Atlantis docked with Mir in June 1995 for about 100 hours

The solar panels convert sunlight to electricity for powering Mir

SPACE STATIONS

Astronauts spend a few days at a time in space aboard the space shuttle. Longer periods, weeks, months or over a year, have been spent on the Mir space station. A space station is a craft permanently in orbit round Earth. It is both a home and a workplace for astronauts. Russian cosmonauts have lived on seven different space stations since 1971. The first were all called Salyut, but Mir was the last and most successful one. Mir was too large to launch in one go. Parts were transported separately and assembled in space. Mir was continuously inhabited by cosmonauts from February 1987 to mid-1999.

The International Space Station (ISS) is now being constructed in orbit around Earth. It will take about 70 flights by Russian Proton rockets and American shuttles to transport all the parts into space for assembly. The first arrived in 1998 and the station should be fully completed by 2004. A cooperative effort by 16 nations, the ISS will provide living-quarters and science laboratories for up to seven astronauts.

▲ This artist's impression shows how the International Space Station (ISS) will look when complete. When built it will be about the size of a football pitch. The living and working modules are in the centre.

In the future this hotel could be the destination of the first space tourists. It is one of a number that have been designed to go into orbit round Earth. The Moon is also being considered as a potential destination for tourists.

SEE ALSO

290–1 Exploring space, 292–3 Rockets and the space shuttle

ARTIFICIAL SATELLITES

Hundreds of artificial satellites have been put into orbit round Earth. They monitor the planet, help us navigate, aid communications and look out into space.

A satellite is an object that orbits a planet. The Moon is Earth's natural satellite, and all the planets except Mercury and Venus have them. Natural satellites, however, are usually referred to as moons, and man-made satellites, often called artificial satellites, are what is usually meant by the word 'satellite'.

There are about 1,000 of them working in space right now. Each one is a scientific package designed to carry out a particular job. They orbit between about 300 and 1,000 km above Earth's surface and are launched into orbit by a rocket or occasionally by a space shuttle.

Once in orbit the satellite is activated and will continue to work for several years. All eventually stop, perhaps through the failure of a part, or of the power supply. The dead satellite will in time fall out of its orbit and burn up as it drops through Earth's atmosphere.

Satellites are placed in particular orbits depending on the job they have to do. Some satellites need to stay above the same part of Earth all the time. These are placed in a geostationary orbit about 36,000 km above Earth's equator. This orbit allows the satellite to remain over the same point on Earth's surface at all times. Many communication satellites are positioned like this and work together as a global network. Other satellites, such as weather satellites work in polar orbit, watching the whole of our planet's surface as Earth turns below.

A satellite's power is supplied by solar cells. The cells are either in panels, which look like wings coming from the sides of the satellite, or are wrapped around the satellite's body. They convert the sunlight that falls onto them into electricity. Panels of cells are folded for launch, and then extended once in space. The panels are positioned so they always face the Sun.

A satellite control centre monitors a satellite during its lifetime, tracking it, receiving its signals and sending it commands. Stick- or dish-shaped antennae on the satellite are kept pointing towards Earth to send and receive data. Tiny rocket thrusters on board can be fired to keep the satellite in the correct position and facing the right way.

Sputnik 1 was the first satellite put in orbit round Earth. The 58 cm diameter aluminium sphere was launched by the Soviets on 4 October 1957. It contained a battery, a radio transmitter, and instruments to test Earth's upper atmosphere.

▼ Satellites occupy one of four main types of orbit. Polar orbits take satellites over the Earth's poles. Geostationary orbits keep a satellite in a fixed position above Earth. A satellite on an eccentric orbit varies its distance from Earth, one on a circular orbit is a constant distance away.

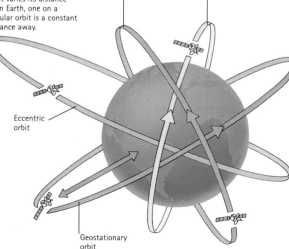

Polar orbit

Circular orbit

Eccentric orbit

Geostationary orbit

▶ A geostationary communications satellite passes telephone calls, TV programmes and computer data between different countries across the world.

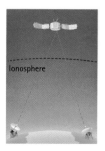

Ionosphere

▼ These satellite dishes on a block of flats in Germany are all pointing toward a communications satellite in space so they can receive TV signals.

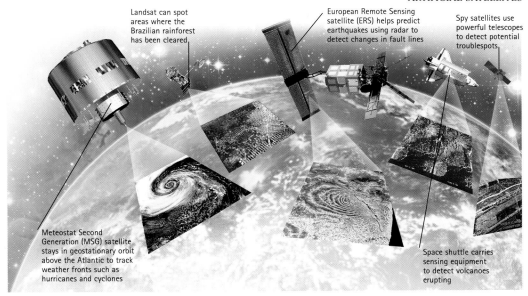

Landsat can spot areas where the Brazilian rainforest has been cleared

European Remote Sensing satellite (ERS) helps predict earthquakes using radar to detect changes in fault lines

Spy satellites use powerful telescopes to detect potential troublespots

Meteostat Second Generation (MSG) satellite stays in geostationary orbit above the Atlantic to track weather fronts such as hurricanes and cyclones

Space shuttle carries sensing equipment to detect volcanoes erupting

SATELLITES AT WORK

Most satellites are made and operated by commercial companies. There are a number of different types; navigation, communication, scientific, military and Earth-monitoring. The first of them started working for us over 40 years ago.

Communication satellites are such a part of everyday life, we use them without realizing it. A sporting event or pop concert happening at one place can be seen by people on the other side of the world at the same time. Cameras film the action, the TV signal is beamed to a satellite above the event, the signal is relayed round Earth by satellites until it reaches one above the opposite side of the planet. The signal is sent down to Earth where it is received and the event watched. Millions of telephone conversations and communications on the Internet, are also handled this way.

Our daily weather forecasts use information from weather satellites that are positioned around the globe. They observe the cloud patterns, monitor Earth's atmosphere, record its temperature range and look out for storms. The whole of Earth's surface has been imaged many times by satellites studying Earth's natural resources such as its forests, icecaps, and oceans. They reveal short- and long-term changes to the planet.

Military satellites can be used for spying on other countries, guiding missiles, or being a weapon themselves. Navigation satellites pinpoint any position on Earth and are an invaluable aid to navigators on land, sea and in the air. Astronomical satellites such as the Hubble Space Telescope look into space and give us spectacular views of the stars and galaxies.

Many of the satellites surrounding Earth are studying our planet. They provide information about the state of the Earth now, and information that forecasts future events, as well as monitoring long- term changes on its surface.

A geologist checks his location using a hand-held GPS (Global Positioning System) receiver. Signals are sent between the handset and up to half of a network of 24 GPS satellites in orbit around the world. Signals from the satellites are used to calculate the geologist's position which is displayed on the handset.

SEE ALSO

14–15 Earth's atmosphere, 274–5 Earth and the Moon, 292–3 Rockets and the space shuttle

MEASURING TIME

The movement of Earth and the Moon around the Sun have always been the basis of time measurement. We have invented increasingly accurate ways to keep time.

In the Middle Ages, candle clocks measured off equal units of time as the flames melted the wax.

Water clocks were used in Ancient Egypt. The amount of water dripping from one bowl to another marked the time passed.

Sundials, used for centuries, show the time by plotting the changing position of the Sun's shadow during the day.

Sand-glasses, popular in the Middle Ages, were used to measure short periods of time.

▶ The world is divided into 24 time zones, starting at the prime meridian (0° longitude) at Greenwich, UK. Each zone west of the meridian is an hour earlier than the last; each zone to the east is one hour later.

Ancient civilizations developed calendars to keep account of the passage of large portions of time such as the days, months and years. The smaller portions of time, the hours and the minutes, were marked by timekeepers. They either used the Sun to directly indicate the time, or they measured the passing of an interval of time.

The earliest time-measuring instruments were those used by the ancient Egyptians. During the day they used a simple form of sundial called a shadow clock. It was made of two wooden rods, one cast a shadow on to the other which had a dial to indicate the hour. At night they observed the position of stars in the sky, and also used the water clock. Water was allowed to flow from one vessel to another. Inside the lower vessel was a scale indicating the passage of the hours as the water flowed.

Sundials and water clocks were later used in Greece and Rome, and then in Europe. These were two of only three timekeepers used up to the early Middle Ages. The third was the sand-glass which measured the passage of time by the flow of sand between two bulbs of glass.

MECHANICAL TIMEKEEPING

The first mechanical clocks were made during the late 13th century. They were clocks intended for general use and placed in a church or other public place where many people could see them. A clock is set in motion in one of two ways – by winding up a spring or by raising a weight. Gear wheels with teeth, or cogs, move the hour, minute and second hands over the face and indicate the time.

Small domestic clocks and the first watches were developed in the 16th century. Cheap, factory-made wristwatches made time available to everyone by the beginning of the 20th century.

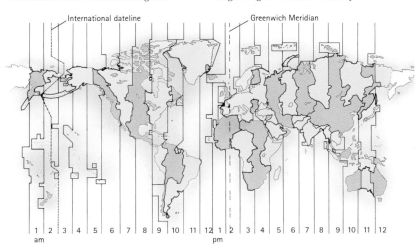

International dateline   Greenwich Meridian

| 1 | 2 | 3 | 4 | 5 | 6 | 7 | 8 | 9 | 10 | 11 | 12 | 1 | 2 | 3 | 4 | 5 | 6 | 7 | 8 | 9 | 10 | 11 | 12 |
am   pm

Timekeeping is needed in every aspect of our lives. However, extremely precise timekeeping is often needed in work and leisure activities. In this athletics stadium, accurate time is displayed on an electronic timing board. The split-second timing of such athletics events often determines the winner.

INTERNATIONAL TIME

Because Earth is spinning, different places face the Sun at different times of day. When it is midday in London, UK, it is dawn in New York, USA, and in Adelaide, Australia, it is still night. If people were to read the time directly from the Sun's position, clocks worldwide would be set to thousands of different times. Chaos would result; for example, it would be impossible to run trains to an accurate timetable.

In 1880, England adopted the time of the Royal Observatory at Greenwich as its standard time, known as Greenwich Mean Time (GMT). By 1884, GMT was adopted as the basis of standard time for the whole world. A system of 24 time zones centred on the Greenwich meridian (0° longitude) was devised and soon became accepted across the world. Time zones to the east are progressively ahead of Greenwich, those to the west, increasingly behind. Tokyo, Japan is nine hours ahead of GMT, so when it is 2:00 am in Greenwich, it is 11:00am in Tokyo. On the opposite side of the world from Greenwich, in the Pacific Ocean, is the international date line, where one day ends and the next begins.

PRECISION TIMEKEEPING

The precise measurement of time has always been important. In the 18th century, astronomers could determine precise time by their observations of the Sun and stars. Navigators wanted to carry that time to sea in a clock so that they could calculate their longitude. The clockmaker John Harrison developed the marine chronometer for this purpose. Today precision timekeeping is still important to navigators, and is used in many other areas of everyday life.

The most accurate clocks are atomic clocks. In fact, they keep time more accurately than the spinning Earth. The time used by everyone today is based on the average rates of a number of atomic clocks around the world. The clocks work by counting the vibrations of light given off by atoms. The most recent, an atomic clock using caesium atoms, is accurate to within one second in 15 million years.

Early mechanical clocks such as this 15th century Italian monastery clock, sounded the hours and indicated them on a dial.

Chronometers were developed in the 18th century for navigation at sea. They were set in motion by a slowly unwinding spring.

The wristwatch became a popular way of telling the time in the 20th century. They were originally powered by a small spring. More recent quartz watches tell the time by recording the vibrations of a quartz crystal. Many types are available, for the astronaut, the deep-sea diver, the schoolchild, or as a fashion item.

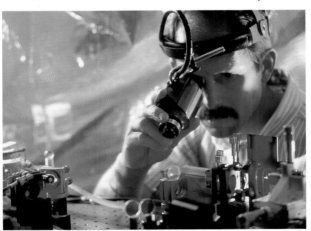

The most accurate clocks are caesium-beam atomic clocks. Here, a scientist uses an infrared detector to look at the laser beam of an atomic clock. Several lasers are used in the clock to measure an atom's vibration. This particular clock has an accuracy of one billionth of a second per day – one second in three million years.

SEE ALSO

9 Earth's rotation,
274–5 Earth and the
Moon

SPACE, TIME AND RELATIVITY

Space, time and gravity are interlinked. Gravity affects everything in space. The gravity of a black hole is so strong it can even change the pace of time.

The scientist Albert Einstein (1879-1955) published his theory of General Relativity, concerning gravity, in 1915. His work led to ideas such as black holes and the Big Bang.

Today, as in the past, scientists work to understand how the Universe works. As we find out more, our new ideas improve, or even replace, old ones. This happened in the early years of the 20th century. Isaac Newton's theory of gravity was improved and in part replaced, by a new theory. In 1915, the German-born scientist Albert Einstein developed a theory called General Relativity which related space and gravity.

In Einstein's theory the gravity of an object distorts space. This is difficult to imagine but start by thinking of a rubber sheet. An object like a marble placed on the sheet will dent it. Now imagine the marble is a massive star and the rubber sheet is space. The star dents the space near it. Other similar massive objects will make a dent, or gravitational well, in space near them.

Astronomers tested this idea during a total solar eclipse in 1919. The light of the Sun was eclipsed and the starlight from a much more distant star could be seen. The starlight travelled close to the Sun on its way to Earth and bent by a predicted amount as it passed the Sun, thus proving that gravity bends space.

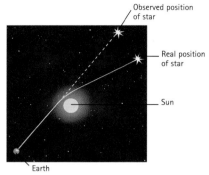

The gravity of the Sun bends space near it. As the light from a distant star passes the Sun on its way to Earth, it too is bent. So astronomers on Earth observe the star away from its true position (marked by continuous line).

A black hole is the remains of a giant star that has blown up. The core that is left after the explosion creates such a powerful gravitational field that any object passing close enough will be pulled into it.

BLACK HOLES

An object such as a black hole makes more than a dent in the space around it. A black hole is what is left over from a massive star that has collapsed at the end of its life. The original star has been super-compressed and now has gravity so strong that it has formed a deep, steep-sided, gravitational well. Light cannot escape so it is 'black'. Nothing at all can escape the well so we think of it as a 'hole' cut off from the rest of space. A hole's gravity not only distorts space around it but also time. The flow of time is disturbed so that it runs slower and slower near the hole.

TIMELINE OF SPACE EXPLORATION

A system of constellations help people understand the night sky.

The first discoveries are made with the newly invented telescope.

It is proved ours is not the only galaxy – there are millions more in the Universe.

Astronomers start to use radio, and later, X-ray and infrared telescopes.

The Space Age begins with the launch of Sputnik, the first satellite.

Soviet cosmonaut Yuri Gagarin becomes the first person in space on 12 April 1961.

| 2000BC | AD1609–10 | 1924 | 1930s–40s | 1957 | 1961 |

WORMHOLE THROUGH SPACE AND TIME

A wormhole is a theoretical tunnel that offers a short cut to a distant place in space. This theory is based on our present understanding of space, time and gravity. The wormhole has two mouths, one at either end of a tunnel. Entry and exit is possible at both ends. One end is at the starting-point of the journey, the other at the destination. Space can be curved, so a distant destination can be brought much closer. Travelling through the wormhole would be considerably quicker than taking the long route through space.

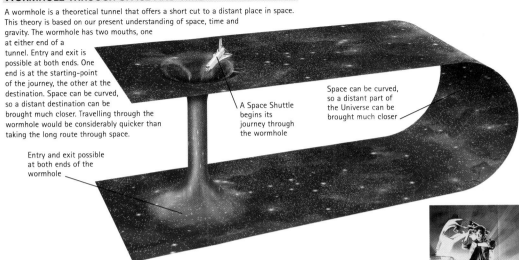

Space can be curved, so a distant part of the Universe can be brought much closer

A Space Shuttle begins its journey through the wormhole

Entry and exit possible at both ends of the wormhole

TIME TRAVEL

All objects, including space, have three dimensions; width, height and breadth. Scientists believe that objects also have a fourth dimension; time. Each day we move through the four dimensions, called space-time. We can move in all directions through space; up and down, side to side, forward and backward. But we can only move forward through time.

Scientists once suggested black holes could offer a way of travelling to very distant places very quickly. Perhaps to another part of our Universe, or even to a different universe. They now know this isn't possible. But a wormhole, a black hole that could be controlled might offer the chance of super-fast, time-beating travel. Anything travelling through a wormhole would move faster than the speed of light. Einstein's theory of relativity states that something travelling faster than light will move backwards through time. But wormholes are only a theory, and they cannot be built. So time travel remains science-fiction.

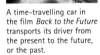

A time-travelling car in the film *Back to the Future* transports its driver from the present to the future, or the past.

SEE ALSO
192 The speed of light, 204–5 Relativity and gravity

Astronauts walk on the Moon. For the first time, they explore a world other than Earth.

Robotic spacecraft explore other planets, and Viking landers look for life on Mars.

The giant outer planets are investigated by the two Voyager spaceprobes.

Telescopes in orbit around Earth make many discoveries and look further into space.

The space shuttle carries parts to build the International Space Station in Earth orbit.

The world's largest and most powerful telescope, the Very Large Telescope, is built in Chile.

1969   1976   1980s   1980s–90s   1999   2000

FACTS AND FIGURES

SOLAR SYSTEM

Sun
Diameter at equator: 1,400,000 km
Spin period at equator: 27 days
Mass (Earth = 1): 330,000
Average surface temperature: 5,500°C

Mercury
Radius at equator (Earth = 1): 0.38
Distance from Sun (Earth = 1): 0.39
Spin period: 58.7 days
Orbit period: 87.9 days (0.24 years)
Mass (Earth = 1): 0.06
Surface temperature: −180°C to 430°C
Moons: 0
Surface gravity (Earth = 1): 0.38

Venus
Radius at equator (Earth = 1): 0.95
Distance from Sun (Earth = 1): 0.72
Spin period: 243 days
Orbit period: 224.7 days (0.62 years)
Mass (Earth = 1): 0.82
Average surface temperature: 460°C
Moons: 0
Surface gravity (Earth = 1): 0.90

Earth
Radius at equator: 6,378 km
Mean distance from Sun: 149,600,000 km
Spin period: 23.93 hours
Orbit period: 365.26 days
Mass (Earth = 1): 1
Average surface temperature: 15°C
Moons: 1
Surface gravity: 9.8 ms⁻²

Mars
Radius at equator (Earth = 1): 0.53
Distance from Sun (Earth = 1): 1.52
Spin period: 24.6 hours
Orbit period: 686.9 days (1.88 years)
Mass (Earth = 1): 0.11
Surface temperature: −87°C to 17°C
Moons: 2
Surface gravity (Earth = 1): 0.38

Jupiter
Radius at equator (Earth = 1): 11.2
Distance from Sun (Earth = 1): 5.2
Spin period: 9.9 hours
Orbit period: 11.9 years
Mass (Earth = 1): 318
Cloud-top temperature: −125°C
Moons: 16
Gravity at cloud tops (Earth = 1): 2.34

Saturn
Radius at equator (Earth = 1): 9.42
Distance from Sun (Earth = 1): 9.54
Spin period: 10.6 hours
Orbit period: 29.5 years
Mass (Earth = 1): 95
Average cloud-top temperature: −140°C
Moons: at least 18
Gravity at cloud tops (Earth = 1): 0.93

Uranus
Radius at equator (Earth = 1): 4.01
Distance from Sun (Earth = 1): 19.2
Spin period: 17.2 hours
Orbit period: 84.0 years
Mass (Earth = 1): 14.5
Average cloud-top temperature: −200°C
Moons: at least 18
Gravity at cloud tops (Earth = 1): 0.90

Neptune
Radius at equator (Earth = 1): 3.88
Distance from Sun (Earth = 1): 30.1
Spin period: 16.1 days
Orbit period: 164.8 years
Mass (Earth = 1): 17.2
Average cloud-top temperature: −200°C
Moons: 8
Gravity at cloud tops (Earth = 1): 1.13

Pluto
Radius (Earth = 1): around 0.18
Distance from Sun (Earth = 1): 29.4
Spin period: 6.4 days
Orbit period: 247.7 years
Mass (Earth = 1): around 0.002
Average surface temperature: −220°C
Moons: 1
Surface gravity (Earth = 1): around 0.07

EARTH'S MOON

Radius at equator: 1,738 km
Average distance from Earth: 384,400 km
Spin period: 27.3 days
Orbit period: 27.3 days
Mass (Earth = 1): 0.012
Surface temperature: −173°C to 127°C
Surface gravity (Earth = 1): 0.17
New moon to new moon: 29.5 days

On 20 July 1969, US astronauts Neil
Armstrong then Edwin Aldrin, became
the first humans to walk on the Moon.

LARGEST MOONS

Moon	Planet	Radius (km)
Ganymede	Jupiter	2,630
Titan	Saturn	2,575
Callisto	Jupiter	2,400
Io	Jupiter	1,815
Moon	Earth	1,738
Europa	Jupiter	1,570

YEARLY METEOR SHOWERS

Quadrantids	1–6 January
Lyrids	19–25 April
Eta Aquarids	24 April–20 May
Delta Aquarids	15 July–20 August
Perseids	23 July–20 August
Orionids	16–27 October
Taurids	20 Oct–30 November
Leonids	15–20 November
Geminids	7–16 December

STARS AND GALAXIES

The ten stars in order of brightness:
Sirius – the Dog Star
Canopus
Alpha Centauri
Arcturus
Vega
Capella
Rigel
Procyon
Achernar
Betelgeuse

CONSTELLATIONS OF THE ZODIAC

Aries – the ram
Taurus – the bull
Gemini – the twins
Cancer – the crab
Leo – the lion
Virgo – the virgin
Libra – the scales
Scorpius – the scorpion
Sagittarius – the archer
Capricornus – the goat
Aquarius – the water carrier
Pisces – the fishes

MILKY WAY

Diameter: 100,000 light years
Disc thickness: 2,000 light years
Centre thickness: 6,000 light years
Mass: 1,000 billion solar masses

LOCAL GROUP – THE TEN NEAREST GALAXIES

Galaxy	Type
Milky Way	Spiral
Sagittarius	Elliptical
Large Magellanic Cloud	Irregular
Small Magellanic Cloud	Irregular
Ursa Minor	Elliptical
Draco	Elliptical
Sculptor	Elliptical
Carina	Elliptical
Sextans	Elliptical
Fornax	Elliptical

TECHNOLOGY

Most powerful telescopes on Earth
(*mirror diameter in brackets*)
Very Large Telescope, 4 x 8 m, Chile
Large Binocular Telescope, 2 x 8.4 m, USA
Hobby–Eberly, 11 m, USA
Keck I, 10m, Hawaii
Keck II, 10m, Hawaii

Key space telescopes

Name	Frequency	Launch date
Oao	UV	1962
Explorer 42	X-ray	1970
Hubble	Visible–UV	1990

READY
REFERENCE

Numbers and Units of Measurement

S.I. Units

The *Système International d'Unités*, or SI, is an international system of units of measurement that came into being in October, 1960. SI units have been adopted by most countries. The system is based on seven principal units:

metre (m)
The metre is the SI unit of length. It is the distance light travels, in a vacuum, in 1/299,792,458 of a second.

kilogram (kg)
The kilogram is the SI unit of mass. It is the mass of an international prototype, which is a cylinder of platinum–iridium alloy kept at Sèvres in France.

second (s)
The second is the SI unit of time. It is the length of time for 9,192,631,770 microwave-frequency oscillations of the caesium-133 atom to occur.

ampere (A)
The ampere is the SI unit of electrical current. When a current flowing through each of a pair of wires, separated by one metre of vacuum, produces a force equal to 0.0000002 (2×10^{-7}) newton per metre, that current is one ampere.

kelvin (K)
The kelvin is the SI unit of temperature. It is 1/273.16 the thermodynamic temperature of the triple point of water.

mole [mol]
The mole is the SI unit of substance. It is the amount of substance that contains as many elementary units as there are atoms in 0.012 kg of carbon–12.

candela [cd]
The candela is the basic unit of luminous intensity. It is the intensity of a source of light, of frequency 520×10^{12} Hz, that produces 1/683 watt per steradian (the steradian is a unit of solid angle).

Metric System of Measurement

Length
10 mm (millimetres) = 1 cm (centimetre)
10 cm = 1 dm (decimetre)
10 dm = 1 m (metre)
10 m = 1 dam (decametre)
10 dam = 1 hm (hectometre)
10 hm = 1 km (kilometre)
1,000 m = 1 km

Area
$100 \text{ mm}^2 = 1 \text{ cm}^2$ (square centimetre)
$10,000 \text{ cm}^2 = 1 \text{ m}^2$
$100 \text{ m}^2 = 1$ are
100 are = 1 hectare
$10,000 \text{ m}^2 = 1$ hectare
$100 \text{ hectare} = 1 \text{ km}^2$
$1,000,000 \text{ m}^2 = 1 \text{ km}^2$

Volume
$1,000 \text{ mm}^3 = 1 \text{ cm}^3$ (cubic centimetre)
$1,000 \text{ cm}^3 = 1 \text{ dm}^3$
$1,000 \text{ dm}^3 = 1 \text{ m}^3$
$1,000,000 \text{ cm}^3 = 1 \text{ m}^3$

Capacity
10 ml (millilitre) = 1 cl (centilitre)
10 cl = 1 dl (decilitre)
10 dl = 1 l (litre)
1,000 ml = 1 l
100 cl = 1 l
1,000 litres = 1 cubic metre (m^3)

Mass
1,000 g (gramme) = 1 kg (kilogram)
1,000 kg = 1 t (tonne; metric ton)

Number Systems

Decimal Base 10	Binary Base 2	Hexadecimal Base 16; Hex
1	1	1
2	10	2
3	11	3
4	100	4
5	101	5
6	110	6
7	111	7
8	1000	8
9	1001	9
10	1010	A
11	1011	B
12	1100	C
13	1101	D
14	1110	E
15	1111	F
16	1000	10
17	10001	11
18	10010	12
19	10011	13
20	10100	14
25	11001	19
30	11110	1E
40	101000	28
50	110010	32
100	1100100	64

Number Prefixes

Number	Prefix
1 trillionth	pico- (p-)
1 billionth	nano- (n-)
1 millionth	micro- (μ-)
1 thousandth	milli- (m-)
1 hundredth	centi- (c-)
1 tenth	deci- (d-)
Ten	deca- (da-)
1 hundred	hecto- (h-)
1 thousand	kilo- (k-)
1 million	mega- (M-)
1 billion	giga- (G-)
1 trillion	tera- (T-)

Fractions, Decimals and Percentages

Common	Percentage	Decimal
½	50.00%	0.500
⅓	33.33%	0.333
¼	25.00%	0.250
⅕	20.00%	0.200
⅙	16.67%	0.167
⅛	12.50%	0.125
⅒	10.00%	0.100
⅔	66.67%	0.667
¾	75.00%	0.750
⅜	37.50%	0.375
⅝	62.50%	0.625
⅞	87.50%	0.875

UK Imperial and US Customary Measurements

Length
12 inches = 1 foot
3 feet = 1 yard
22 yards = 1 chain
10 chains = 1 furlong
8 furlongs = 1 mile
5,280 feet = 1 mile
1,760 yards = 1 mile

Mass
437.5 grains = 1 ounce (1 oz)
16 oz = 1 pound (1 lb; 7,000 grains)
14 lb = 1 stone
8 stones = 1 hundredweight (1 cwt)
20 cwt = 1 ton (long ton; 2,240 pounds)

Area
144 square inches = 1 square foot
9 square feet = 1 square yard
4840 square yards = 1 acre
640 acres = 1 square mile

Capacity (UK Imperial)
20 fluid ounces = 1 pint
4 gills = 1 pint
2 pints = 1 quart
4 quarts = 1 gallon (8 pints)

Volume
1,728 cubic inches = 1 cubic foot
27 cubic feet = 1 cubic yard

Capacity (US Customary – dry)
2 pints = 1 quart
8 quarts = 1 peck
4 pecks = 1 bushel

Capacity (US Customary – liquid)
16 fluid ounces = 1 pint
4 gills = 1 pint
2 pints = 1 quart
4 quarts = 1 gallon (8 pints)
1 US gallon = 0.8 UK gallon

Mass (US Customary)
2,000 lb = 1 ton (short ton)
1 short ton = 1.12 long ton

GEOMETRICAL SHAPES

POLYGONS

A polygon is a flat shape with three or more straight sides.

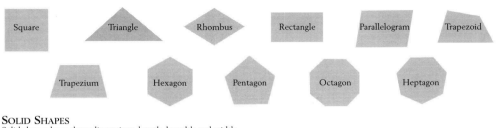

Square · Triangle · Rhombus · Rectangle · Parallelogram · Trapezoid

Trapezium · Hexagon · Pentagon · Octagon · Heptagon

SOLID SHAPES

Solid shapes have three dimensions, length, breadth and width.

Sphere · Cube · Cuboid · Cone · Cylinder · Prism · Pyramid

PARTS OF A CIRCLE

A **circle** is a curved line on which all points are equally distant from the centre. The **circumference** of a circle is the length around its outside. An **arc** is a part of the circumference. The **radius** of a circle is a straight line between a point on the circumference and the centre of the circle. A **chord** is a straight line between two points on the circumference. The **diameter** is a chord that passes through the centre of a circle; its length is twice the radius. A **sector** is an area bounded by two radii and an arc of the circumference. A **segment** is an area bounded by a chord and an arc of the circumference.

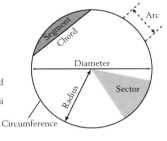

Segment
Chord
Arc
Diameter
Sector
Radius
Circumference

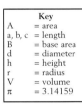

	Key
A	= area
a, b, c	= length
B	= base area
d	= diameter
h	= height
r	= radius
V	= volume
π	= 3.14159

AREAS OF SHAPES

Circle · Triangle · Rhombus · Square · Rectangle · Parallelogram

$A = \pi r^2$ $A = \frac{1}{2}ah$ $A = \frac{1}{2}bh$ $A = a^2$ $A = ab$ $A = ah$

VOLUMES OF SOLIDS

Sphere · Pyramid · Cube · Cuboid · Rhomboid

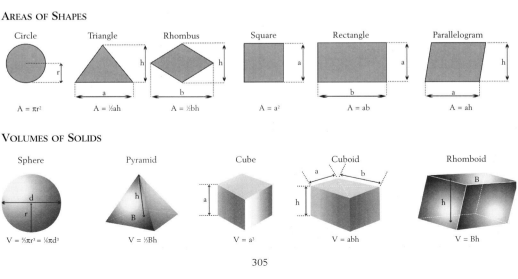

$V = \frac{4}{3}\pi r^3 = \frac{1}{6}\pi d^3$ $V = \frac{1}{3}Bh$ $V = a^3$ $V = abh$ $V = Bh$

FAMOUS SCIENTISTS

Archimedes (c.287–212BC)
A Greek mathematician and inventor, Archimedes invented a spiral pump for raising water. The pump, called an Archimedean screw is still used. He discovered differential calculus and the formula for the volume of a sphere. He is best known for his work on the principles of buoyancy, which he is alleged to have discovered when sitting in a bath and noticing how his body displaced the water. He was killed by a Roman soldier during an invasion of his home city of Syracuse.

Nicolaus Copernicus (AD1473–1543)
Born in Poland, Copernicus studied mathematics and optics at Kraców. After 30 years of work, he put forward a theory that Earth rotates daily about its own axis, and that Earth and the other planets orbit the Sun with years of different lengths. This challenged the ancient belief that Earth was the centre of the Universe. Copernicus was reluctant to publish his controversial theory, but other leading astronomers developed and expanded his theory after his death. These astronomers included Kepler, Galileo and Newton.

Tycho Brahe (1546–1601)
Brahe was born in southern Sweden, which at the time was under Danish rule. Studying astronomy before the invention of the telescope, Brahe discovered serious errors in the astronomical tables of the time. He started a project to correct those errors that would last until his death. He proved that comets were heavenly bodies and calculated the length of a year on Earth to within one second. Working from Brahe's observations, his assistant Johannes Kepler calculated that Mars travels in an elliptical orbit.

Galileo Galilei (1564–1642)
An Italian astronomer and mathematician, Galileo improved the refracting telescope. He observed the phases of Venus and was one of the first people to study sunspots. He found that a pendulum swings at a constant rate and discovered the laws of falling bodies. A believer in Copernicus' theory of the Universe, he fell foul of the religious authorities and had to renounce his beliefs. He continued his research while under house arrest in Florence. By 1637, Galileo had become totally blind.

Blaise Pascal (1623–1662)
Born in France, Blaise Pascal invented a calculating machine (1647), and later the barometer, the hydraulic press and the syringe. He studied fluid pressure, showing that pressure in a liquid acts equally in all directions and that changes in pressure are transmitted instantly.

Robert Boyle (1627–1691)
An Irish-born scientist, Boyle performed experiments on air, vacuum, combustion and respiration. In 1662, he observed that the pressure and volume of a gas at a fixed temperature are inversely proportional. This relationship is now called Boyle's law. Boyle also studied acids and alkalis, density, crystallography, and refraction.

Isaac Newton (1642–1727)
British physicist and mathematician Isaac Newton is well known for his work concerning gravity, but this was just part of a host of important laws and discoveries he made. He developed his three laws of motion and discovered that white light is made up of rays of light of different colours. He also built the first reflecting telescope (1868). Newton was involved in many disputes throughout his life, most notably a conflict with Karl Leibniz over who first discovered the branch of mathematics known as calculus.

Benjamin Franklin (1706–1790)
Born in Boston, Massachusetts, Franklin was a printer, statesman, prolific writer and inventor. He installed street lighting and re-organized the American postal system. He also invented the fuel-efficient Franklin stove, the lightning rod, bifocal lenses for glasses, the first copying machine and the harmonica. Franklin's most famous scientific work concerned his experiments with electricity. He developed the fluid model of electric current, understanding that it consists of the movement of charged microscopic particles. In a famous kite-flying experiment he proved that lightning was a form of electricity.

Joseph Priestley (1733–1804)
British chemist Joseph Priestley discovered oxygen in 1774. He also identified many other gases, including ammonia, carbon monoxide, nitrous oxide and sulphur dioxide. He discovered that green plants give off oxygen and require sunlight. A schoolteacher, writer and politician, Priestley supported the French Revolution and opposed the slave trade, emigrating to the United States in 1794.

Antoine Lavoisier (1743–1794)
Regarded as the founder of modern chemistry, French scientist Lavoisier showed air to be a mixture of gases, which he named oxygen and nitrogen, and proved that water contains hydrogen and oxygen. He devised a method of naming chemical compounds and was a member of the commission that devised the metric system. A critic of the French revolution, Lavoisier was guillotined in Paris in 1794.

John Dalton (1766–1844)

British chemist Dalton's scientific work started in the fields of colour blindness and meteorology – from 1787 he kept a weather journal recording some 200,000 observations. He advanced a theory of atoms in 1803, proposing that molecules are made from atoms combined in simple ratios. In 1808, Dalton published the first table of comparative atomic weights. He also researched the force of steam, expansion of gases by heat and developed his law of partial pressures.

Michael Faraday (1791–1867)

This British physicist and chemist discovered benzene in 1825 and was the founder of electrochemistry. He discovered electromagnetic induction, which led to the dynamo and electric motor. His work contributed greatly to modern understanding of electricity, electrolysis and the development of the battery. Faraday was the first person to use pressure to turn a gas into a liquid.

Charles Darwin (1809–1892)

Darwin studied first medicine and then biology at university before becoming the naturalist aboard HMS *Beagle*, a scientific survey ship bound for South America in 1831. Returning to Britain in 1836, he wrote many works on plants and animals, but he is most famous for his theories regarding evolution, particularly his works, *On the Origin of Species by Means of Natural Selection* (1859) and *The Descent of Man* (1871). Darwin believed that species were not created individually but developed over a long period of time from other species in a struggle for existence that created survival of the fittest.

Gregor Mendel (1822–1884)

Born in Austria, Mendel was ordained as a priest in 1847. He trained as a science teacher while growing plants in a series of experiments. In researching inheritance in plants – especially edible peas – Mendel discovered and developed a number of the key principles that govern genetics, including the law that traits are inherited independently of each other – the basis for recessive and dominant gene composition. The importance of his work was not recognized until it was rediscovered in the early 20th century.

Louis Pasteur (1822–1895)

A French chemist, Pasteur was the first person to show that microbes cause fermentation and disease. He developed the process of using heat to kill germs (pasteurization) and popularized the sterilization of medical equipment, which saved many lives. Pasteur discovered vaccines for rabies and anthrax. In 1888, he founded the Institut Pasteur in Paris for the treatment of contagious diseases. He worked there until his death.

James Clerk Maxwell (1831–1879)

This British physicist was the first person to write down the laws of magnetism and electricity in mathematical form. In 1864, he proved that electromagnetic waves are combinations of oscillating electric and magnetic fields. Maxwell identified light to be a form of electromagnetic radiation, and he increased understanding of the movement of gases, showing that the velocity of molecules in a gas depends on their temperature.

Joseph John Thomson (1856–1940)

Born in Manchester, Thomson studied at Cambridge, where he became Professor of Experimental Physics in 1884. His work led to the discovery of the electron. Thomson discovered that gases are able to conduct electricity and was a pioneer of nuclear physics. Thomson won the 1906 Nobel Prize in physics.

Marie Curie (1867–1934)

Polish-born Marie Curie, and her husband, Pierre, worked in Paris. Along with Henri Becquerel, they were awarded the 1903 Nobel Prize in physics for their work in discovering radioactivity. In 1906, Marie Curie became Professor of Physics at the Sorbonne, Paris, and isolated the elements polonium and radium, as well as discovering plutonium. Curie received the 1911 Nobel Prize in chemistry. Curie died of pernicious anaemia caused by many years of exposure to radiation.

Ernest Rutherford (1871–1937)

This New Zealand-born physicist worked with J.J. Thomson at Cambridge. Rutherford discovered the presence of nuclei in atoms. Along with Frederick Soddy, he proposed that radioactivity results from the disintegration of atoms and was the first person to split the atom. He won the 1908 Nobel Prize in chemistry for his studies of the different types of radiation.

Albert Einstein (1879–1955)

Born in Germany, Einstein is most famously associated with his 1905 Special Theory of Relativity, which related matter and energy in the equation $E = mc^2$. This and his 1915 General Theory of Relativity redefined the way scientists view the Universe. His other achievements included a photoelectric theory, for which he was awarded the 1921 Nobel Prize in physics.

Dorothy Hodgkin (1910–1994)

Born in Egypt, Hodgkin studied at Oxford and Cambridge before becoming a research fellow at Oxford from 1936 until 1977. A researcher in the field of crystallography, she was awarded the 1964 Nobel Prize in chemistry for her discovery, using X-ray techniques, of the structure of vitamin B_{12} and other complex molecules, including penicillin and insulin.

INVENTIONS AND DISCOVERIES

Aircraft

The first heavier-than-air craft was a glider built by British inventor George Cayley in 1808. The concept was developed further by German glider builder Otto Lilienthal, who invented an arched-wing glider in 1877. The first controlled powered flight was in 1903, when US brothers Orville and Wilbur Wright built and flew their Wright *Flyer 1*. The Wrights continued to develop more controllable aircraft throughout the first decade of the 20th century and inspired many others to develop the potential of air transport.

Antiseptic

In 1865, French chemist Louis Pasteur's proposal that airborne microbes caused infection was read by British surgeon Joseph Lister with great interest. Through his experiences with infections around bone fractures, Lister devised a series of successful trials using a solution of carbolic acid, or phenol, to protect wounds from microbes in the air. Today, antiseptics are widely used in surgery and healthcare, as well as to sterilize food preparation areas.

Deoxyribonucleic acid (DNA)

Nucleic acids were first discovered in 1869 by Swiss physiologist Friedrich Miescher. In 1944, a US team led by Canadian-born Oswald T. Avery showed that DNA transmits genetic information. In 1953, DNA's structure was detailed as a double helix by James Watson and Francis Crick. Their findings were based on research by two crystallographers, Maurice Wilkins and Rosalind Franklin. Wilkins was jointly awarded the Nobel Prize with Watson and Crick in 1962.

Dynamite

In the early 1860s, Swedish chemist Alfred Nobel worked to develop a way to safely use the powerful but unstable explosive nitroglycerine. In 1867, he patented a mixture, based on nitroglycerine and a type of clay called kieselguhr, that was stable until it was ignited by a detonator. Nobel called this mixture dynamite. Dynamite soon became widely used for blasting tunnels and quarries

Electron microscope

In 1932, German scientists, Max Knoll and Ernst Ruska produced a microscope that used beams of electrons rather than light to produce images. Knoll and Ruska's microscope could magnify by 17 times. Modern electron microscopes magnify by up to 2 million times – much more than is possible with an optical microscope. They make it possible to investigate materials and organisms that could not be viewed through an optical microscope.

Helicopter

It took more than four centuries for Leonardo da Vinci's designs for a vertical-takeoff, heavier-than-air craft to become reality. The first fully-controlled helicopter was the twin-rotored Focke–Wulf *Fw 61* helicopter, designed by German aircraft designer Heinrich Focke in 1936. By 1938, the *Fw 61* had established a number of records, including altitude and distance.

Internal-combustion engine

Etienne Lenoir is credited with producing the first internal-combustion engine to operate successfully. His two-stroke engine was built in 1860. In 1867, German engineers Nikolaus Otto and Eugene Langen developed a basic internal-combustion engine, and in 1876, Otto patented the first practical four-stroke internal-combustion engine – the forerunner of the engines that now power motor vehicles all over the world.

Jet engine

In 1930, a British Royal Air Force cadet named Frank Whittle patented his design for a jet engine. Whittle tested a complete jet engine on the ground in 1937. Around the same time, German engineer Hans von Ohain built a similar engine that would power the first-ever jet aircraft, the Heinkel *He 178*, in 1939. The first British jet aircraft was the Gloster *E28/39*, fitted with a Whittle-designed *W-1* jet that provided a maximum thrust of 3,800 KN. The Gloster first flew on 15 May 1941.

Laser

The principle of laser action was detailed as early as 1917 by Albert Einstein. It was only in the 1950s that theory was put into practice: US physicists Gordon Gould, Theodore Maiman, Leonard Schawlow and Charles Townes all developed lasers at this time, as did Soviet physicists Nikolay Basov and Aleksandr Prokhorov. In 1964, Maiman's ruby laser was used to treat lesions of the retina – the first practical application of a laser. Since then, many types of lasers have found uses in medicine, industry and communications.

Light bulb

In 1801, British chemist Humphry Davy demonstrated incandescence, which is the ability of substances to give off light when heated to high temperatures. In 1878, US scientist Joseph Wilson Swan made the first light bulb using an incandescent carbon filament in a glass bulb from which the air had been pumped. US inventor Thomas Alva Edison produced a similar bulb one year later. Edison went on to develop the equipment needed to produce the first practical lighting systems.

Microchip

In 1947, US scientists John Bardeen, Walter Brattain and William Shockley invented the semiconducting transistor. By 1971, a microchip was launched that had 2,300 transistors etched on a single chip. Much more powerful microchips are now available. They are used as the processors of computers and the microprocessors that control the operations of household appliances such as washing machines.

Photography

In 1826, French inventor Joseph Niépce made the first permanent photographic image on bitumen-coated metal plates. From 1829 until his death in 1833, Niépce worked with French artist and inventor Louis Daguerre. By 1837, Daguerre had developed the first practical techniques for printing photographs. In 1841, British scientist William Talbot patented a process that used a negative image to make several copies, or prints, of a single image.

Printing

The Chinese used engraved blocks of wood to print on paper before AD200; moulded metal type may have been developed by the Koreans as early as the 14th century. The first European printing press to use movable type is credited to German printer Johann Gutenberg around 1447. William Caxton used a similar press in England from 1476, while Juan Pablos set up a press in Mexico City in 1539.

Radio

German physicist Heinrich Hertz was the first to demonstrate the existence of radio waves in 1888. His discovery was applied by Italian inventor Guglielmo Marconi, who built the first radio system in 1895. In 1901, Marconi transmitted a radio signal across the Atlantic Ocean. In 1906, the first voice broadcast was made by Canadian physicist Reginald Fessenden.

Satellite

The first artificial satellite, Sputnik 1, was built by the Soviet Union and launched on 4 October 1957. Sputnik 1 carried a radio beacon and a thermometer. Since then, hundreds of satellites have been launched for a variety of applications, including global surveying, communications and military surveillance. TIROS 1, the first weather satellite, was launched in 1960.

Sound recording

In 1855, French inventor Leon Scott built a phonautograph that recorded sound vibrations on paper. In 1877, Thomas Edison developed the phonograph, which recorded sound as a groove in tinfoil and could then play back the sound. The first gramophone to use discs was invented by Emile Berliner in 1887. Dutch company Philips launched the first tape cassette in 1961 and, with Sony, the CD in 1980.

Space rocket

Russian physicist Konstantin Tsiolovsky first suggested the use of rockets for space research in the 19th century. US engineer Robert Goddard launched the first liquid-fuel rocket in 1926. German-born rocket scientist Wernher von Braun developed the V-2 long-range rocket in the 1940s and went on to lead the development of NASA's Saturn V rocket, which launched the Apollo missions to the Moon.

Spinning jenny

In 1764, British inventor James Hargreaves developed one of the first machines to increase the efficiency of the textiles industry. Named after his daughter, the hand-operated spinning jenny teased out a number of threads from a single bundle of fibres onto a number of individual spindles. This enabled yarn to be made faster.

Steam engine

The French physicist Denis Papin detailed the principles of steam power in 1679. In 1712, Thomas Newcomen and Thomas Savery built the first practical steam engine which was widely used to pump water out of mines. In 1768, British engineer James Watt built a steam engine, patented the following year, which played a major role in the Industrial Revolution.

Telephone

Based on what he had learned from his work to improve the telegraph, Scottish-born inventor Alexander Graham Bell filed a patent application for his telephone device at noon on the 14 January 1876 – just two hours before US inventor Elisha Gray did the same. In March 1876, Bell transmitted intelligible words over a prototype telephone for the first time.

Television

In 1884, German scientist Paul Nipkow invented a disc with a spiral of lenses that split an image into lines as it rotated. British inventor John Logie Baird used a Nipkow disc in his mechanical television system, which he first demonstrated in 1926. The BBC used Baird's system to broadcast news and shows from 1929. Electronic television was developed in the 1920s and 1930s by Russian scientist Vladimir Zworykin, and US inventors Philo Farnworth and Allen DuMont.

X-ray

In November 1895, German physicist Wilhelm Conrad Roentgen was working in his laboratory with a cathode-ray tube and a fluorescent screen. He noticed that some form of radiation was lighting up a screen that was shielded from the cathode rays. He named the radiation 'X-ray'. A month later he took the first X-ray photograph, or radiograph. It showed the bones in his wife's hand. Roentgen won the 1901 Nobel Prize in physics for his work.

GLOSSARY

acceleration The rate of change in the velocity of a moving object.

acid A chemical compound, whose formula usually includes hydrogen, that dissolves in water to produce hydrogen ions.

aerofoil A teardrop-shaped structure that produces force as it cuts through air. Aircraft wings and control surfaces are aerofoils.

alkali A base that dissolves in water to form hydroxide ions, and forms salt with acids.

alloy A mixture of two or more metals, or a mixture of a metal and a non-metal.

astronomical unit (AU) A unit of measure equal to the average distance between Earth and the Sun – 149, 597,870 km.

atmosphere The gases and clouds that surround a planet, star or moon.

atom The smallest particle of matter that has the chemical properties of an element.

bacteria Microscopic living organisms, some of which can cause diseases.

base A substance that can react with an acid to form a salt and water.

biodegradable Able to decompose in the natural environment by biological means in a moderate amount of time.

biome A climatic region, made up of different habitats, inhabited by distinct forms of plant and animal life.

biotechnology The use of living organisms to generate useful products.

black hole A dark region of space that is so massive that not even light can escape its gravitational pull.

catalyst A substance that increases the rate of a chemical reaction but is unchanged at the end of the reaction.

cell The basic unit from which living all things – plants and animals – are composed.

cloning The creation of genetically identical organisms by natural or artificial means.

colour spectrum The range of frequencies of visible light; the colours of the rainbow arranged in order of frequency.

combustion The scientific term used to describe all forms of burning.

compound A substance that consists of two or more elements chemically bonded together.

conductor A substance through which heat or an electrical current can flow.

constellation A region of the night sky containing a group of stars. There are 88 constellations.

density mass per unit volume; a measure of how tightly packed the mass is in a substance.

DNA Deoxyribonucleic acid. DNA is present in every cell and carries all the genetic information of an organism.

eclipse When one star, planet or satellite passes in front of another, blocking the light from a celestial body such as the Sun.

ecosystem A self-contained community of plants and animals and their environment. A rainforest is an example of an ecosystem.

electrochemistry A branch of chemistry that interrelates chemical reactions and electrical currents.

electromagnetic waves Waves of constantly changing electric and magnetic fields, such as light, radio and X-rays.

electron A subatomic particle that has a negative electrical charge and that usually orbits the nucleus of an atom.

element A substance that cannot be transformed into simpler substances by chemical reactions.

enzyme A chemical substance that occurs in living organisms and that acts as a catalyst for chemical reactions in those organisms.

equator An imaginary circle on the surface of a planet or star at equal distances from its two poles.

erosion The process by which material is worn down and transported by water, wind, gravity or ice.

escape velocity The velocity an object must reach to escape the gravity of an astronomical object such as a planet.

evolution Fundamental changes in the genetic makeup of a species over a series of generations.

fault A break or fracture in rocks along which movement has taken place; joint between two tectonic plates – often a source of earthquakes.

force A form of pushing or pulling influence that can change the velocity of an object or modify its shape.

fossil fuel Energy-containing substance formed from remains of prehistoric plants or animals; coal, oil or gas.

friction A force that resists motion between surfaces that are in contact.

galaxies A collection of millions or billions of stars, planets, gas and dust bound together by gravity.

gas A low-density form of matter that will fill a container of any size or shape.

gene A section of DNA that passes from parents to offspring and determines physical characteristics in an organism.

genetic engineering The science of altering or transplanting genes to create new organisms or to produce useful substances.

gestation In mammals, the time during which a foetus develops and is nourished inside its mother's body.

gravity A force of attraction between objects that is due to their mass.

greenhouse effect The retention of heat by Earth's atmosphere as a result of gases such as methane and carbon dioxide.

habitat The place where an animal or plant lives or grows.

hormones Chemical substance produced by an organism to control processes of the body, such as growth.

hydraulics The use of the pressure in a liquid to transfer power.

igneous rocks Rocks formed under Earth's surface when molten material solidifies.

immunization A medical procedure that primes the body's immune system to fight specific infections.

inertia The tendency of objects to resist a change in their velocity.

infrared radiation Electromagnetic waves just beyond the low-frequency limit of the visible spectrum.

insulator A substance that does not conduct electricity or heat well.

invertebrate Creature that does not have a backbone.

isotopes Atoms of an element that have the same number of protons, but different numbers of neutrons in their nuclei.

kinetic energy The energy an object has due to its speed.

laser (light amplification by stimulated emission of radiation) A device that produces a narrow, intense beam of light of a single frequency. Lasers are used in medicine, communications and industry.

lightning A flash of light caused by a large discharge of static electricity through gases in the atmosphere.

light year The distance light travels in one Earth year; 9,465,000,000,000 kilometres.

liquid A fluid state of matter that collects at the bottom of its container.

luminous The term used to describe any object that gives off light.

magnet An object that attracts iron, nickel and cobalt, and attracts or repels other magnets.

magnetic field A region of space in which a magnet experiences a force.

metamorphic rock Rock that has been modified by heat, pressure, or both.

mineral Naturally occurring substance that is usually found in rocks.

mixture Two or more substances that occupy the same volume but are not chemically bonded together.

molten A substance that has melted and is in a liquid state.

nanotechnology The science and engineering of nanometre-scale objects.

neutron A subatomic particle that has the mass of a proton but no electrical charge.

nucleus (*plural* nuclei) A mass of protons and neutrons at the core of an atom.

orbit The path of one body around another, such as the Moon's path around Earth.

ore A mineral from which useful products, such as metals, can be extracted.

oxidation The addition of oxygen to, or the removal of hydrogen from, a compound; the loss of electrons from a substance; the opposite of reduction.

oxide A compound of one or more elements bonded with oxygen.

ozone layer A layer of the atmosphere that contains a high concentration of ozone (O_3), a gas that absorbs harmful ultraviolet radiation from the Sun's rays.

permafrost Soil or rock that is constantly below the freezing point of water.

planet A large sphere of matter that orbits around the Sun or another star.

pneumatics The use of the pressure in a gas, usually air, to transfer power.

polymers A substance that consists of huge molecules made by a chemical reaction that links together many small molecules, called monomers, in a chain or network.

potential energy The energy that an object has as a result of its position or state.

pressure The force exerted over a specific area by a solid, liquid or gas.

proton A subatomic particle that has the mass of a neutron and a positive electrical charge equal in size to the charge of an electron; the nucleus of a hydrogen atom.

radiation Electromagnetic energy that travels in waves; matter and energy released by the decay of a radioactive substance.

radioactivity The tendency of the unstable nuclei of certain isotopes to release radiation.

recycling The collection and reuse of materials to conserve energy and resources.

reduction The removal of oxygen from, or the addition of hydrogen to, a compound; the addition of electrons to a substance; the opposite of oxidation.

reflection A process in which a wave, often of light, bounces off of another medium.

refraction The bending of light waves as they cross the boundary between transparent substances of different densities.

renewable energy Sources of energy, such as wind or tides, that can be harnessed to generate power without being used up.

reproduction A process in which living things make more organisms similar to themselves in order for a species to survive.

resistance The ability of a substance to prevent or reduce the flow of an electrical current through it.

respiration The absorption of oxygen, in order to release energy from food, and the emission of carbon dioxide and moisture.

satellite An object that orbits a larger object. The Moon is a natural satellite of Earth. Many artificial satellites orbit Earth to provide telecommunications links and photographic images of Earth's surface.

sedimentary rock Type of rock formed from the sediment – stones, sand, dead organisms and mud – that gathers under a body of water and hardens over time.

semiconductor A substance that conducts electricity better than an insulator, but not as well as a conductor.

solid A state of matter that has a definite shape due to the forces that hold it together.

solution A mixture that consists of a solid or gas dissolved in a liquid.

species A group of similar organisms that can produce fertile offspring.

star A large sphere of gases, mostly hydrogen and helium, in which nuclear fusion produces heat and light. The Sun is a star.

superconductivity The ability to conduct an electrical current without losses due to resistance. Certain materials become superconductors at very low temperatures.

synthetic Chemical compounds and other substances artificially made by chemical reactions in a factory or chemicals plant.

tectonics The study of the plate structures that form Earth's crust, and the forces and processes that change it.

telecommunications The transmission of information over long distances by electrical, optical or radio signals.

tissue The cells that make up a particular part of a plant or animal.

troposphere The lower 13.5 km of Earth's atmosphere in which most clouds are found.

ultraviolet radiation Electromagnetic radiation beyond the high-frequency limit of the visible spectrum.

vacuum A space that contains no matter. A partial vacuum contains a greatly reduced concentration of matter at low pressure.

velocity The distance travelled by a moving object in a particular direction in a given amount of time, usually a second.

vertebrate Creature that has a backbone.

vibrations Regularly repeated movements that are described in terms of the frequency and the amplitude of the movement.

virus In biology, a microscopic organism – smaller than bacteria – that can carry or cause disease in another organism.

wavelength The distance between two similar points in a wave.

weight The strength of the pull of gravity on an object due to its mass.

X-rays High-frequency radiation that can pass through some opaque objects.

Computer terms

algorithm A set of rules for calculations.

ASCII System of codes used to represent letters, numbers and other characters.

bit Binary digit: the smallest piece of data.

bug Error in the computer hardware or software that causes it to fail.

byte Eight bits.

CD-ROM Type of compact disc that holds computer programs and information.

data General term for the information handled by a computer.

database Organized collection of data that can be sorted, added to and retrieved.

disk A data-storage device. Hard disks are fixed in computers; removable disks are portable data stores.

file Collection of information, such as a document or an image, stored with a specific name.

hardware The physical equipment which makes up a computer system.

input devices Equipment that puts information into a computer, such as a mouse, keyboard or scanner.

Internet Communications network that links computers throughout the world.

mainframe Powerful computer that can be used by many people at the same time.

microprocessor Powerful set of miniature electronic circuits, etched on a silicon chip, that processes data in a device.

network Group of computers linked to share information and resources.

operating system The programs that run a computer and its many parts. The operating system starts up every time a computer is switched on.

peripherals A part of a computer system that is outside the computer housing, such as a keyboard or printer.

pixel A picture element. Pixels are the dots that can light up to form an image on a computer monitor or television screen.

program A set of instructions, in order, that are used carry out a particular job.

RAM (Random Access Memory) Memory where data are held as they wait to be processed. The contents of the RAM are lost when the computer is switched off.

ROM (Read Only Memory) Where essential programs and data are held. The ROM keeps its contents when the power is off.

software Name given to all the programs used to run a computer.

virus Program that is designed to cause damage such as deleting files.

Visual Display Unit (VDU) The display screen used with a computer.

word processor Program that allows users to store, edit and print documents.

INDEX

ACKNOWLEDGEMENTS

The publishers wish to thank the following for their contribution to the book:

Photographs *(t = top; b = bottom; m = middle; l = left; r = right)*
Page 8 *b* Photo Library International/ESA/Science Photo Library; 10/11 *t* Dr B. Booth/GeoScience Features; 11 *tr* Dr B. Booth/GeoScience Features; *mr* John Reader/Science Photo Library; 12 *tl* Dr B. Booth/GeoScience Features 13 *b* Pekka Parviainen/Science Photo Library; 14 *tl* Nils Jorgensen/Rex Features; 15 *bl* Simon Fraser/Mauna Loa Observatory/Science Photo Library; *br* DRA/Still Pictures; 17 *tl* Aaron Chang/The Stock Market; *br* PowerStock Photo Library; *bl* GSFC/Science Photo Library; 18/19 *t* NASA/Science Photo Library; 19 *tr* David Parker/Science Photo Library; *mr* M. Hobbs/GeoScience Features; 20 *tr* D. Decobecq/Still Pictures; 21 *mr* Alan Watson/Still Pictures; *bl* Dr B. Booth/GeoScience Features; *br* Prof. Stewart Lowther/Science Photo Library; 22/23 *t* Paul X. Scott/Sygma;21 *mr* Rex Features; 24 *tl* James L. Amos, Peter Arnold Inc./Science Photo Library; *ml* Claude Nuridsany & Marie Perennou/Science Photo Library; *l* Adrienne Hart-Davis/Science Photo Library; *bl* Sinclair Stammers/Science Photo Library; *br* Tony Craddock/Science Photo Library; *tl* The Stock Market; 25 *ml* Dr B. Booth/GeoScience Features; *bl* Dr B. Booth/GeoScience Features; *tr* Mike Jackson/Still Pictures; 26 *tl* U.S. Dept. of Energy/Science Photo Library; *l* Dr B. Booth/GeoScience Features; *tr* Dr B. Booth/GeoScience Features; 26/27 *b* Sinclair Stammers/Science Photo Library; 27 *tr* Adam Jones/Science Photo Library; *r* Astrid & Hans-Frieder Michler/Science Photo Library; *br* Simon Fraser/Science Photo Library; 31 *bl* Dr B. Booth/GeoScience Features; 32 *bl* Keith Kent/Still Pictures; 33 *tr* Stewart Cook/Rex Features; *r* Bob Evans/Still Pictures; 35 *tr* NASA/Science Photo Library; *br* JHC Wilson/Robert Harding Picture Library; 39 *mr* Tom Pantages/Phototake NYC/Robert Harding; 40 *tl* P. Morris/Ardea; 42 *tr* Peter Parks/Oxford Scientific Films; *bl* Barry Dowsett/Science Photo Library; *br* Microfield Scientific Ltd/Science Photo Library; 43 *br* Klein/Hubert/Still Pictures; 45 *r* Peter O'Toole/Oxford Scientific Films; *b* John Brown/Oxford Scientific Films; 47 *br* E. A. Janes/NHPA; 48 *tl* Nigel Cattlin/Holt Studios; *b* Nigel Cattlin/Holt Studios; 50 *br* Kenneth W. Fink/Ardea; 51 *b* Scott Camazine/Oxford Scientific Films; 53 *br* John Mead/Science Photo Library; 55 *br* Laurence Gould/Oxford Scientific Films; 56 *ml* Valerie Taylor/Ardea; *tr* Peter Parks/Oxford Scientific Films; *br* Frederik Ehrenstrom/Oxford Scientific Films; 57 *tr* A.N.T./NHPA; 58 *tr* Still Pictures; 59 *tr* Zig Leszczynski/Oxford Scientific Films; 62 *br* Hans D. Dossenbach/Ardea; 63 *bl* P. Robert/Sygma; 65 *b* David B. Fleetham/Oxford Scientific Films; 66 *br* Stephen Dalton/NHPA; 69 *bl* Francois Gohier/Ardea; 70 *br* Hans & Judy Beste/Ardea; 72 *tl* Martyn Colbeck/Oxford Scientific Films; *b* Mark Carwardine/Still Pictures; 75 *mr* Matthew Polak/Sygma; *br* Ferrero-Labat/Ardea; *tr* Konrad Wothe/Oxford Scientific Films; *m* Rich Kirchner/NHPA; *b* Stephen Dalton/NHPA; 78 *tr* V Rocher, Jerrican/Science Photo Library; 79 *tr* BSIP Meullemiestre/Science Photo Library; *bl* Quest/Science Photo Library; 101 *tr* Richard Wehr/Custom Medical Stock Photo/Science Photo Library; 81 *br* CNRI/Science Photo Library; 82 *tl* Scott Camazine/Science Photo Library; 83 *tm* Philippe Plailly/Science Photo Library; *tr* Quest/Science Photo Library; 84 *tl* Laura Bosco/Sygma; 85 *bl* Don Fawcett/Science Photo Library; 87 *ml* CNRI/Science Photo Library; 88 *tr* Larry Mulvehill/Science Photo Library; 89 *tr* Prof. P. Motta/G. Franchitto/University "La Sapienza", Rome/Science Photo Library; *br* Secchi-Lecague/Roussel-UCLAF/CNRI/Science Photo Library; 90 *tl* Kingfisher; *tr* CNRI/Science Photo Library; 93 *tl* BSIP GEMS EUROPE/Science Photo Library; *tr* Prof. P. Motta/Dept of Anatomy/University "La Sapienza", Rome/Science Photo Library; 95 *t* Rex Features; *b* Mark Clarke/Science Photo Library; 96 *ml* Quest/Science Photo Library; 97 *bl* CNRI/Science Photo Library; 98 *tr* National Cancer Institute/Science Photo Library; 100 *bl* Science Photo Library; 101 *br* CNRI/Science Photo Library; 102 *tl* Julian Holland; *tr* Julian Holland/Rainbows End Cafe, Glastonbury; *b* John Kelly/Tony Stone; 103 *bl* Dr C. Liguory/CNRI/Science Photo Library; 105 *t* Dept of Clinical Radiology, Salisbury District Hospital/Science Photo Library; 105 *r* V. Clement, Jerrican/Science Photo Library; 106 *tr* Kingfisher; 107 *br* Biophoto Associates/Science Photo Library; 111 *tl* Mary Evans Picture Library; *br* Photo Researchers/Science Photo Library; 113 *tr* Philippe Plailly/Science Photo Library; *tr* Philippe Plailly/Science Photo Library; 114 *tr* Martyn F. Chillmaid/Science Photo Library; 115 *tr* David Stewart-Smith/Katz Pictures; *mr* Dylan Garcia/Still Pictures; *br* Rex Features; 116/117 *t* Photo Library International/Science Photo Library; 118/119 *t* Pekka Parviainen/Science Photo Library; 118 *b* Nigel Cattlin/Holt Studios; 119 *br* Richard Megna/Fundamental/Science Photo Library; 120 *br* Patrick Barth/Rex Features; 121 *tr* John Mead/Science Photo Library; *bl* Mills Tandy/Oxford Scientific Films; 122 *tl* James Holmes/Zedcor/Science Photo Library; 123 *tr* David Taylor/Science Photo Library; *br* Sygma; 124 *b* Charles D. Winters/Science Photo Library; 125 *b* Rosenfeld Images Ltd/Science Photo Library; 127 *tl* Rex Features; *tr* Richard Folwell/Science Photo Library; 128 *tl* D. Giry/Sygma; *ml* Clive Freeman, The Royal Institution/Science Photo Library; 129 *tr* NASA/Science Photo Library; *l* Charles D. Winters/Science Photo Library; 130 *tl* Gianni Giansanti/Sygma; 131 *ml* J. P. Delobelle/Still Pictures; *tr* Stephen Dalton/NHPA; 132 *tl* Charles D. Winters/Science Photo Library; 133 *tr* Michael Freeman/Phototake NYC/Robert Harding; *r* Professor Max Perutz, MRC Laboratory of Mollecular Biology/Science Photo Library; 134 *tl* Ullstein Bilderdienst; 135 *bl* Stephen Dalton/NHPA; *br* Philip Moore/Rex Features; 136 *b* Milepost 92; 137 *b* F. Pitchal/Sygma; 138 *tl* Robert Harding; *tr* R. Bossu/Sygma; 139 *tr* Cornu, Publiphoto Diffusion/Science Photo Library; 140 *b* David Taylor/Science Photo Library; 141 *tr* Martin Bond/Science Photo Library; *b* Holt Studios/Willem Harinck; 144/145 *t* Jon Jones/Sygma; 144 *bl* David Parker/Science Photo Library; *br* Andrew Syred/Science Photo Library; 145 *b* Rover Group; 146 *tr* Sygma; 147 *tr* Rex Features; *br* John Meek; 148 *bl* Klaus Guldbrandsen/Science Photo Library; 148/149 *b* National Railway Museum/Science & Society Picture Library; 151 *tr* Claude Nuridsany & Marie Perennou/Science Photo Library; *bl* Robert Harding; 152 *tl* Alex Bartel/Science Photo Library; *tr* Harry Nor-Hansen/Science Photo Library; *tl* Pekka Parviainen/Science Photo Library; 153 *tr* Robert Harding; *b* Simon Fraser/Science Photo Library; 154 *tr* Fred Dott/Still Pictures; *b* Allen Green/Science Photo Library; 156 *tr* Eye of Science/Science Photo Library; *br* I. Vanderharst/Robert Harding; 157 *tr* Craig Prentis/Allsport; 161 *b* Milepost 92; 162 *l* Corbis; *b* Rosenfeld Images/Science Photo Library; 162/163 *t* Doug Martin/Science Photo Library; 163 *tr* Hartmut Schwarzbach/Still Pictures; *bl* Bernard Sidler/Sygma; *br* Bernard Sidler/Sygma; 166 *tl* Bluestone Productions/TCL Stock Directory; *bl* NASA/Science Photo Library; *tr* Bavaria-Bildagentur/Telegraph Colour Library; 167 *tl* V.C.L./Tipp Howell/TCL Stock Directory; *tr* Werner Gartung/Katz Pictures; *mr* Simon Fraser/Science Photo Library; *b* U.S. Dept. of Energy/Science Photo Library; 168 Jeff Jacobson/Katz Pictures; 169 *mr* Dick Luria/Science Photo Library; 172 *ml* Conmet/Telegraph Colour Library; 173 *tr* Rural History Centre, University of Reading; 174/175 *b* Gerard Soury/Oxford Scientific Films; 175 *mr* J. T. Turner/Telegraph Colour Library; *br* Schuster/Robert Harding; 176 *br* J. T. Turner/Telegraph Colour Library; 177 *br* Jamie Baker/Telegraph Colour Library; 178 *tl* Ralf Schultheiss/Tony Stone; *b* Clint Clemens/International Stock/Robert Harding; 179 *tr* Adam Hart-Davis/Science Photo Library; *mr* John Cocking/TCL Stock Directory; *b* David Nunuk/Science Photo Library; 180 *b* Jerome Wexler/Science Photo Library; 180/181 *b* Belinda Banks/Tony Stone; 181 *mr* Pekka Parviainen/Science Photo Library; *br* The Stock Market; 182 *tr* Tony Hopewell/Telegraph Colour Library; *bl* Martyn F. Chillmaid/Robert Harding; 183 *bl* NASA/Science Photo Library; 185 *tr* Dr Jeremy Burgess/Science Photo Library; *r* Donald Cooper/Photostage; *b* J. Bright/Robert Harding; 186 *ml* Masterfile/Telegraph Colour Library; 187 *tr* Shehzad Nooran/Still Pictures; *bl* GeoScience Features; *br* Vaughan Fleming/Science Photo Library; *l*-Anthony Bannister/NHPA; 189 *tr* Patrick Durand/Sygma; *mr* Fuji Photo Fil (UK) Ltd; *b* Yves Forestier/Sygma; 190 *tl* Philippe Plailly/Science Photo Library; *bl* John McGrail/Telegraph Colour Library; *tr* Adam Hart-Davis/Science Photo Library; 191 *r* World View/Bert Blockhuis/Science Photo Library; *b* Julian Baum/Science Photo Library; 192 *bl* Space Frontiers/Telegraph Colour Library; 193 *tr* Philippe Plailly/Science Photo Library; *bl* Bildagentur/Bramaz/Robert Harding; 196 *tl* Ferrero-Labat/Ardea; *br* John Massis/Rex Features; 197 *tr* Alan Abramowitz/Tony Stone; 198 *tl* Jerome Prevost/Tony Stone Images; *br* Dr. Gary S. Settles & Stephen S. McIntyre/Science Photo Library; 199 *tr* Rex Features; *br* Paul Van Riel/Robert Harding; *bl* John Meek; 200 *tl* Gray Mortimore/Allsport; *tr* Jean-Paul Ferrero/Ardea; 202 *b* Mark Thompson/Allsport; 203 *t* Shaun Botterill/Allsport USA; 204 *tl* AKG London/IMS; *l* The Kobal Collection; 204/205 *b* Ken Fisher/Tony Stone; 205 *t* NASA/Science Photo Library; *br* Julian Baum/Science Photo Library; 206 *b* Robert Harding; 207 *b* Stewart Cohen/Tony Stone; 208 *bl* David Madison/Tony Stone; 209 Royer PH./Explorer/Robert Harding; 210 *b* Nick Vedros/Tony Stone; 211 *b* Daikusan/Tony Stone; 212 *tl* Tom Till/Tony Stone; *tr* Oscar Burriel/Science Photo Library; 214 *br* Kevin Schafer/NHPA; 215 *mr* Yves Baulieu, Publiphoto Diffusion/Science Photo Library; 217 *tr* Nicholas DeVore/Tony Stone; *r* Tony Stone; 218/219 *t* Jean Becker-Sygma; 222/223 *t* Dr Gary Settles/Science Photo Library; 223 *r* Tony Craddock/Science Photo Library; *b* Keith Kent/Science Photo Library; 226/227 *b* Martin Bond/Science Photo Library; 227 *r* Science Museum/Science & Society Picture Library; 233 *tr* Robert Harding; 235 *tr* Volker Steger/Sandia National Laboratory/Science Photo Library; *bl* Hornby Hobbies Ltd; 236/237 *t* Mark Edwards/Still Pictures; 236 *b* Collections/Nick Oakes; 238/239 *b* Pat & Tome Leeson/Science Photo Library; 239 *b* Massimo Lupidi/Still Pictures; 241 *br* Catherine Poedras/Science Photo Library; 242 *b* Simon Fraser/Northumbria Circuits/Science Photo Library; 243 *tl* Comsat/Science Photo Library; 244 *bl* Robert Harding; 244/245 *b* David Parker/IMI/University of Birmingham High TC Consortium/Science Photo Library; 246 *tr* US Dept of Energy/Science Photo Library; *b* Science Photo Library; 248 *tr* Conor Caffrey/Science Photo Library; 249 *r* Julian Holland; *br* Julian Holland; 250 *tl* Illustrated London News; 251 *tr* Digital Studio; *mr* Jeremy Young/Rex Features; *br* J.L.Atlan/Sygma; 252 *tr* Charles Falco/Science Photo Library; 253 *tr* Peter Menzel/Science Photo Library; 254 *tl* Los Alamos National Laboratory/Science Photo Library; *tr* Science Museum/Science & Society Picture Library; 255 *tr* Rex Features; *mr* Psion; 256 *tr* Hulton Getty Picture Library; 257 *tr* NCSA, University of Illinois/Science Photo Library; *bl* Weiss, Jerrican/Science Photo Library; *br* Hartmut Schwarzbach/Still Pictures; 260 *tr* X-ray Astronomy Group, Leicester University/Science Photo Library; 261 *mr* Space Telescope Science Institute/NASA/Science Photo Library; *br* Hale Observatories/Science Photo Library; 262 *bl* Royal Observatory, Edinburgh/Science Photo Library; 265 *mr* Space Telescope Science Institute/NASA/Science Photo Library; 266 *tr* Celestial Image Co./Science Photo Library; *ml* Royal Observatory, Edinburgh/AATB/Science Photo Library; 270 *b* David Nunuk/Science Photo Library; 271 Victor Habbick Visions/Science Photo Library; 273 *mr* NASA/Science Photo Library; *br* NASA/Science Photo Library; 275 *bl* Sipa-Press/Rex Features; *br* C. Simonpietri/Sygma; 276 *tl* Jason Ware/Science Photo Library; *m* David Nunuk/Science Photo Library; 278 *tr* David P. Anderson, SMU/NASA/Science Photo Library; 279 *b* NASA/Science Photo Library; 280 *tr* NASA/Science Photo Library; 282 *tr* Chris Butler/Science Photo Library; 285 *mr* NASA/Science Photo Library; 286 *m* Detlev Van Ravenswaay/Science Photo Library; 287 *b* David Parker/Science Photo Library; 288 *bl* Richard J. Wainscoat/Still Pictures; 289 *br* Space Telescope Science Institute/NASA/Science Photo Library; 291 *tl* European Space Agency/Science Photo Library; *br* Science Photo Library; 292 *tr* European Space Agency/Science Photo Library; 293 *tr* David Baker; 294 *tl* Novosti/Science Photo Library; *ml* Novosti/Science Photo Library; *b* NASA/Science Photo Library; 295 *t* John Frassanito, NASA/Science Photo Library; *mr* NASA/Science Photo Library; 296 *br* Michael Klinec/Environmental Images; 297 *b* Ken M. Johns/Science Photo Library; 298/299 *t* Mark Thompson/Allsport; 299 *br* Alexander Tsiaras/Science Photo Library; 301 *mr* The Ronald Grant Archive.

Additional artwork: Jonathan Adams, Marianne Appleton, N. Ardley, Mike Atkinson, Graham Austen, Craig Austin, Richard Berridge, Louise Bolton, Simone Boni, Richard Bonson, Peter Bull, Julian Burgess, Robin Carter, Jim Channell, Kuo Kang Chen, Jeanne Colville, Tom Connell, Richard Coombes, Sandra Doyle, Richard Draper, David Eddington, Brin Edwards, David Etchell, Jeff Farrow, D. Fletcher, Eugene Fleury, Chris Forsey, Mark Franklin, Garden Studio, Sally Goodman, Jeremy Gower, Ray Grinaway, Terence Gubby, Terry Hadler, Alan Hancocks, A. Hardcastle, Hardlines, Ron Hayward, G. Hinks, Karen Hiscock, Christa Hooke, Lisa Horstman, Ian Howatson, Industrial Artists, Ian Jackson, Bridgette James, John James, E. Jenner, Ron Jobson, Kevin Jones, Felicity Kayes, Roger Kent, Elly and Christopher King, Terence Lambert, Adrian Lascom, Steve Latibeaudiere, Ruth Lindsay, Rachel Lockwood, Bernard Long, Mike Long, Chris Lyon, Kevin Madison, Mainline Design – Guy Smith, Alan Male, Maltings Partnership, Janos Marffy, Shane Marsh, Josephine Martin, Eve Melavish, Simon Mendez, Carol Merryman, Ian Moores, Patrick Murey, William Oliver, R. Payne, Bruce Pearson, David Phipps, Jonathan Potter, Malcolm Porter, Sebastian Quigley, E. Rice, Paul Richardson, John Ridyard, Steve Roberts, B. Robinson, Eric Robson, Mike Rolfe, David Russell, Valerie Sangster, Mike Saunders, Nick Shewning, Chris Shields, Mike Stacey, Roger Stewart, Lucy Su, Mike Taylor, Simon Teg, George Thompson, Ian Thompson, Linda Thursby, Guy Troughton, Simon Turvey, T.K.Uayte, Ross Watton, Phil Wear, Paul Weare, David Webb, Steve Weston, Graham White, Peter Wilkinson, Ann Winterbotham, John Woodcock, David Wright.